The Army Spouse Handbook

The Army Spouse Handbook

21st Century Guide for the 21st Century Spouse

Ann Crossley

with

Ginger Perkins

ABI Press

Cover: The emblem on the front cover represents the United States Armed Forces insignia, patterned after the Great Seal of the United States. Adopted on June 20, 1782, the Great Seal symbolizes the sovereignty of the United States. The American eagle, with his right talons holding an olive branch and his left talons holding arrows, represents self-reliance. This combination of olive branch and arrows is meant to convey a desire for peace but the ability to wage war. The number of leaves on the olive branch and number of arrows symbolize our nation's thirteen original states.

The Army Spouse Handbook
By Ann Crossley with Ginger Perkins
Copyright © 2019 ABI Press

All rights reserved. No part of this book may be reproduced in any form or by any means, except for the inclusion of brief quotations.

This publication is intended to provide accurate and authoritative information with regard to the subject matter covered. It should not be considered as reflecting the official views of the United States Army.

ISBN 978-0-578-52294-4

To order this book, please go to sales@armyspousehandbook.com. Price is $27 per book, plus shipping and handling. Reduced prices are available for spouse clubs when buying in bulk.

To all the Army spouses

Foreword

I would first like to thank Ann Crossley for the opportunity to help her update *The Army Wife Handbook* to the new *The Army Spouse Handbook*. I love sharing Army protocol, customs, and traditions with our Army Spouses. It has been my passion to empower, educate, and inspire Army Spouses.

I would highly encourage every Army Spouse to start a reference library. Mrs. Crossley's book, *The Army Wife Handbook,* has been the backbone of my library for many years. Our new book, *The Army Spouse Handbook,* has been updated for today's spouses.

Traditions allow us to remember and celebrate the rich heritage of the US Army, an organization older than the Nation it serves. Army Spouses have always played a key role in that history. To go through the journey of an Army Spouse and not celebrate legacies and traditions is a missed opportunity. This book gives you that opportunity. You can modify the custom or tradition to reflect your personality and your circumstances but remember to share the historical background as you mentor Army Spouses. We hope you embrace this way of life as our Soldiers continue to fight to keep our freedom.

I hope this handbook becomes a part of your library as you continue your Army walk of life!

Hooah hugs,
Ginger

TABLE OF CONTENTS

Foreword vii
Preface xv
Introduction xvi
Prologue: Legacies xviii

Section One: First Impressions

1. **Meeting People** 3
Standard Steps; Shake Hands and Make Eye Contact; How to Remember Names; Making Conversation; Active Listening; Compliments

2. **Introductions** 10
Introducing Two People; Groups; Yourself; Family Members; Responding to Introductions; Standing; Standing to Talk

3. **Names, Ranks, and Official Titles** 15
Introducing Others; Yourself; Children, Introducing and Addressing Military; Diplomatic and Civil Dignitaries; Using First Names; Sir and Ma'am; All About Name Tags

Section Two: The Art of Communication

4. **Correspondence** 25
Stationery; Personalized Stationery; Informals; Social and Business Correspondence; Social Titles; Salutations, Signatures, and Suffixes; Change of Address Cards; How to Change your Address with the Postal Service; Greeting Card Etiquette; Holiday Cards and Letters; Form Letters

5. **Business and Social Cards** 37
Calling/Social Cards; Address Cards; Business Cards; Ordering Guidelines

6. **Telephone Manners** 46
The Landline—Just Say "Hello"; For the Caller; For the Answerer; Children and Phones; Taking Messages; Interruptions; Ending Phone Calls; Telephone Time; Phone Use by Guests; How Long to Let the Phone Ring; Voicemail and Answering Machines; Call Waiting; The Business Phone; Texting/Messaging

Section Three: Invitations
7. **Extending Invitations**   55
   Early Planning; What to Include; Cancellation or Postponement; Formal Invitation Guidelines; Informal Invitation Guidelines; Examples of Invitations; Invitation Reminders

8. **Addressing Envelopes**   73
   Official and Social Correspondence; Rank and Initials; Examples; Foreign Titles and Names; Return Address; Hand Delivery; Holiday Invitations

9. **Responding to Invitations**   83
   Response Indicators; Oral/Email Response; Written Response; Response Time; Failure to Respond; Oral Invitations and Responses; Response Complications; Keeping Track of Your Invitations; Two Invitations for the Same Time; When No Response Is Received; Educate Politely; An Invitation to the White House

Section Four: Social Graces
10. **Dress for the Occasion**   93
    Dress Terms; Formal; Informal; Casual; Dress Term Chart; Wearing Special Awards; Travel; Sock Color for Civilian Clothes; Tying a Bow Tie

11. **Guests' Responsibilities**   102
    Party Guests; Houseguests

12. **Table Manners**   109
    Good Table Manners; Which Fork to Use; American and European Styles; Special Challenges

13. **Expressing Thanks**   115
    Written Thanks; Basic Guidelines; How Soon to Write; Thank-You Note Formula

Section Five: Entertaining
14. **Entertaining with Ease**   123
    Easy Steps to Becoming a Successful Host/Hostess; Don't Overlook Spontaneity

15. **Party Preparations** 131
    Pre-Party Schedule; Party Checklist; Entertainment Record

16. **Buffet Dinners** 138
    Menu Selection; Buffet Table Location; Room Setup; Traffic Pattern; Table Organization; Lap Trays; Silverware and Napkins; Centerpiece; Candles; Appetizers; Beverages; Multiple Rooms; Host May Serve; Moving to the Buffet; Second Servings; Ask for No Help; Clearing the Table; Dessert and Coffee

17. **Seated Dinner** 145
    Seating Plan; Place Cards; Course Order; Methods of Service; Place Setting; Place Plates/Chargers and Service Plates; Table Linens; Napkins; Centerpiece; Candles; Table Appearance; Chilled Water; Pouring Beverages; Dinner Partners; Entering the Dining Room; Hostess's Manners; Portions; Dessert at the Table; Last-Minute Cancellation; Late Guests

18. **Seating Arrangements** 158
    Seating Plan; Places of Honor; Multiples of Four; Other Considerations; Round Table Seating; Head Table Seating; Order of Precedence; Guest of Honor

19. **Entertaining Tips** 167
    Entertaining Equipment; Ironing Linens; Treating Spills; Silver-Polishing Gloves; Candle Wicks; Cooking with Wines; Serving Wines; Bamboo Lap Trays; Bread-Basket Liners; Gifts of Flowers; Tub for Silverware; Bartender; Back-to-Back Entertaining; New Ideas; Entertaining Outside the Home

Section Six: Social Functions
20. **Coffees** 177
    The Unit Coffee; Hosts and Hostesses; Guidelines; For Hosts and Hostesses; For Attendees; For Coffee Group Leader; Attendance/Inclusion; Welcoming Committee; Newsletters; Welcome Coffee; Farewell Coffee; Family Readiness Groups (FRG)

21. **Showers** 192
    Baby Showers; Wedding, Birth, and Adoption Announcements

22. **Spouse Clubs** 197
    Spouse Club History and Traditions; Army Spouse Clubs; Membership; Activities; Spouse Club Courtesies; Spouse Club Board;

Elections; Advisors and Honorary President

23. **Teas** 205
Guidelines for Guests; Guidelines for Pourers; Planning a Tea; Escorts; Pinning on a Corsage

24. **Receptions and Receiving Lines** 223
Attire and Arrival; The Reception Line; Going Through the Receiving Line; Sitting; Setting Up a Receiving Line; Receiving Line Guidance; Reception/Receiving Line Diagram; New Year's Reception

25. **Formal Functions** 231
Formal balls; Formal Farewells; The Dining-Out; Toasting

Section Seven: Military Functions

26. **Parades, Changes of Command, Changes of Responsibility, and Retirement** 241
On the Field; Seating; General Guidance; Invited Guests; Changes of Command/Changes of Responsibility; Guidelines for the Spouse of Incoming and Outgoing Commander, Command Sergeant Major, and Senior Enlisted Advisor; Retirements

27. **Military Weddings** 256
Uniforms; Military Chapel; Arch of Sabers; Invitations; Addressing the Envelopes; Gifts; Children; Rehearsal and Dinner; Wedding Photographs/Videos; Usher and Saber Bearer Duties; Guest' Participation; Reception

28. **Military Funerals** 271
Honors; Chapel Service; Funeral Procession; Without Chapel Service; Graveside; Cremation; Symbolism; Funeral Courtesies; Memorial Service; How Can You Help?; What to Say and Not Say; Helpful Books; Helpful Resources

29. **Military Courtesies** 282
Flag Etiquette; Promotion and Award Ceremonies; Staff Car Seating; When to Stand; No Rank Among Spouses; The Honor of Being First; The Honor of Leaving First; Approach Senior People; Call Senior Officers by Name; New Year's Reception—A Command Performance; Babies at Adult Functions; Carrying the Umbrella; Revolving Door Courtesy

Section Eight: Military Roles

30. Commanders' Spouses   291
It's Your Choice; Preparing for the Role; Leadership Skills; Leadlrship Role; Practical Advice; When the Command Comes to an End; Command Reflections

31. NCOs' Spouses   306
Long Hours; Opportunities for Involvement; Leadership Roles; Ladder of Communication and Support; Who's in Charge?; Keeping the Proper Perspective; Working Spouses; Heritage of Service

32. General Officers' Spouses   317
Increased Opportunities; Pitfalls to Guard Against; Setting the Standard; Social (Official) Calendar; Aides; Stewardship of Quarters; Coping Skills for New Spouses

Section Nine: Military Living

33. Community Life   329
Living on Post/Off Post/Camp/Station; On-Post Courtesies; Being a Good Neighbor; Being a Good Sponsor; Military Time; Support Services and Resources; Army Family Team Building (AFTB); Duty in Washington, D.C.; Keeping Informed

34. Overseas Assignments   346
Overseas Living; Host Nation Courtesy and Customs; Language Training; Home Link

35. Joint Assignments: Service Traditions   351
Duty with a Unified Combatant Command; Duty with Another Service; Sister Services; Service Customs; Attaché Duty; Rank Insignia Chart

36. The United States Army   364
Chain of Command; Army Structure; Unit Structure; Rank Structure; Military Education; Acronyms and Abbreviations; Army Family

37. Lessons Learned   377
Friends Are Essential; Children Also Serve; Avoid Family Separations When Possible; Promotion Lists Require Kindness; When a Commander Is Relieved; Divorce Happens; The Army Is Not a Democracy; Nurture Your Spiritual Health; Dealing with the Media; Moving Smart; Travel Smart; Decorating Tips; Army Spouses Care

in Ways that Count; Life After the Army; Letting Go Gracefully; The Military Spouse—A Noble Creation

38. Suggestions for Further Reading   **392**
Military Wives' History; Social Guides for Military Spouses; Service Etiquette and Official Protocol; General Etiquette and Manners; Entertaining; Support Resources; U.S. Department of the Army Publications; Family/Parents' Guides; Videos; The Changing World of Manners and Customs

Appendix: Army Spouses' Historical Chronology	399
Index	401
About the Authors	405

Preface

It's been almost thirty years since *The Army Wife Handbook* was first sold at an AWAG conference, American Women's Activities in Germany. From there, many copies of the book were sold on Ways and Means tables and in post exchanges at Army facilities around the world. Two years later, with the help of my co-author, Carol Keller, who brought her Air Force-wife experience and history background, we wrote and published *The Air Force Wife Handbook*. In 1994 we merged many features of the two books to create the second edition of *The Army Wife Handbook*. At the time, a few people urged us to replace "Wife" with "Spouse," but it seemed too early. Of course, I knew that there were already some male spouses accompanying their female soldiers and, in fact, I made reference to that in the Introduction of the last Army Wife book. But at that time Army wives still made up the vast majority of the spouses, and I wanted to address them. Now society and, by extension, the Army have changed, so it's time for this book to change as well.

The question in my mind was, "How could I update the Army book since my experience as an Army wife ended in 1988?" And "How could I even find someone with active or recent Army-spouse experience to help?" Well, it must have been Providence that guided me, because I found the perfect person to lead the update, Ginger Perkins. And she, in turn, put together a wonderful committee of Army spouses and active-duty members to help. Those willing contributors were Carol Brooks, Brad Combs, Rochelle Crockett, Holly Dailey, Claudia Davenport, Jennine Duelge, Gretchen Escribano, Sheila Hairr, Lynda MacFarland, Angel Mangum, Shand Mayville, Kathleen Palmer, Joelle Pekala, Cassandra Perkins, Maggie Phillips, Aimee Randazzo, Connie Roy, Cindy Scaparrotti, Linda Tarsa, and Linda Via.

Thanks are inadequate to Ginger and this committee for the time and expertise they have contributed to this effort, but I think I can speak for them when I say it's done as a labor of love for today's Army spouses and Army Family.

Ann Crossley

Introduction

Since I first wrote *The Army Wife Handbook*, way back in 1990, American society has changed and our Army has changed to reflect this new normal. Now we have not only the typical couples with male soldiers and civilian female wives, but we also have couples with both serving as soldiers, female soldiers with civilian male spouses, and same-gender couples. Hence, this handbook has been written for the broader audience of Army spouses and titled to reflect this change.

Writing a book in any manner is not easy, especially when writing before the advent of personal computers. The first book I wrote, *The Army Wife Handbook*, was before I had a computer; I handwrote the entire book on yellow legal tablets, later transcribed into a computer. My next two books, *The Air Force Wife Handbook*, and the 2nd edition of *The Army Wife Handbook* were written with a co-author, Carol Keller. You will see that we have kept her magnificent "Prologue: Legacies" in this book. However, our work together was before email, so all of our co-writing had to be done by "snail mail." Carol lived out west and I lived in Florida; to coordinate a single chapter took weeks. In fact, it wasn't until the Air Force book was written and printed that we first met. Now we are into the age of email, and I have learned that working with a committee is unbelievably easy; all it takes to transfer text is just a click of the computer key. What advancements lie ahead?

The committee had hoped at first to prepare this book in eBook format to make it quickly accessible to all military spouses. Unfortunately, we have learned that digital rights management (DRM) software can easily be overcome. That doesn't seem fair to the hard-working committee that has put many hours into updating this book. Therefore, we have made the difficult decision to print this book. As with my other books, we will sell this book to spouse clubs at a discount so they can earn a profit for their clubs' ways and means. We hope that all Army spouses will find this book useful, and that it will enrich their experience by retaining the traditions of our proud Army!

Preparing any manuscript, whether for a print book or an eBook requires significant effort and technological expertise. I turned to both Ed "Tex" MacMillan and his sister, Katie Egan, for their invaluable help and guidance in this endeavor. However, any errors that you might find are mine alone.

You will notice a new feature in this book: the inclusion of BLUFs—Bottom Line Up Front. BLUFs are intended as a quick summary of the point we are trying to make in each particular section. They are short, meaningful, and easy to remember.

As with the earlier handbooks, the bottom line is often the most important. And the bottom line for any social guide is to use the information with grace; it's better to be thoughtful to others than to be socially correct. That's the BLUF for this book: Follow the Golden Rule of Good Manners; Treat Others the Way You Would Like to be Treated.

Prologue: Legacies

"Few of us have the greatness to bend history, but
each of us can change a small portion of events."
Anonymous

The early morning sun has just begun to take the chill from the air on this Saturday morning in October 1980, as two women, one tall and soft-spoken, the other shorter and more gregarious, enter the foyer of the Sheraton Washington Hotel. They cross the posh lobby and move down a long hall to the now-silent central meeting room. Round cloth-covered tables, chairs, easels with thick pads of blank paper attached, and a raised podium have been set up; everything appears in order. In a few hours, the room will be crowded with over two hundred and fifty women who have come from all over the world. They are Army wives, delegates, who have come to represent themselves and other Army spouses of their communities at the first Army Family Symposium. This is the moment Joyce Oft and Betty Steiger, Symposium co-leaders, have anticipated and planned for over four years. By nine o'clock, every seat is filled and the room is again quiet. The delegates listen to Steiger's welcoming remarks: "This is a safe platform for the expression of your ideas, your enthusiasm, your anger, your viewpoints.... Be sure your ideas are realistic.... We are talking about a peacetime Army.... We are here by choice. Your spouse has chosen to be in the Army." For the next two days, these 1980s women will help the United States Army understand and define its relationship, commitment, and responsibilities to Army wives and families. Their ideas and vision of the future will shape the development of Army family policy into the decade of the 1990s, and be their legacy to the next generation of Army wives.

The modern U.S. Army wife shares a heritage with the past. She carries with her a trunk load of memories—a legacy. She is given this inheritance by those who have gone before, along the wagon trails of the American West and across the jet contrails of the contemporary world. From the grueling life of the earliest Revolutionary War camp followers, the companionship of the frontier and constabulary forces, the harsh realities of two World Wars, the seemingly endless Vietnam decade for those waiting wives, to the new Total Force era of today—each generation has experienced the trials and joys of service life and added to the contents of the Army wife's metaphoric trunk of legacies. As a framework for understanding the social character of this life, let us unpack that trunk.

THE FORMATIVE YEARS: 1775 - CIVIL WAR

The first generation of American military wives followed the Continental Army, in spite of concerted efforts by its leaders to prevent them from doing so. Many wives actually accompanied their soldier husbands in the small regular Army and the citizen militia. They provided aid, comfort, and whatever social life was possible, and their actions set a pattern that continues today. The best known of these women was Martha Washington, who joined the General in winter quarters throughout the war. She arrived at Valley Forge in a carriage well stocked with provisions for her husband and his men. Once there, she dispensed the food and medicines, and gathered daily with the other officers' wives to knit, sew, and visit with and tend the sick. She also gave small dinners and held open house twice a week for her husband's officers. Other officers' wives followed Mrs. Washington's example and did their share of entertaining to ensure life continued as normally as possible. Martha Washington set the standard. Writing in later years, she recalled, "[I] learned from experience that the greater part of our happiness or misery depends upon our dispositions and not upon our circumstances."

Throughout the war, and in spite of the harsh conditions, the Washingtons took every opportunity to demonstrate good manners and civility. Martha had married the one man in the American colonies who had written his own guide to good manners at the age of fifteen. Though Washington's "Rules of Civility" were borrowed from English guides, he made them uniquely American and filled with common sense. However, it was Martha who became the principal guide to common courtesy. Fittingly, the first American-authored guide to manners written after the Revolution was dedicated to her.

Life in the new Republic was characterized by social change. With new opportunities for a higher standard of living came a renewed interest in social decorum. In the decades before the Civil War, an average of three new books on manners and good conduct appeared annually, more than at any earlier time. Most were written by Americans; they were remarkably egalitarian, crisply written, and sold for only twenty-five cents. The new books granted women a surprising amount of new freedom. Girls were encouraged to acquire some knowledge or skill so that they might become self-supporting if necessary. Ladies could converse with "respectful" strangers and undertake long journeys alone. For the first time, manners were not only for the rich and wellborn—they were for everyone.

The regular Army, during these years, reflected the needs of the Republic. Small in size, except for two periods of war (with England in 1812 and Mexico in 1846), its role was to protect the frontier and the coasts. Since the establishment of the United States Military Academy at West Point in 1802 and the Naval Academy at Annapolis in the 1840s, the officer corps was professionally trained. Officers married the socially prominent daughters of the well-to-do who carried the social customs of the antebellum period with them to frontier garrisons. Wives shared totally the life of the Army and, in small garrisons, created a society closely resembling that which they had left "back home"—dances, picnics, socials, and afternoon calls—all meant to give their lives pleasant intimacy. But wives resented the frequent moves and separations of Army life. The indomitable Lydia Lane made twenty moves in three years and crossed the great plains four times in an eight-year period. The Grants' first assignment was to Detroit where they arrived with high expectations, only to discover that Ulysses' orders had been changed to another post. Julia wrote her parents not only of the tiresomeness of the additional journey, but of its considerable cost and of her efforts to keep a home in her mother's custom.

Far less is known of the lives of soldiers' wives, since most enlisted men were forbidden marriage. During the Revolutionary War, many of these wives, like the artillery heroine, Molly Pitcher, followed their men on campaign. The early frontier Army was different. In 1847, Congress declared "no man with a wife or a child may enlist." The policy was frequently ignored and many soldiers did marry the only women with official Army status—the laundresses. These women were critical to camp life for, in addition to doing the laundry, they also served as nurses and midwives.

The coming of the Civil War changed the Army and disrupted the pattern of frontier garrison life. This war divided colleagues, friends, and families. One-half of the officer corps resigned to fight for the Confederacy. The war involved greater numbers of men, and that meant more camp followers in all ranks. These women's contributions during that war, especially in nursing, were remarkable.

FRONTIER WIVES: 1865 - World War I

Post Civil War industrialism and prosperity created a new model for American manners that lasted until World War I. Great wealth helped to create an American aristocracy, Victorian in style and customs. Mark Twain called this time in our history the "Gilded Age." Immigrants poured into the country and the American Army went west again, to pacify the Indians and ensure that the railroad opened the frontier. Women's magazines enjoyed great popularity, and etiquette books reached their zenith—five or six new ones appeared each year. Among them was a book by a Navy wife, Mrs. John A. Dahlgren, thought to be the first etiquette book written by a military wife. The stated purpose of these books was to teach a form of social decorum and gracious living that was in keeping with the rising fortunes of the middle class. Instructions were given for dress, conversation, table manners, smoking etiquette (men only), the intricacies of the proper use of calling cards, ballroom and cotillion (a complicated dance involving partner changes and the presentation of expensive favors) decorum, the debut, and chaperones. All of this resulted in one of the most complicated structures of etiquette in our nation's history. These were the guides and the image of society that Army brides took with them on their way west.

Garrison life with the post-war frontier Army began to change. The size of the Army was reduced to pre-war levels; it never exceeded 2,100 officers. The Army became primarily an Indian-pacification force. As the railroad was completed and permanent housing replaced the dugouts and tents of earlier times, life grew easier. The brick residences built on some posts in the 1870s and 1880s meant that garrison life could indeed be much like life back East.

We probably have General William T. Sherman to thank for our extensive knowledge of this period. He met with the men going west and urged them to take their wives. He encouraged the ladies to keep diaries and to take with them "all the needed comforts" for their lives in the "newly opened country." Sherman didn't need to encourage Libbie Custer, who had already followed her dashing husband throughout the war. She traveled with him on summer campaigns and made a home in the large, new, Victorian frame house at Fort Lincoln, South Dakota, in the winter of 1876. Afternoon musicals, book reviews, amateur theatricals, and sewing with "Godey's Lady's Book" dress patterns formed the rituals of ladies' social life in garrison. The bonds of friendship were strong and necessary, for the women also grieved together when word came the following July of the disaster at Little Big Horn.

Martha Summerhayes called her life in the Arizona territory "glittering misery," but would not change it. Army officers' wives learned to ride, spent leisure time reading, and looked forward to entertaining their husbands' superior officer. They were not unaware of the need to make an impression; one wife wrote proudly of serving a five-course meal for Generals Sherman and Sheridan in her tented quarters, borrowing from the other garrison wives to ensure that only the best linen and china were used. Frances Roe described garrison life as delightful, with frequent dinner and card parties that were elegant affairs graced with silver and china. Wives also internalized Army life, as depicted in a wonderful photograph of Frances in her husband's West Point Dress blouse with his Kepie jauntily perched on her head.

The War Department did not officially recognize officers' wives and their families, but it did provide for them. Soldiers' wives and families were often less fortunate. Living conditions varied from post to post, but overcrowding prevailed. Quarters were normally provided for senior NCOs and laundresses, but in 1878 the Army barred further enlistment of laundresses. This deprived the Army of skillfully laundered uniforms, and also deprived garrison wives of critical mid-wife services. The rigid social structure of the time also meant there was little interaction between soldiers' wives and officers' wives. Each group lived with their husbands in different areas of the post, and these came to be known as "soap-suds row" and "officers row." Yet, Army wives of the frontier era are remembered as women who succeeded in making the best of the difficulties of garrison life and helped to shape the destiny of the American West. By 1890, the Indian wars were over, troop strength reduced, and the frontier "closed." Forts across the West were deactivated; at Fort Lincoln, Libbie Custer's house was dismantled, along with the rest of the buildings, and the timber sold in nearby Pierre.

Less than a decade later, America had become an empire. The nation's expanded horizons became those of military wives as well. They boarded ships with their husbands, who were bound for constabulary duty in the Philippines. Once there, the wives continued to maintain their tradition of helping out wherever there was a need. They volunteered in orphanages, cared for and educated their children, and tried to maintain a sense of humor throughout. Army wives in Manila also found time and the necessary ingredients for developing an elegant social life. Domestic help was plentiful and the occupation villas spacious. Shopping filled leisure hours, and they were able to purchase beautifully embroidered linens to grace their tables at elegant dinner parties. Army life abroad mirrored that of affluent society back home. Officers' wives became the "Army's aristocracy." They patterned their lifestyles by the glitter and display of the "American society" of their era.

During these years, the women's club movement was born in America. The clubs supported philanthropic and educational endeavors, temperance, and suffrage reform. All across the nation, women began to reshape their genteel social views in order to have more influence in public affairs. Military wives paralleled the efforts of their civilian sisters, and expanded the focus of their literary societies and music clubs to include the social needs of their post communities. They sought to increase widows' pensions and soldiers' families' stipends. However, a few military wives were not convinced of the virtues of women's suffrage. Both Ellen Sherman and Madeline Dahlgren, wives of Army and Navy flag officers, are known to have signed an anti-suffrage petition to the United States Senate.

New thinking moved within the armed forces as well, resulting from a new concept of military professionalism and new weapons—the battleship, machine gun, tank, and airplane. Each was tested in World War I, and their success soon led to a revolution in twentieth-century warfare.

RENEWAL: EARLY TWENTIETH CENTURY

While World War I initiated major changes in the armed forces, it also expedited the transition to a new set of social manners for society. Prohibition, automobiles, the vote, and movies precipitated demands for a relaxation of past social structures. Outmoded Victorian manners were replaced. Etiquette books maintained their popularity; an average of five or more new ones appeared yearly, some with phenomenal sales. The most successful writer was Emily Post, whose first book was written to support herself and her children after divorce from her wealthy, socially prominent, banker husband.

While Emily Post initially wanted to preserve the manners of her own social heritage, within five years she announced that each generation had the right to interpret etiquette for itself. Her next edition "vanished" the chaperone, "countenanced" women smoking, and announced that one of the least important "rules" of etiquette was that concerning which fork to use. Lillian Eichler, whose books sold over a million copies, went even further with the new etiquette; she advised readers that what was once regarded as highly "ill-bred" might be acceptable before they finished reading her book. The etiquette of courtship relaxed, applying lipstick in public was to be preferred to appearing "wan-lipped," entertaining became less formal with the introduction of the cocktail party, and six courses replaced the standard twelve. The cotillion and grand ball lost their glamour, and debuts were less frequent. Business etiquette books also began to appear, the first for the businesswoman in 1924. Such was the shifting pattern of society, which was reflected in the armed forces during the interwar years.

How did this social revolution impact on the society of the "Old Army" that had grown to maturity during the more rigid social structure of an earlier era? The entire Army now numbered less than 140,000. Senior officers continued to be mostly Academy graduates and almost everyone knew one another. The pay was poor and promotions came very slowly. Though the Army was smaller in size, the number of Army wives increased when, in 1925, the War Department permitted enlisted men to marry. However, they could only do so with the permission of their superior officer; "marriage must be for some good reason in the public interest," and "the efficiency of the service is to be the first consideration." Military wives, in this brief period of peacetime, were more isolated from the civilian community and developed a sense of uniqueness, dedication, and closeness.

In these isolated communities during the interwar years, the roles of officers' wives were formalized. They learned to ride, played golf and tennis, and entertained; they also spent time with their husbands making calls and watching polo matches. (Some officers received special allowances for polo.) Their lives centered on home and children. Yet, we must not assume idleness and self-interest, for they also raised funds for the Red Cross, the Army and Navy League, and followed their husbands in duty all over the world. The traditions and social protocols, refined in years spent at remote garrisons and overseas, remained central to the thinking of "Old Army" wives.

These halcyon days of the peace-time Army ended with World War II. The largest mobilization in American history began—nearly nine million men and women. Civilians who knew little of the Army were drafted and Reserve officers were called up. The dramatic increase in dependents led to the passage of Public Law 490, providing dependency benefits; the dependent ID card was born! Wives who were unable to leave small children to work in war industries sought meaningful volunteer work. The Army Relief Society became the Army Emergency Relief Society, staffed by Army-wife volunteers. The "spotter" program was established by Army wives to help new wives coming into the Washington, D.C. area find their "spot" in needed war work.

The war also put American manners to the test abroad. The government prepared "pocket guides" for men and women in uniform to explain the customs of other nations, and issued warnings not to give offense. For those going to England, the advice was, "It isn't a good idea to say 'bloody' in mixed company," and for Egypt, "Don't offer a Moslem alcohol or pork in any form." Back home, Army brides learned what they knew of service etiquette from other wives. Manners and customs acquired in the war years were to sustain them through the turbulence of demobilization, the build-up for the long years of the Cold War, and a time when they, too, would become the "old guard."

THE ERA OF CHANGE: LATE TWENTIETH CENTURY

The pace of change again accelerated dramatically in the decade after the war. Rapid demobilization reduced the armed forces to one-half million. Returning soldiers, sailors, and airmen, happy to have survived, wanted homes and families; the birth rate soared, creating the "baby boom." Seventy million babies were born in America between 1946-1961, touching off a demographic revolution that continues to be felt. Women left their factory jobs, but not the workforce. Americans began an active pursuit of success, defined primarily in financial terms. Once again, they turned to etiquette books to guide them through the "minefields" of a new, upwardly mobile society. Millicent Fenwick's popular *Vogue's Book of Etiquette* (1948) contained extensive notes on dress and stressed the civic value of etiquette. That same year, a new expert on manners, Amy Vanderbilt, began the book that was to dominate the manners scene for the next twenty years and make her the successor to Emily Post.

Four years in the writing, *Amy Vanderbilt's Complete Book of Etiquette: A Guide to Gracious Living* continued the tradition of a member of America's genteel aristocracy providing guidance on public manners. Five printings were issued in quick succession, and it became the best-selling guide of the decade. Vanderbilt promised to address "every possible social problem one is likely to encounter in modern social living." Like Post before her, Vanderbilt wrote with encyclopedic detail; however, she went beyond Post to address basic social issues for "baby boom" parents. An entire section was devoted to "The Family and Social Education of the Children." Calling-card use was still described, but so too was the proper dress for women in public office and for an appearance on television. Post and Vanderbilt were standard reference books in the American middle-class household in the 1950s.

During this time, the passion for proper etiquette found a voice for military wives in the writings of Nancy Shea. The precepts of military courtesy shaped by military wives in garrison life and constabulary duty, refined during the interwar years as "old Army," were now to be shared with a new generation of wives. Nancy Shea, an Army Air Corps wife, wrote *The Army Wife* in 1941. More than a guide to military etiquette, Shea wanted to explain "what the Army wife may expect from the Service and what the Service expects of her." Shea followed the next year with *The Navy Wife* and, when the Air Force became a separate service in 1947, she responded with the publication of her third "service bible," *The Air Force Wife*. Military wives traveled to Germany and Japan, ready to provide "all the needed comforts," and Shea's book went with them.

Nancy Shea was sensitive to the impact the reduced size and restructuring of the post-war Army had on Army wives. In the 1948 edition of *The Army Wife*, she cautioned her readers that they might "need to make loose-leaf for the many changes to come in the next few years." But it was no sooner off the press than the

Korean War dictated Army expansion. Shea's books enjoyed great popularity and success with the new arrivals to military life. Her service books went through several editions and revisions in the 1950s and sold for only three dollars and fifty cents. They were especially suited to the mood and style of the 1950s, offering not only guidance for young wives on the customs of the service, but also "the management of an Army household." Establishing a congenial home, assuming her Army responsibilities, being a helpmate in her husbands' career, and following him wherever the Army sent him—this was the role of the Army wife Nancy Shea described.

Back at home, Americans enjoyed the prosperity and the "consensus society" of the fabulous '50s. Television, youth culture, the growth of suburbia, and the upsurge in automobile production which put America on wheels—each contributed to changes in American manners. Women's clubs flourished, especially in the Army, where posts were expanding at home and abroad. New post housing provided more opportunity for socialization and recreation. Informally organized wives' clubs, which varied from post to post before the war, were now more equitable and organized as officers' wives' clubs, NCO wives' clubs, and enlisted wives' clubs. Customs and courtesies from the "old Army" days were updated. These social changes within the services led the Navy to publish in 1959 the first edition of *Service Etiquette* for officers in the naval service.

The 1950s "hat and glove" era is remembered not only for the lovely formal teas, card parties, and gala balls. During that time, Army wives and wives' clubs also created and staffed the first post nurseries; opened and operated floral shops, thrift shops, and youth centers; and raised funds for scholarships and a host of community welfare activities. They actively supported the Junior Army Navy Guild Organization (JANGOS), which offered service daughters valuable hospital training and volunteer opportunities. Wives' clubs sponsored balls so that young Army daughters and sons would have the same social opportunity as those in the civilian community.

Books written specifically for the Army wife remained popular. Shea's format was replicated when Ester Wier continued the tradition of companion works for each of the services. Her first book, *Army Social Customs,* was followed by comparable books for Navy and Air Force wives. The guides written by service wives continued to be more than etiquette books. They described proper social decorum, dress, and behavior; they also offered help, "how to's," and general "survival" advice.

As the "fabulous '50s" passed into the 1960s, the American dream seemed alive. Proper etiquette and respect had forged a bond among Army wives that fostered a unique social system dedicated not only to common courtesy, but community building. This translated in 1961 to a commitment by officers' wives' clubs to support a home for widows of Army officers at Distaff Hall. Army wives also helped establish and staff Army Community Services (ACS). First organized in the early 1960s at Ft. Benning to function as a liaison between the Army and Army families, ACS became officially recognized Army-wide in 1964. All wives' clubs flourished, and NCO and enlisted wives' clubs grew in number. Their purpose was to be social, welfare, and community focused. In 1969, Mary Preston Gross dedicated her *Mrs. NCO* to all the NCO wives who had contributed so greatly in giving social status to the families of noncommissioned officers.

At the same time, the world and the Army were changing. First, the Cold War escalated; then, the Cuban missile crisis erupted; next, the war in Vietnam precipitated dramatic service expansion. The decade that began with President Kennedy's image of Camelot and President Johnson's high hopes for a "Great Society" ended in social, political, and military turbulence—the civil rights movement, the women's movement, youth counter-culture, and the longest war in America's history. During the Vietnam War, most "waiting wives" joined informal groups or groups sponsored by wives' clubs and ACS; they bonded together and they grieved together. To meet the needs of those military families living beyond the reach of post medical facilities, Congress created the Civilian Health and Medical Program of the Uniform Services (CHAMPUS).

The social revolution of the 1960s and '70s had a momentous impact on American manners and customs. The classic guides by Post and Vanderbilt remained available, but were printed less frequently; few new guides appeared. Many people waited to see what was to replace the old order, as America turned inward. The changing structure of American society was reflected in significant changes within the services: creation of the all-volunteer force and great influx of women into the peacetime military. These changes were reflected in the all-new 1977 edition of *Service Etiquette* by Oretha D. Swartz, the recognized authority on protocol for naval officers. This expanded version, written for all services, focused on both male and female officers and their families. Changes in the traditional service etiquette and customs reflected in this book were an indication of the far greater social restructuring taking place within America's armed forces.

The increasingly intense national dialogue on women's roles, their place in the workplace and individual aspirations, had a profound influence on the services. Service wives, during the decade of the '70s, took a critical look at military families, their needs, and their relationship to the institution. Research contributions by service-wife professionals helped to focus attention in each of the branches of the armed forces on the need for an affirmative family policy. Changing family patterns within the Army, indicated by increased numbers of Army husbands and dual-career couples, influenced joint-spouse assignment concerns. Issues of mission readiness and retention and their relationship to family life provided an additional sense of urgency.

Changes in the structure of the American workforce had a great impact on the military's family programs that were dependent upon volunteers. As more Army officers' wives entered the workforce, "work for pay," frowned upon in the 1960s, became acceptable and necessary by the mid-1970s. Employment was frequently a financial imperative for soldiers' wives. Army wives had less time for total involvement in what had evolved as traditional aspects of Army community life. Additionally, they were now more independent and wanted more organizational support and recognition of their own needs and those of their families. With the decline in volunteerism, agencies once fully staffed by Army-wife volunteers and regarded as "nice-to-have," were now recognized as critical to the quality of Army life.

The decade of the 1980s was characterized by significant changes in both the way the armed forces viewed the value of family members and how family members viewed themselves. In 1978, the Army established the Army Quality of Life Action Planning Committee and, a year later, a Quality of Life Office on the Army Staff. In order to address voluntary programs from the point of view of the participant, an Army wife was appointed volunteer consultant to the Department of the Army. Her task was to help provide structure to volunteer programs and to find experienced volunteer leaders to serve as counselors to the major Army commands (MACOM's). However, the real vitality and issues-problem-solving came from the "grass-roots" level, the Army wives themselves. Determined to take a role in the decision-making process that affected their lives, they responded in October 1980, to the invitation of the Army Officers' Wives' Club of the Greater Washington Area and the co-sponsor, the Association of the United States Army (AUSA), to attend the first Army Family Symposium. The women listened intently when Army Chief of Staff General Edward C. Meyer said, "We recruit soldiers, but we retain families," and went on record as recognizing changes in society and the Army family.

The Symposium report generated more than a hundred recommendations, and the Army listened. As a result, the Army Family Liaison Office (FLO) was established. The term "dependent," now considered demeaning and inaccurate, was replaced with "family member." Spouse-employment opportunities were expanded with the Department of Defense authorization for spouse-hiring preference. Commanders' wives were invited to accompany their husbands to the pre-command courses for "command team" training, and Army wives wrote *The Commander's Link* (later revised and titled *The Leader's Link*) in preparation for these new seminars. Thus, the crucial initiatives of Army wives helped to shape family programs and policies during the 1980s, and led to official recognition of the important role of Army wives—both as individuals and as "mission multipliers."

Along with role recognition, came choices. The Army acknowledged, as a result of a policy recommendation of the family symposium movement, that Army wives had the right to pursue any job or activity they wished, without fear of their actions impacting on their husbands' assignments or careers. Thus, the Army was already in compliance when the Department of Defense issued Directive 1400.33 stating, "No DOD official shall, directly or indirectly, impede or otherwise interfere with the right of a spouse of a military member to pursue and hold a job, attend school, or perform volunteer services on or off a military installation." This directive confirmed the revolution in social attitudes within all the armed services and helped to define the relationship between each branch of the service and service spouses.

The increased openness of American civilian society during the '80s led to a renewed interest in manners, and new authors emerged who promised a "new look" that reflected the recent social revolution. Continuing in the tradition of socially prominent women interpreting society, Charlotte Ford's *Book of Modern Manners* was the first guide. Ford added to traditional guidelines the new topics of teenagers, divorce, marriage contracts, and the social decorum of living together when not married. In this same style, Letitia Baldrige, who was already known for revising Vanderbilt's encyclopedic guide in 1978, wrote her own guide to the new manners in 1990, and included a discussion on "The New Manners Concerning Sex." In a separate category all her own, Judith Martin, under the copyrighted name "Miss Manners," continues to bring wit and clever satiric commentary on America's ever-changing social landscape.

Military etiquette also went through a transition period as a result of the turbulence of the '80s and the self-evaluation by the Army wives of that decade. Preoccupation with concerns perceived to be more central to their daily lives had led service wives to a benign neglect of traditional common courtesies once thought to be an automatic extension of military life. However, a resurgence of interest in this subject had begun. By the mid-eighties, protocol sessions at the American Women's Activities Germany Conference (AWAG) were among the best attended, primarily by young wives who wanted to know something about the strange world of "military manners." Wives' club magazines included etiquette columns, and programs featured guest speakers on the subject. But no new full-length guides, which recognized just how real the revolution in societal roles for Army wives had been, appeared on the scene to fill the need. There was not to be a fresh look at service traditions until 1990 when the first edition of Ann Crossley's *The Army Wife Handbook: A Complete Social Guide* appeared. In the forward, Alma Powell (Mrs. Colin Powell) commented on the many changes since Nancy Shea: "Our lives are much more relaxed, the moves of society are constantly changing and evolving, but we as military wives face many social and official occasions that are structured by long traditions."

As a framework for understanding the character of contemporary service etiquette, we have unpacked the trunk filled with the customs, traditions, and history of military wives who have gone before us. Whether visualized in a battered leather trunk, dented metal footlocker, or sleek new briefcase, this heritage provides the legacy that has helped to frame our society in the 1990s. It remains to be seen if today's military manners will stabilize for a time or continue in a state of transition into the twenty-first century. That's up to you, the Army spouses of today. You will write the next chapter in our history. It's time for you to ponder, "What will be our legacy?"

1993, Austin, Texas   CAROL A. KELLER

[Updated 2019 by Ginger Perkins]

THE NEXT CHAPTER: THE TWENTY-FIRST CENTURY

Change has continued at an accelerated pace since the last edition. Our intent for this new book is to ensure the newest information is available in an easy-to-digest form. *The Army Wife Handbook*, now *The Army Spouse Handbook*, has been and will continue to be the bedrock of many Army-spouse libraries. It is the go-to book to use as a reference on the customs and courtesies of our Army. Starting from military weddings (the saber send-off) to the solemn and beautiful Arlington cemetery funeral processions, Army spouses can look to this book for guidance before embarking on their military experience or at any point during their journey.

One of the most notable and obvious changes is the name of this handbook from "wives" to "spouses." It is the *21st Century Guide for the 21st Century Spouse*. This handbook is for all spouses—men, women, active duty, reserve, National Guard, retirees, civilians, career-oriented spouses, stay-at-home spouses, working-from-home spouses, same-gender spouses, fallen soldiers' spouses, and professional volunteer spouses. We all share the name and role of Army spouse.

The Army recognizes the sacrifices that Army spouses have and continue to make. Whether it is following their soldiers, like Martha Summerhayes: "I cast my lot with a soldier and where he was, was home to me" (1866 Army Wife), or spouses that homestead for their personal career, family considerations, or senior-year high school stabilization. The Stabilization for Soldiers program began in April 2001 allowing soldiers to extend their tour of duty at their current duty station for an additional year so a family member can graduate from their current high school. Army spouses continue to embrace "Home is where the Army sends you" as they wait for orders, housing, promotions and assignments. There's always an adventure waiting, for you're married to a Soldier!

Our nation has been repeatedly sending soldiers to war since 9-11. Never before has our military had so many deployments for so many years away from home. During times of crisis in a unit, whether it's during a deployment or an accident at home base, Army spouses step up and help each other and their families. Army spouses started their own care teams during Operation Iraqi Freedom. Posts started their own memorial walls and tree-lined streets in honor of their Fallen Heroes.

During this time, Family Support Groups and FSG Liaisons morphed into Family Readiness Groups. Virtual FRGs for remote posts or national guard units, and the family readiness support assistant (FRSA) were created. In 2019 the FRG is being renamed the Soldier and Family Readiness Group. While the unit family readiness is a commander's responsibility, the FRSA is there to assist the commander and FRG leaders (most often spouses) in the execution of that responsibility throughout all phases of a unit's life, particularly during the deployment cycle. The FRSAs in deploying units are the continuity, updating chain of command rosters, chain of concern rosters, and care team rosters—enabling spouses to be the heart of the unit. FRG newsletters became email newsletters sent to family members: spouses, parents, grandparents. Spouses are encouraged to prepare family notebooks of important documents—birth certificates, marriage certificates, powers of attorney—as they take on the sole responsibility of the family during deployment. Spouses may need to file taxes, sell homes, cars, and move while their soldiers are deployed.

Designed to provide a link between the military unit and Army Community Service, the Unit Service Coordinator Program was provided, utilizing ACS staff to assist commanders and rear-detachment commanders and Family Readiness Group leader-spouses. The Army created job descriptions for the FRG volunteer positions, and ACS added more classes to assist spouses in their more formal volunteer jobs, including evening classes for the many working spouses. The Command Team, meaning the commander and command sergeant major and their spouses, ideally work together for the welfare of the unit's soldiers and families. Their spouses are "Battle Buddies"; they work together for the betterment and enrichment of all members of the unit. (Battle Buddies can also refer to the close friend or friends who support one another through Army life's challenges. Most often they are stationed with us, but some are only a text or phone call away for those truly rough days that no one else will understand.)

During the height of large unit deployments, families learned to value quality time over quantity. The Performance Triad was introduced. It focuses on sleep, activity and nutrition to increase physical and mental performance. Soldier, spouses and family members can participate in this program for a more balanced lifestyle. They also learned that a healthy body helps keep a healthy mind. FRGs created events like "Walk to Iraq (then Afghanistan) & Back," a tracker that helped keep families fit as they marked milestones throughout long deployments. Children were born during deployments and war and were viewed for the first time via FaceTime or Skype. Many families during Operation Iraqi Freedom/Operation Enduring Freedom spent a Saturday afternoon waiting in line to speak with their soldier over video teleconferencing. Luckily today, Army families communicate daily through text messaging and social media.

The Army realized that spouse volunteers move every year or two, so the Army created the Volunteer Management Information System through MyArmyOneSource to track and document volunteer hours. Spouses can include these verified hours and volunteer positions on their civilian resumés to document time spent away from the workforce. The Army also awards spouses for their exemplary service and dedication. These awards consist of certificates, pins and medals; all can be transferred to civilian resumés.

Chaplains are wonderfully equipped both emotionally and with encouraging reading material for Army spouses and family members during deployments. Chaplains and ACS hold helpful reintegration classes for units returning from combat or peacekeeping deployments. Civilian companies have come up with wonderful, memorable items such as "deployment dolls" and imprintable pillowcases to help families cope with long separations. Many communities gather to send holiday stockings to soldiers, as well as care packages throughout the year.

During deployments, spouses attended many emotional unit memorials and funerals with the support of other spouses. Communities stepped in to embrace these families. Motorcycle clubs came to support the units as well. Spouses continued to take care of one another by providing meals, love and support for families whose soldiers were killed in action or injured. All of those still occur when a soldier dies while serving on active duty. Additionally Survivor Outreach Services was formed for the Fallen Heroes Families and Gold Star Families, along with Honor and Remember groups that help those families whose loved ones made the ultimate sacrifice. The Silver Star Service Banner was approved in 2010 for those wounded, ill and injured.

In spite of the challenges of war and deployment, Army spouses have continued the traditions of coffees, teas, balls, formals, holiday and New Year's receptions. They have adapted these traditions to fit into their busy lifestyles and their unit's rotations. Some groups alternate having unit coffees at different homes while others have coffees at restaurants or coffee houses; some substitute wine tastings. Wherever they are held, coffees are important as ever; they are networking venues for personal, professional and military growth, as well as a time to create lasting friendships. Traditional terms may change, such as "calling cards" to "business cards," but the intent is the same. Army spouses use these business cards to share their contact information. This networking is key among Army spouses. Some traditional posts and units continue to host a welcome or farewell tea for a spouse of a senior leader or NCO. Balls and formals have become beautiful unit send-offs and welcome-home celebrations from deployments. Due to deployment cycles, holiday and New Year's receptions are held perhaps only once during a two-year command. "Organizational days" continue to be a great source of fun, networking and camaraderie. Dining-Ins and Dining-Outs may be less frequent, but a new phenomenon started with the Dining Inside-Out, a twist that spouse clubs or units may host with spouses only.

The Army families have had many strains placed on them during these years of frequent deployments. Single parents and working spouses are an increasingly large part of the Army landscape. Posts increased their child care facilities, and they remain a priority. The Army Family Covenant, succeeded by Total Army Strong, was instituted with many programs to assist the soldiers, families, and civilians. For example, the Exceptional Family Member Program has made new strides in assisting special needs families with their moves and individualized education programs. Military Interstate Children's Compact Commission was implemented and continues to assist with graduation waiver requirements, exit exams and credit transfers during senior year. The Military Child Education Coalition founded in 1999 assists military families transition from one school system to another through student-to-student and parent-to-parent programs as well as school liaisons. Army Brats meet the challenges of moving every couple of years, making new friends and entering new schools. Many Army families have children who enter the military. When led to join the Army, they become resilient and formidable soldiers.

Spouse employment and spouse education were key topics for the Army Family Action Plan. The Army heard the battle cry and authorized a spouse professional weight allowance for moving their professional, work-related books and equipment. The Army especially values spouse involvement and success. Many spouse courses are offered through the professional military education courses at the pre-command course, general officer course, sergeants major academy course, and nominative leader course for nominative command sergeants major. ACS continues to offer many new classes to spouses including classes for new daddies, healthy cooking to relaxation classes and, as always, Army Family Team Building classes.

As the social gap between officers and NCOs closed, wives' clubs transitioned to all-rank spouse clubs. Spouse clubs continue to foster post and community networking and to raise funds for community grants to worthy recipients. And with the new GI benefits bill, education is within grasp for any soldier, spouse or child. Many thrift shops and post museums, formerly run by volunteers, became private organizations. AWAG, American Women's Activities – Germany, was renamed Americans Working Around the Globe and continues to educate and professionally develop volunteers.

Advisorship and mentorship are still key ingredients to continuing our Army spouse legacy. Nothing takes the place of personal interaction and engaged leadership. Knowledge needs to be shared at every level. Seasoned spouses invest their time and talents because they know that Army families are special. Their soldiers are service members—those that "serve" and protect the American Constitution, something larger than themselves.

Social graces and etiquette never go out of style; Army traditions need to be passed on to keep them alive. Our traditions make us special in a modern world that does not always see the value and purpose of such things. We continue to unpack our trunk filled with customs and traditions, altering the design as required. Continue to keep the "5 F's: Faith, Family, Fitness and Finding a Fabulous Battle Buddy" and ponder, "What will our legacy be?"

GINGER PERKINS
New Hampshire
2019

SECTION ONE: FIRST IMPRESSIONS

Chapter 1

Meeting People

"You never get a second chance to make a first impression."
Will Rogers

    The friends and acquaintances we make during our life with the Army are what make it special. Ask any Army spouse whose soldier has retired what they miss most about the military, and they'll say, "the people." Of course, living and traveling in foreign countries and meeting the challenge of making a home for one's family in different locations can be very exciting, but the people are what make any location a real pleasure. Thus, learning to meet people with ease and develop friendships quickly is vital for every Army spouse.
    Some people seem to have no difficulty meeting someone new. They appear relaxed and confident right from the start. They always seem to know the correct things to say, and move from one interesting topic to the next without embarrassing lulls in the conversation. No doubt, they will even remember everyone's name, where they met, and the conversations they had. We all wish we could be like that. The good news is, with practice, we can!

STANDARD STEPS
    Because first impressions are often lasting, it's important to try to do and say the right things every time you meet another person. There are a few, simple steps that are the accepted standard for every meeting. Follow these and you will gain confidence that you are presenting yourself well. Practice them and they will become a natural part of your behavior.

- *Smile and appear friendly.* A friendly smile says you are approachable and conveys a sense of goodwill.
- *Shake hands.* To reach out and touch someone in a handshake is more friendly and sincere than simply saying hello.
- *Look the person in the eyes when you greet them.* This is one of the most important keys to appearing sincere, but the most often overlooked. It says, "I'm really thinking about you at this moment."
- *Call the person by name.* When you've just been introduced, use this

technique to help you remember the name. When you've met before, use the person's name to reinforce your memory. If you can't remember the person's name, don't hesitate to ask; doing so implies that you are genuinely interested in that person and in learning his or her name.

- *Stay relaxed and natural.* Trying to be someone you're not is never as successful as just being yourself.
- *Listen to what the other person says.* Show your interest and attention by looking directly at the person speaking. Glancing around signals that you're allowing yourself to be distracted and not really listening.
- *Say something nice when you part.* "It was so nice to meet you," and "I hope to see you again soon" may be overused, but they convey the right message.

These steps are simple and basic. They are the keys to appearing friendly and sincere—and making a good impression.

SHAKE HANDS AND MAKE EYE CONTACT

> BLUF
>
> "Shake hands as you would hold a dove—not too strongly that you would kill the bird and not too lightly that it would fly away."
> *MAJ Cassandra Perkins*

Shaking hands has long been the traditional method of greeting people and bidding them farewell. It is usually well received by most people you are meeting for the first time. It has always been a part of the world of business, politics, and diplomacy. Shaking hands should be regarded as more than a formality; it is a gesture of friendship.

Making eye contact is the complement of shaking hands; the two should always go together. A handshake without eye contact equates to an insincere hello; it says I'm shaking hands simply because it is expected. Most of us look hastily at peoples' faces when we greet them, but seldom directly at their eyes. Yet, this adds immeasurably to the sincerity of our greeting. Try it; you'll like it! It just takes a little practice to remember to make eye contact every time you shake hands with someone.

Guidelines for Shaking Hands

Who Offers First - It doesn't matter who offers to shake hands first. In the old-fashioned American tradition, a woman offered her hand to a gentleman upon their meeting or being introduced *if* she cared to shake his hand. It was considered bad manners for the gentleman to offer his hand first. This stilted form of etiquette is no longer practiced. Modern handshaking etiquette doesn't distinguish between sex, age, or position; it makes everyone equal. Even very young Army spouses should not hesitate to extend their hands first when meeting older or more senior people. What's important is the gesture of friendliness that the offered hand conveys. However, be considerate of international customs that may be different from ours (see below).

When - Shake hands at every appropriate opportunity; when you are introduced to someone, meet a friend or acquaintance, and when you say goodbye. A host and hostess should always shake hands with their guests, both as they arrive and depart. Parents should teach their children that it is polite and proper to stand and shake hands when being introduced to adults.

How - The manner in which you shake hands is important because your handshake seems to reflect your character, at least in the other person's mind. If you give a limp, boneless handshake, people may consider you a "shrinking violet." On the other extreme, those who give a bone-crushing handshake will come across as trying to impress people with their strength. A nice, firm handshake—neither too hard nor too weak—sends the signal that you are a self-assured and friendly person.

The Hand Kiss - A few words of caution to the ladies when shaking hands with European gentlemen: Be prepared to have your hand kissed, instead of shaken. This gallant custom is still frequently observed in European and diplomatic circles. It's a delightful greeting, but it can be embarrassing to the lady if she has already started shaking hands. Approach these situations prepared to go either way.

International Greetings - Be aware of other countries customs. In some countries, such as Italy and Germany, they greet each other with air kisses, usually two times alternating cheeks or three times alternating cheeks starting right cheek to right cheek, such as in the Netherlands and Russia. In some countries, the opposite sex do not touch when greeting. In Japan and Korea, they greet one another with a bow. Many Muslim women do not shake hands with men due to religious considerations regarding a woman being touched by a man outside of her family. If you are female, do not reach for the hand of a practicing Muslim male unless he initiates it; this is the same with Orthodox Jewish males.

Gloves - If wearing gloves when you're introduced or you meet someone, the general rule is to remove your right glove before shaking hands. When unexpectedly introduced to someone who is already extending a hand, slip your glove off quickly—if you can. Otherwise simply say, "Please excuse my glove," and shake hands with it on. Alternatively, if your arms are full and you can't get your glove off easily, or you're outside in the freezing cold, it's understandable that you would keep your glove on. Of course, you could slip your glove off and say, "Excuse my cold hand; remember it's cold hand, warm heart." Nevertheless, the *normal situation* calls for men and women to shake hands without gloves—to feel the warmth of the hand, rather than the cool cloth or leather of the glove.

Receiving Line - Those standing in receiving lines should consider not wearing rings on their right hand, or at least consider removing them while shaking hands. It only takes a few bone-crushing handshakes to understand why.

HOW TO REMEMBER NAMES

> **BLUF**
>
> "Mnemonic devices are easy and fun ways to remember names, i.e., She had on a blue dress, blue like the sky, S - her name was Sarah."
> *MAJ Cassandra Perkins*

Remembering names doesn't come easily; it takes concentration and effort, but it's not impossible for anyone. To young spouses, it sometimes seems that older, more experienced Army spouses are able to remember the names of everyone they meet. While this isn't necessarily true, there are people who have learned how to aid their memory. Anyone can master these techniques, as they are not difficult. Of course, they only work if you put them into practice.

1. *Listen carefully.* You can't remember the name if you haven't really heard it. For a variety of reasons, listening carefully is the hardest skill to master. During introductions, human nature pushes us to concentrate more on the sound of our own name, the appearance of the other person, and how we are coming across to the other person. We often spend valuable time during the introduction trying to think of something appropriate or witty to say. Instead, *use this time to concentrate* on listening for that name. If you don't hear the name clearly during the introduction, ask for it to be repeated. Don't let that embarrass you. In fact, the other person will be *flattered* that you care. If it's an unusual name, you may even have to ask how it is spelled. Whatever it takes, get that name!

2. *Remember the name briefly.* Do that by using one or more of the following techniques:

- Visualize the name written; let your mind see it.
- Make a quick association. Immediately try to think of someone or something to associate with some part of this new name. For example, if you know someone with the same first name, make the association between the two people. Think to yourself, "She has the same first name as my best friend back home."
- Remember the sound, with special attention to the first syllable. Sometimes recalling the first letter or syllable is enough to help you remember the whole name. This is especially useful when trying to remember unusual names.

- Tell yourself to concentrate on that name until you can write it down. That may mean you will have to re-visualize it or say it silently several times. Let your mind keep repeating the name, even while the conversation continues, so that you won't forget it.
- If possible, say the name aloud. For example, during the introduction, you can use the name in your greeting. Later, try to include it in your conversation. By saying the name and hearing it repeated, you are reinforcing your memory.

3. *Write it down.* As soon as possible, write down the name. It also helps to note where you met, something about the individual's appearance, and anything significant from your conversation. Use one or two words for each fact. These notes will help you put the person and the name together, and give you topics for conversation the next time you meet. For example, think how good you would both feel if, after being introduced to a wife and learning that she has a new baby, you could mention them both by name the next time you meet. Without having taken notes after your first meeting, this probably wouldn't be possible.

A small notebook in your purse or pocket is a handy place to keep your notes. You could also enter it into your smartphone if time allows. You will find this is especially useful at a new assignment when you are meeting so many new people at once. Alternatively, someone may give you his or her business card. When that happens, keep the card as your reminder and jot on it the date and occasion when you met. This then substitutes for your notebook or phone. Either way, recording the information is the key.

4. *Refresh your memory.* Periodically re-read your notes, say the name aloud, and visualize the person you are thinking about. If you expect to see the person at a function, re-read your notes just before you leave home.

5. *Use the name.* Once you know a person's name, use it. It makes the individual feel special to be called by name and has the added benefit of reinforcing your memory.

Large Groups - These techniques for remembering names obviously work best in small groups or one-on-one situations. However, don't give up and stop trying when confronted with larger groups. Instead, decide to concentrate on remembering *just a few names*. If you do that every time you are with this new group, soon you will know them all by name, and everyone will marvel at your memory!

Helping Others - You can help others learn your name by saying it slowly and distinctly when introducing yourself. Also, should you meet them again and suspect that they don't remember your name, tell them who you are and where you met. This helps them place you and, hopefully, remember your name the next time.

MAKING CONVERSATION

Conversing with people you've just met or known only a short time may seem difficult; yet, it need not be—especially with military people. There are so many topics of conversations and questions to ask of someone who moves around that you will surely run out of time before you run out of questions. How long have you been here? How do you like it? Where did you live before? What was it like there? Ad infinitum. Isn't it boring to ask and hear those same questions so many times? Yes, a little bit. Nevertheless, the answers are truly interesting, and they allow you to get to know a person.

Certainly, we should all be able to discuss deeper subjects than just a person's last assignment, current unit, the Army in general, or the weather. Other suggestions are: Keep up with what's happening in the world and discuss a recent news item; bring up post or community events; talk about your interests and talents. These are all good topics of conversation about which others can either contribute or express interest. It's better to avoid long stories and potentially controversial topics that might result in an argument or hard feelings. However, any other subject is fair game for a conversation. The easiest way to start a discussion is simply to ask, "What do you think about...?"

ACTIVE LISTENING

Remember that it takes two to create a conversation, the one speaking and the one listening. Don't forget to let the other person do half of the talking. Sometimes in our eagerness to appear interesting, we dominate the conversation. Guard against that, as well as saying too little, which forces the other person to keep the conversation going alone. Pleasing conversation is a balance of give and take.

Being a good listener means that you *really hear* what is being said. Too often we talk with people who only half listen. Their eyes glance around the room instead of looking directly at us. An active listener cares about the other person's viewpoint or concern. It's not necessary to agree with them, but care enough to listen. This requires that you look at the person speaking, appear interested, and concentrate on the conversation.

Occasionally, asking appropriate questions helps the other person know that you are interested and that you want to hear more. This is an especially useful approach when the topic of conversation is one about which you know very little. However, don't let asking questions substitute for contributing ideas and opinions of your own.

Shy and insecure people often think that what they have to say wouldn't be of interest to others. They may be good listeners but are reluctant to join into conversations. If you are in that group, know that it isn't hard to change. Start by being an active listener, then join in at an appropriate time with a humorous or personal experience of your own. You will soon see that others enjoy your contributions just as much as you do theirs.

COMPLIMENTS

There is an art to giving and receiving compliments.

Giving Compliments - Compliments should always be sincere. This is best accomplished by being specific; this makes your words seem more meaningful and genuine. For example, "What a lovely dress! That shade of blue looks wonderful on you." When complimenting someone on long hours of volunteer work, you might say, "Bob, your work with the Family Readiness Group has been invaluable." If nothing specific comes to mind, don't let that deter you. A compliment is a good way to put a newcomer at ease.

Receiving Compliments - Compliments should always be accepted gracefully. A simple smile and an acknowledging comment about the compliment are all that are required. A response such as, "Thank you, how nice of you to notice," or "I really appreciate your saying that—the children are very special to me," or, "You've made my day!" makes the person who gave the compliment feel good in return. Never feel embarrassed by a sincere compliment, even if you don't think you deserve it. If you protest too much, you embarrass the person giving the compliment. Simply enjoy the compliment and acknowledge it with a smile and a few words of thanks.

* * * * * * * * * * * * *

Everyone can become poised and self-confident when meeting new people – because the required skills are learned, not natural-born.

Chapter 2

Introductions

"Social customs are to society what habits are to the individual."
Dorothy Stratton

Almost as often as we meet new people in our military way of life, we are confronted with having to make introductions. Making those introductions in the correct manner is a simple courtesy to those involved. Yet, most of us find it difficult to remember the proper procedure for introducing people. Compounding that is the frustration of trying to remember everyone's name in the haste of the moment. Believe it or not, it isn't hard to make introductions correctly, and it's a skill that each of us can and should master.

INTRODUCING TWO PEOPLE

The rules governing introductions have evolved in order to show respect to women, maturity, and position. When introducing two individuals to each other, one person is usually considered to take precedence over the other, and this *honored person's name is stated first.*

Sex—State the woman's name before the man's name.

Age—State an older person's name before a younger person's name.

Position (or rank)—State a senior person's name before a junior person's name.

There are two forms of introduction in common use today: the standard form and a simplified version. The standard form uses the traditional, double-introduction format—"Mary, may I present John Jones? John, this is Mary Smith." The simplified form uses a single-introduction format, with a few preliminary words—"I don't believe you two have met. Mary Smith, John Jones."

Decide which method of introduction you feel more comfortable with, the standard or the simplified form, and put it into practice. In formal or diplomatic situations, it is more appropriate to use the standard form. However, if this double-introduction format confuses you or makes you feel uncomfortable, use the simplified form. What's important is the principle of showing respect for the sex, age, or position of those involved by *always* naming the honored person first.

Complications can occur when the two people you are introducing are both in the same category, or when there are several considerations. For example, suppose you want to introduce a man and woman, but they are both in the military and he outranks her. The answer to this and all complicated situations is simple—apply common sense. If the situation just described occurs in a military setting, the importance of rank prevails; the senior male is named first in the introduction. However, with the same individuals in a purely social setting, the importance of the female prevails; the young woman is named first (unless she's very young and he's very senior).

Sometimes it helps to add a bit of identifying information when you're introducing people. For example, "Mary is my next-door neighbor." This background can provide a ready topic for the ensuing conversation. However, it's better not to introduce someone as "my friend," as this might imply that the other individual is not your friend.

GROUPS

Small Groups - Forget all the considerations of sex, age, and position when introducing one or two people to a small group. The sensible procedure in such a situation is: (1st) name all the people of the group, logically progressing from one side to the other, (2nd) say the name of the newcomer. The rationale for this approach is that by first saying the names of the individuals in the group, you get their attention so they will be listening when you say the name of the newcomer.

Large Groups - Introducing one or two people to a large group generally arises when you're having a big party and new guests arrive. It isn't necessary to introduce the newcomers to everyone, certainly not right away. Simply take your new guests to a small group and make the introductions, following the rule for making introductions to a "small group." Later, if you have time or see the need, you can take these guests to another group and introduce them there. However, most guests will circulate and introduce themselves. If a large group arrives at the same time, simply bring them into the party, smile pleasantly and say, "I'll let you introduce yourselves."

YOURSELF

Nothing could be easier than introducing yourself because, unlike other introduction situations, you're not likely to forget your own name. Also, there are no rules of introduction to worry about. So don't be shy—step right up, smile, look the person in the eyes, say "I don't believe we've met," extend your hand for a handshake and introduce yourself, "I'm Ann Crossley." You'll discover that shyness melts away as you concentrate on the other person, make a new friend, and continue that wonderful military tradition of friendliness.

If you ever find yourself in the potentially embarrassing situation of having someone say that they've already met you just after you have introduced yourself, simply smile and say, "How nice of you to remember me." This is the perfect lead-in to a conversation about where you first met this person who has such a wonderful memory for names and faces.

Conversely, if you get a puzzled look from someone whom you've met before when you say "Hello," help that person remember you by mentioning where you met, or perhaps a mutual friend that you both have. If you have difficulty placing a name with a vaguely familiar face, simply say, "I'm sorry—I'm sure we've met, but I can't remember where." For those occasions when you remember where you met but can't remember the name, you might say something like, "I believe we met at the parade last month. I'm Carol Keller."

FAMILY MEMBERS

The three considerations in making introductions (sex, age, position) are ignored when introducing a member of your own family. Courtesy demands that the other person is always given precedence. Therefore, the other person's name is mentioned first, before your family member's name, as in "General Jones—my mother, Mrs. Brown."

When introducing your husband or child to someone who already knows your last name, it isn't necessary to say the last name (unless you go by your maiden name). Simply say, "Mrs. Smith, I'd like to introduce my husband, Bill," or "Sergeant Major Green, this is my son, Jack."

RESPONDING TO INTRODUCTIONS

Respond immediately to an introduction with a smile and a handshake. Though you need not repeat the person's name, it may help you remember it. It's easy to include the name by saying, "How do you do, Betty," or "I'm pleased to meet you, Betty," or just "Hello, Betty." When someone introduces himself or herself to you, respond similarly and include your own name. If you neglect to give your name, they are perfectly correct to follow up courteously with, "I don't believe I know your name."

STANDING

During Introductions - "Should I stand or not?" is often a question in an individual's mind when being introduced, because etiquette on this point is changing. American protocol of who should stand and who should not stand once followed much the same rules as introductions. The determinants were sex, age, and position. Women remained seated except for greeting older women or much more prominent people. Men usually stood to greet everyone. However, in today's American society, both civilian and military, we are beginning to follow a different guideline.

Today, it is considered good manners for everyone—regardless of age, rank, or position—to stand when being introduced. Body language is important; if you remain seated when a person is introduced to you, the message sent is that you don't care. European custom, equality of the sexes, and our more-relaxed approach to manners have combined to bring about this change. However, as it would be awkward for everyone in a large group to stand when a newcomer is introduced, this new etiquette usually applies only when in a small group.

Whichever tradition you prefer and follow, there are a few situations in which the old-style manners should always be observed. They are:

Men should stand when being introduced to anyone.

Men and women should stand when being introduced to older women or very prominent people.

Children should stand when being introduced to adults.

Host and/or hostess should stand to greet and bid farewell to a guest.

Everyone should stand when being introduced by a speaker from the podium. (In very informal settings, it is acceptable to remain seated and simply acknowledge being introduced with a wave.)

At Official Gatherings - Army life presents several situations that require everyone to rise:

When the senior officer walks into the room of an *official gathering and is announced* (e.g., promotion ceremony, awards ceremony, or auditorium where he or she is to speak), all those present should rise. This includes not only the military members, but also family members, civilians, and even the spouse. This is done as a courtesy to the officer's rank and position.

At official ceremonies such as parades, everyone (military and civilian) should stand for the following (listed in order of occurrence): "Ruffles and Flourishes" and "The General's March," all national anthems and "To the Colors," any invocation, as the American flag and those of other nations pass in front of you, and "The Army Song."

At informal ceremonies (such as promotions, retirements, awards, or changes of command held indoors), military members must stand, and civilians should stand out of courtesy, if the announcement, "Attention to orders," prefaces the reading of the orders or award.

STANDING TO TALK

Another situation that calls for an Army spouse to stand is when a very senior spouse or a much older person walks up to talk while you are seated. In the past, this courtesy used to be carried to the extreme; when the senior person entered the door at a coffee, everyone would stand up. However, today, we simply follow the courtesy of standing when such a person approaches and begins a conversation with us. In fact, with the new equality approach to etiquette, it is considered gracious to rise when anyone—male or female, young or old—approaches to talk with you and you are seated. This courtesy results in everyone being on the same level, without one having to bend over to talk. Thoughtfulness toward others is the key to good manners.

* * * * * * * * * * * * *

Every introduction is the beginning of a potential friendship;
do it correctly and you'll create the perfect beginning.

Chapter 3

Names, Ranks, and Official Titles

"I've learned that people will forget what you said, people will forget what you did, but people will never forget how you made them feel."
Maya Angelou

An important aspect of making introductions correctly and addressing others in an appropriate manner is the proper use of names, ranks, and titles. However, since some occasions require more formality than others, what is correct for one situation may not be correct for another. The guidelines for different situations discussed in this chapter may be helpful.

INTRODUCING OTHERS

- Use the names the newly introduced people are likely to use with one another. If you know "Mrs. Brigadier" well and call her by her first name, it's fine to introduce her to other generals' spouses as Betty Brigadier. However, if you introduce "Linda Lieutenant" to her, then she is introduced as Mrs. Brigadier. If she wants to say to Linda, "Please call me Betty," that's her prerogative.
- Always use complete names when introducing people. It should either be Mrs. Smith or Sally Smith, but never just Sally.
- In normal social settings, it is not appropriate to use your spouse's rank or title when introducing or referring to him or her. If you feel that some clarification might be necessary in order to avoid embarrassment for the other person, mention your spouse's job or position. For example, "This is my husband, Charlie. He's the battalion command sergeant major."
  Certainly, you should *never* refer to your spouse in conversation by his or her rank, or rank and name. For a spouse to say about their soldier, "The Colonel said ..." or "Sergeant Major Stripes said ...," is considered pompous. Use the name you normally call your spouse or, if that's too casual (or personal) for the situation, simply refer to him or her as "my husband or wife."
- When introducing your spouse or child to someone who already knows your last name, it isn't necessary to repeat it. Simply say, "This is my wife, Suzy," or "This is my daughter, Margaret." Logical exceptions would be: If you kept your maiden name, it would be important when introducing your spouse to say the last name distinctly; or if your daughter is married, you would want to introduce her using her first and last name.

YOURSELF
- As a general rule, you should introduce yourself with the name you want the other person to call you.
- On the other hand, when a senior person introduces himself or herself to you by using a first name, you aren't automatically entitled to use that name. Courtesy demands that you wait until a senior person invites you to call him or her by the first name before doing so.

CHILDREN
- When introducing a child to an adult, use the adult's title (or rank) and last name. If the adult and/or child's parents prefer for the child to be on a first-name basis with the adult, that is their choice.
- When introducing yourself to a youngster, give the name you want the child to use in addressing you. An exception might occur with an older teen or young adult. For example, if you think they should call you Mrs. Brown, but feel that introducing yourself that way is too formal, give your first and last name. Hopefully, the young person will know enough not to take this as a signal to call you by your first name, unless you specifically offer.
- "Miss (first name) or Mr. (first name)" is another respectful way an older teenager or college student may address an adult if their parents are family friends and all agree.

INTRODUCING AND ADDRESSING MILITARY
- When introducing a military person in his or her *official capacity*, the rule is to use the full rank, as it is written. However, Army spouses normally make introductions in social situations, and the rule changes.
- When introducing a military person in most *social situations:*
  - First and second lieutenants are introduced and addressed as "Lieutenant," lieutenant colonels as "Colonel," and all general officers as "General."
  - Warrant officers are introduced and addressed as "Mr." or "Miss" or "Chief." If a female warrant officer is married, she is introduced and addressed as "Mrs." or "Ms." or "Chief."
  - Command sergeants major and sergeants major are introduced and addressed as "Sergeant Major." (Note that the plural of sergeant major is sergeants major, not sergeant majors.)
  - A complete list of ranks is provided in Chapter 36, "The United States Army."
- When introducing military medical doctors, it is proper to use their military rank. However, in general conversation they may be called "Doctor," unless

they are assigned as hospital commanders or are general officers. In those cases, they are always referred to by their rank.
- Military chaplains are introduced by their rank; then their position should be mentioned. Once introduced, chaplains are always referred to as "Chaplain."
- Retired military are introduced using their military rank. It's appropriate to add a comment telling from which branch of service the individual retired. Even in civilian settings, a retired military person should be introduced using his or her rank, because military rank is a professional title that is retained, just as a civilian doctor retains his or her title after retirement. Other titles that are retained are governor, ambassador, and justice of the Supreme Court. Retired judges are also introduced using their former title as a courtesy.

DIPLOMATIC AND CIVIL DIGNITARIES

Occasionally, we are privileged to meet high-ranking diplomatic and civil dignitaries. On those occasions, we need to know the proper form of address customarily accorded those officials, both to honor their positions and to feel confident we are using their correct titles. In the following chart, formal introductions have been omitted, because Army spouses do not typically make such introductions of senior officials. (Written forms of address are discussed in Chapter 8, "Addressing Envelopes.")

Position	Introduction (other than formal)	Spoken Address
American Ambassador	The American Ambassador,[1] Mr. Doe	Mr. (or Madame) Ambassador (on leave, Mr. or Mrs. Doe)
U.S. or State Senator	Senator Doe	Senator Doe or Senator
Governor	Governor Doe or The Governor	Governor Doe or Governor
U.S. Representative	Congressman Doe Congresswoman Doe	Congressman Doe[2] or Mr., Miss, Mrs. Doe
Secretary of the Army (Air Force, Navy)	The Secretary of the Army Mr. Doe	Mr. Secretary or Mr. Doe
American Consul	Mr. Doe	Mr. Doe
Mayor	Mayor Doe	Mayor Doe, Madame Mayor, or Mr., Miss, Mrs. Mayor

[1] In Latin American countries, it is preferable to use the more correct title of The Ambassador of the United States of America.
[2] It is not incorrect to introduce or speak to a congresswoman using the title of Congressman.

USING FIRST NAMES

Knowing when it is permissible to call someone by his or her first name seems fairly easy at first glance. As a general rule, an Army spouse automatically calls by their first name: (1) all contemporaries, meaning all military with the same rank as our soldier and their spouses, and (2) all spouses whose soldiers have a lower rank than our soldier. Senior military should be called by their rank and last name. Junior military should also be addressed by the proper title of rank and last name, unless you get to know them well on a social basis. The spouse of someone senior should be addressed by the title of "Mr." or "Mrs." and their last name, unless the spouse invites you to call him or her by their first name.

Complications arise as follows:

Promotion of Contemporaries - One of your spouse's contemporaries, with whom you've been on a first-name basis, gets promoted. What do you call the couple now? The recommended solution is for you and your soldier to continue calling the spouse by his or her first name, but call the newly promoted soldier by the new rank and last name in public. If he or she says, "Why aren't you calling me (first name) anymore?"—then you know it's appropriate to return to a first-name basis. (The exceptions would be family members and very close friends, with whom first names would continue to be appropriate.)

What to Call "the Boss" - Your soldier's commander insists that the spouses use his or her first name. All of the soldiers, of course, are calling the commander by rank and last name, or "Sir/Ma'am." You don't like the idea of calling the boss by his/her first name when your soldier doesn't. What do you do? You can thank the commander, then explain that you feel uncomfortable using his or her first name. Hopefully, this will be sufficient. However, if the commander still insists on first names, but you feel that you can't address him or her so informally, try to simply avoid using the first name.

What to Call the Senior Lady/Gentleman - You're a young spouse who was brought up to call your elders by their titles. The senior spouse, Mary (John) Doe, says they want everyone to call them Mary (John), but you really can't bring yourself to do that. My goodness—they are as old as your mother (father)! What can you do? Continue to call her Mrs. (Mr.) Doe. If they ask you in private to please call her Mary (John), explain that you just don't feel comfortable calling them anything but Mrs. (Mr.) Doe. They'll understand.

What to Call the Senior NCO's Spouse - What do you call a senior NCO's spouse who is considerably older than you, if your spouse outranks the senior NCO? Respect for the senior NCO's spouse's age and position in the community should take precedence over the soldier's rank. Call him or her "Mr." or "Mrs." unless you are invited to do otherwise.

First Names in Private Only - When very senior military members or their spouses ask you to call them by their first names and you do, it doesn't mean that you should do so in all situations. Reserve this informality for private occasions and those in which everyone is on a first-name basis. When in public together or when you are referring to them in public, revert to the more formal rank or title and last name.

Speaking about Someone - Speaking with (or writing to) a senior spouse about their soldier (for example in a thank-you note), when you're not on a first-name basis *with their soldier,* can seem awkward. However, by referring to the soldier sometimes as "your spouse or your husband/wife" and other times by the rank and last name, you can avoid the repetitiveness of using only one.

SIR AND MA'AM

Many Southerners are taught as children to say "Sir" and "Ma'am" to their elders; in the Army environment, military members use these terms of respect to address any officer who outranks them. So when does an Army spouse stop using "Sir" and "Ma'am," or is there ever a situation when that can happen? Certainly, as a very young person, you can use these terms when speaking with someone much older than you. However as you mature, there comes a time to drop these terms. Don't be misled into thinking that you should continue to address senior officers as "Sir" and "Ma'am" just because the military members do. Senior officers recognize these as terms of respect from lower-ranking military members, but do not expect anyone else to call them "Sir" or "Ma'am." In place of these terms, a spouse should use the individual's name instead. For example, rather than saying "Good morning, Sir/Ma'am," as your soldier might say, you should say "Good morning, General Smith."

ALL ABOUT NAME TAGS

How nice it is to walk into a room and know everyone's name! But that doesn't happen very often to Army spouses. It seems as though every time we just get to know the names of all those in our coffee group, our soldiers come home with orders. So what can we do to learn one another's names faster and ease the embarrassment of forgetting someone's name? Use name tags!

Name tags come in all shapes and sizes: some homemade, some purchased. They are all wonderful—because of the useful purpose they serve. The main thing to remember about name tags is that the name must be printed large and clearly enough to be read at arm's length. Suggestions: (1) Print! Don't write in longhand; it's harder to read. (2) Use large, dark letters that can be read from a distance. A small felt-tipped pen is perfect for writing on name tags if it's a casual event; a broad-tipped ink pen can be used to make more graceful lines for formal occasions. (3) Printers can make printing name tags for large groups a breeze. You may even be able to use a software program that will create a calligraphy script; just be sure to use a very large, bold font.

Press-on or clip-on name tags are preferred to the pin-on type because they are easier to put on. If you are presented with a pin-on name tag, you may find it better to pin it into a seam for greater support, if that's possible. When attending a function where you know you will be wearing a name tag, it might help to plan your apparel accordingly. Normally, name tags are not worn for formal evening functions.

What name should be used on the name tag? That depends on the formality of the event. Starting with the most formal function and progressing to the most casual, a married person's name tag could read Mr. or Mrs. John Doe, Mr. or Mrs. Doe, John/Mary Doe, or simply John/Mary. If titles are being used on the name tags, use Ms. for a married woman who has retained her maiden name. With today's informality, name tags with only first and last names are appropriate for most spouse social functions. For military members, use the person's rank (abbreviated), preferred first name, and last name on the name tag.

Susan Davis *LTC Derek Davis*

All things being equal, the name tag is better worn on your right side. The rationale for this advice is that this placement allows others to steal a quick glance at your name tag as their eyes follow the path from your handshake to your eyes.

* * * * * * * * * * * *

Introductions are made to bring people together,
not to put them in their place.

SECTION TWO: THE ART OF COMMUNICATION

Chapter 4

Correspondence

"Think of letter writing as conversation put on paper."
Amy Vanderbilt

America remains a nation of letter writers—even with today's more frequent use of telephones, e-mail, texts, and instant messaging. In 2017 alone, the United States Postal Service handled 18.5 billion pieces of first-class, single-piece, stamped mail that includes personal correspondence, letters, and cards.

Developing your skills as a correspondent is important for both personal and practical reasons. When you put something into written form, you are sharing your thoughts with someone in a lasting way. You want the result to convey what's on your mind as accurately and effectively as if you were there to speak to the person face to face. The appearance of your correspondence is also important, for it is a reflection of you as a person. Therefore, not only should you choose your words with care, but also your stationery, the color of ink, and the format of your correspondence. The information in this chapter is provided to help you with those choices.

STATIONERY

Stationery comes in a wide assortment of colors, designs, sizes, and paper quality. You will probably want a variety to meet your different correspondence needs. Everyone needs some conservative stationery—white, beige, or pale gray—for business correspondence. This is usually the larger size stationery, approximately 7 by 11 inches. If you happen to be a conservative person, this size might also double for your personal letters as well, although social stationery is usually smaller, approximately 5 by 7 inches. You may prefer something a little more interesting for social correspondence, and there are a multitude of colors and designs from which to choose. Additionally, you will find it useful to keep note cards on hand. These, too, come in a wide variety of colors and designs, and are perfect when you only need to write a brief message.

The ink color you use is equally as important as the stationery itself. For business letters, write in black or dark blue ink, type, or use a laser-quality computer printer. For social letters and brief notes, the color of ink used may be influenced by the purpose of the correspondence. For the light-hearted social letter or note, select any color that matches or complements the stationery. However, for more serious social letters, black or dark blue would be more appropriate. For letters of condolence and written responses to formal invitations, black ink should always be used.

PERSONALIZED STATIONERY

Ordering personalized stationery or note cards was once a luxury few could afford. However today personalized stationery is both fashionable and affordable. In the event you want to purchase personalized items, these guidelines for the proper use of names and initials will be helpful.

- *Full name:* Either Mr./Mrs. John Doe or simply your first and last name is appropriate. You may prefer to have small amounts of both types printed to provide additional flexibility to your social correspondence, since the first is more formal than the second.
- *Single initial:* Traditionally, women use the initial of their first name (or the name they go by); men use the initial of their last name.
- *Two-initial monogram:* Use the initials of your first name (or the name you go by) and last name.
- *Three-initial monogram*: First, middle, and last name initials in order, the same font size. By tradition, a married woman usually uses the first initial of her first name, maiden (or middle) name, and married name. They also appear in the normal order, when the initials are printed all the same size.
- *Three-initial monogram with large center initial:* The large center initial should be that of the last name.

INFORMALS

Informals are small cards that can be used for a wide variety of purposes: invitations, reminders, postponement or cancellation of invitations, written acceptances or regrets, gift enclosures, thank-you notes, and other brief notes. These multipurpose cards are usually fold-over, but can be flat. Informals usually measure from 3 by 4 inches to 4½ by 6½ inches. They can be ordered personalized with your monogram, name, or both your and your soldier's names (joint informal). You can also select from the vast assortment of plain, embossed, or decorated informals sold at department stores, stationery stores, or online. For a summary of uses and correct wording of informals personalized with your name, see the last two sections of Chapter 5, "Business and Social Cards."

Flat informals are made of heavier card stock paper than fold-over informals and are usually slightly larger. Because of their larger size, they are not used as gift enclosures.

Informals, like stationery, are always sold with envelopes. This makes them very convenient for mailing; however, be aware that some of the small informals may not be appropriate for mailing. Postal regulations require envelopes to measure at least 3½ by 5 inches.

Where you begin writing on fold-over informals varies. On personalized fold-overs, the invitation or message is always written on the inside because of the monogram or name centered on the front page. However, on nonpersonalized fold-overs, the appearance of the front page determines where to begin writing. Write on the front page if it is perfectly plain, or has a small border, or an initial or pattern in one corner. Some people mistakenly think they should never use the front page of fold-over cards, but it is perfectly correct to do so—if there is room to write. Only when a name or decoration is centered on the front, or the decoration is so large that it's impossible to write there, should you begin writing on the inside. Flat informals are always designed so that there is room to write on the front.

When writing a note on fold-over informals, the normal sequence of pages is as follows:

When you can't write on the front

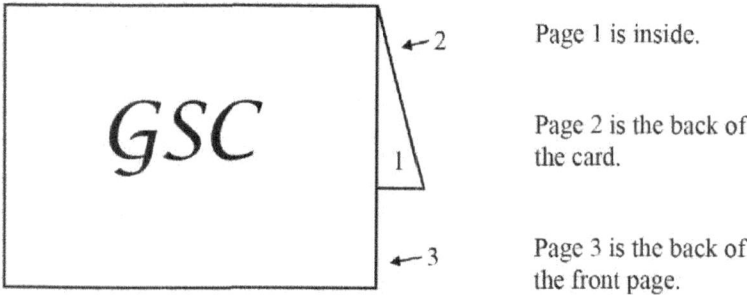

Page 1 is inside.

Page 2 is the back of the card.

Page 3 is the back of the front page.

When you can write on the front

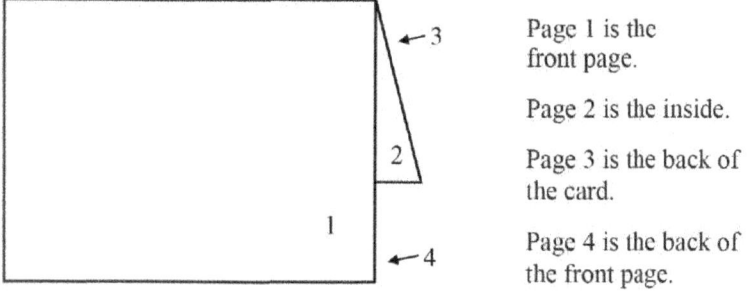

Page 1 is the front page.

Page 2 is the inside.

Page 3 is the back of the card.

Page 4 is the back of the front page.

This rather convoluted page order is followed because informals are normally used for very *brief* notes or invitations, and seldom does the writing spill over to the fourth page. For those occasions when you know in advance that you plan to write a lot, you may write on the front (if the design of the informal allows), then open the note flat and write straight down from the top to the bottom, thus making the back of the note the last page. If you can't write on the front and know you plan to write a lot, you can open the note flat and write straight down from the top, again making the back of the note the last page.

Insert the informal into the envelope so the card will be right-side up and facing the reader when the envelope is opened. (The reader is looking at the back of the envelope.) This same rule, right-side up and facing the reader, also applies when placing letters and greeting cards into envelopes.

SOCIAL AND BUSINESS CORRESPONDENCE

Personal correspondence can be divided into two categories: social and business. Social correspondence includes personal letters, notes, and everything that deals with entertaining, from invitations to thank-yous. Business correspondence is just what the name implies. The differences between the two, besides the nature of the correspondence itself, are: the format, names used, and signature.

Social Correspondence

Social correspondence may include a return address on the upper right corner of the letter, but usually it is omitted. The date can appear on the top right corner of the first page or on the left side of the last page, just below the level of the signature. The month and day can be completely stated—January 13—or simply the day— Monday, or Monday, 13th. A prompt thank-you note dated "Monday," for an event you attended on Sunday, needs no further clarification. Brief notes often have paragraphs consisting of only one or two sentences. Names used on the envelope of social correspondence do not need to include the middle name or initial of either the sender or addressee. Some social correspondence may be typed; however, informal invitations, replies, thank-yous, and condolence notes should be handwritten. When social correspondence is typed, the envelope should be typed as well.

May 20

Dear Sally,

Bob just came home with the good news that we're moving to Washington next month. I can hardly wait to see you and Tom again.

Maybe we can even have Thanksgiving dinner together, just like last year – only this time you cook the turkey!

We don't know exactly when we'll arrive, but I'll call as soon as the date is set.

*Love,
Mary*

*Mrs. Robert Smith
180 A Avenue
Armytown, GA 55555*

*Mrs. Thomas Turner
909 Beltway Road
Alexandria, VA 88888*

Business Correspondence

Business correspondence can be divided into two categories: personal business and regular business. Personal business correspondence usually relates to family business matters, social activities, school/unit fundraising, or other community work for which you are more likely to be identified as someone's spouse or parent. Regular business correspondence is normally conducted for your own professional business reasons, such as employment or legal matters, in which you wish to be identified as an individual. These two types of business correspondence usually differ in format and in the signature block. Of course, the content of any letter is of primary importance.

Writing a Good Business Letter

Whether you are writing a personal business or regular business letter, your correspondence needs to be professional in both content and appearance. The following guidelines will help you organize your thoughts and write in a clear, concise manner:
- Before writing the rough draft of the letter, list the major points you want to make.
- When answering a letter, keep it and your list of major points in front of you as you write.
- State clearly in the first paragraph why you are writing.
- Write in the active voice (e.g., I will, we can), give specifics, and make it sound natural.
- The last paragraph should remind the reader what you want him or her to do, or what you're going to do.
- Include contact information: i.e., email address and telephone number.
- Edit ruthlessly.

Personal Business Correspondence Format

The usual format for the personal business letter is the modified block or the modified semi-block. In this, the return address, date, complimentary closing, and signature block appear lined up on the right side of the page. The name and address of the addressee, salutation, and any enclosure notation appear on the left margin. The difference between the modified block and modified semi-block is whether or not the first line of each paragraph in the letter is indented. In this type of correspondence, a married woman, if she wishes to be identified by her title of Mrs., should use in the signature block her first and last name (and sign the letter in this manner).

Example of personal business correspondence
(modified semi-block)

                                                      1408 Post Road
                                                      Fort Lewis, WA 66666
                                                      May 10, 2018

Best Home Realty Company
Attn: Mr. Samuel Smith
801 Main Street
Atlanta, GA 22222

    Re: 7602 Rolling Hills Road, Apt. 4

Dear Mr. Smith:

    Thank you for your prompt reply concerning the status of the lease on the property listed above.

    We are pleased to learn that the present tenants are interested in renewing their lease for another year. My husband and I do not expect to return to the Atlanta area to live until next July, so a one-year lease extension is perfect for us.

    Please prepare the necessary lease extension agreement for our signatures, keeping the rent at the same level it has been for the past year.

    Your assistance in this matter is greatly appreciated. Your property management fee for June is enclosed.

                                              Sincerely,

                                              Janet Jones

                                              (Mrs. Jason E. Jones)

1 Encl.

    Mrs. Jason E. Jones
    1408 Post Road
    Fort Lewis, WA 66666

                                          Best Home Realty Company
                                          Attn: Mr. Samuel Smith
                                          801 Main Street
                                          Atlanta, GA 22222

Example of regular business correspondence
(block format)

1408 Post Road
Fort Lewis, WA 66666
January 13, 2018

Marietta School Board
Hiring Committee
1075 Peachtree Road
Marietta, GA 22223

To Whom It May Concern:

This letter is a request for information on the availability of openings for reading teachers in the Marietta elementary schools for the upcoming school year, 2019-20.

My résumé is enclosed for your review. As you will see, I have ten years' experience in the field of elementary education and have specialized in reading for the last three years.

Although presently living in Washington state, I will be moving with my family to the Atlanta area next summer. My expected arrival date is 1 July, 2019, and I could be available for an interview any time after that date.

Your prompt reply wi~~ll be greatly appreciated.~~

Sincerely,

Janet Jones

Janet S. Jones
206-xxx-xxxx
janetj@gmail.com

Ms. Janet S. Jones
1408 Post Road
Fort Lewis, WA 66666

Marietta School Board
Hiring Committee
1075 Peachtree Road
Marietta, GA 22223

Regular Business Correspondence Format
The format for the regular business letter follows that of the personal business letter except that, for convenience, it is normally written in the block format with *everything* beginning at the left margin. In this type of correspondence, use in the signature block your first name, middle or maiden name or initial, and last name. On the envelope, the title of "Mr." or "Ms." is added in the return address, unless you have a professional title. See the preceding example.

SOCIAL TITLES
Ms.
"Ms." can be used by or for any woman (single, married, or divorced) in conjunction with her own name. For example, in the personal and regular business letters just shown, Mrs. Jason E. Jones is also Ms. Janet S. Jones. When you don't know how a woman prefers to be addressed, "Ms." remains a safe way to address any correspondence.

Miss and Master
Traditionally, the proper titles to use when addressing social correspondence to young children and teens depended on their gender and, for boys, their age. Today, a young girl can be a "Miss" from the time she is born. Since "Miss" is not an abbreviation, it is not written with a period. A young boy used to be addressed as "Master" until he was between six and eight years old. From then until after high school, he was addressed without a title. Today, a young man of 18 and older may be addressed as "Mister," which is always written as "Mr."

SALUTATIONS, SIGNATURES, AND SUFFIXES
Salutations
Use the same name in a letter's salutation as the one you use when speaking to that person. If you're on a first-name basis, then "Dear Margaret" is appropriate. For someone you have just met or with whom you are more formal, "Dear Mrs. Smith," "Dear Ms. Smith," or "Dear Margaret Smith," would be better. If the person to whom you are writing a business letter has a professional title, it should be used, e.g., "Dr. Smith."

For those occasions when you do not know the name of the individual who will read your letter or when it is being addressed to a group, the two safe bets are "To Whom It May Concern," or name the group in the salutation, such as "Dear Board Members."

Signatures
Married women often have problems knowing how to sign their names. Should they use their own name without a title, or use Mrs. followed by their spouse's name? The rule is really quite simple. A married woman signs her name *without a title,* using her first name, possibly her middle or maiden name (or initial), and spouse's last name (as in Jane Doe, Jane Marie Doe, or Jane Jones Doe). This

signature is used in almost all circumstances, even for signing personal checks.

A working woman who established a business reputation when she was single may prefer to continue working under her maiden name after she marries. In that case, she may find it easier to be consistent and use that name for all of her business, banking, and legal matters. It also makes sense because women in business need to establish credit under their business name.

Joint Signatures - Use joint signatures only for greeting cards, postcards, and gift enclosures. (The general rule for all other types of correspondence is: The person who writes the note, signs the note.) When joint signatures are appropriate, the person doing the writing courteously signs his or her name last.

Suffixes

Junior, II - A man's name may end in *junior* if his name is exactly the same as his father's. Properly written, junior is not capitalized unless it is abbreviated; use either "Mr. David Doe, junior," or "Mr. David Doe, Jr." The abbreviated form is used when it's necessary to shorten the length of the name. A man can become the II only when some close male in his family, other than his father, has the same name—possibly a grandfather or uncle. The younger man usually drops the junior or II suffix when the person originally having the name dies, unless it would create confusion for the widow. Note that any suffix added to a man's name is separated from the name by a comma.

CHANGE-OF-ADDRESS CARDS

Our mobile military lifestyle often results in frequent changes of address, which require us to notify our business and personal correspondents. The postal service has free change-of-address cards that can be used to notify your business and magazine correspondents. To notify family, friends, and other personal correspondents, here are several suggestions:

- Purchase "we've moved" greeting cards and fill in the necessary information.
- Purchase blank postcards from the post office, use return-address stickers with the new address and add a short message.
- Use an online greeting card service (i.e., Paperless Post) to email your new address to friends and family.
- If it's near the holidays, send out holiday cards early and ask your friends to be sure to note your new address.

HOW TO CHANGE YOUR ADDRESS WITH THE POSTAL SERVICE

To let the post office know you are going to change your address and want your mail forwarded to your new location, you have two options:
- Go to USPS.com/move to change your address online.
  - This is the fastest and easiest way, and you immediately get an email confirming the change.

- o   There is a $1 charge to change your address online. You will need a credit or debit card and a valid email address. The $1 charge to your card is an identity verification fee to prevent fraud and make sure you're the one making the change.
- Go to your local post office and request a Mover's Guide packet.
- o   Inside the packet is PS Form 3575. Fill out this change of address form and give it to a postal worker behind the counter or drop it into the letter mail slot inside the post office.
- o   You should receive a confirmation letter at your new address in five business days.

GREETING CARD ETIQUETTE

Most of us enjoy sending and receiving greeting cards for special occasions. However, even with such a common practice, there are accepted rules of etiquette.

- Greeting cards should always include a handwritten message, however brief, in addition to your signature.
- It's nice to personalize the card by writing the recipient's name above the printed message.
- When signing a card for two people, courtesy requires the writer to sign his or her name last.
- When signing a greeting card for the entire family, you may sign everyone's name (in which case the order is parents, then children), or sign "The Smiths" or "From the Smiths."
- Traditionally, greeting cards printed with a couple's names, though seldom used today, should list the wife's name before the husband's, and always include the last name, e.g., Jane and John Doe.
- When sending a greeting card to an entire family, address the envelope to the parents only, not "and Family." Greetings to the children should be expressed inside the card.
- Acknowledge birthday and anniversary cards that you receive. Thank local well-wishers in person, and send at least a brief note or text of thanks for out-of-town cards.

HOLIDAY CARDS AND LETTERS

Because we make so many friends as we move around from one Army post to another, sending holiday cards to all of them becomes difficult. So make the holidays as easy on yourself as possible, and don't try to send cards to all of your local friends. Mail cards only to out-of-town friends and, perhaps, to local friends whom you don't expect to see. Usually, we extend our holiday greetings to local friends and neighbors when we see them during the holiday season or, if we don't see them, we can give them a call or send a text.

It is not appropriate to send religious cards to non-religious friends or friends of a different faith. But you might want to consider sending a simple "holiday greetings" card to those friends. One of the things that gives military life its rich texture is its diversity, and respecting those differences can go a long way.

When a holiday card is received from someone to whom you did not send a card, it is nice to acknowledge it in some way. You can thank them the next time you see them or, if there's time, send a card in return. If neither is possible because of time or distance, you can write a short note of thanks on a New Year's card or note paper.

FORM LETTERS

If you like to send a form letter at holiday time to family members and close friends in place of individually written cards, that's fine. The letter will have the advantage of catching everyone up on your family's activities and achievements. It's especially interesting if you can add pictures of the family so friends will see how you look and how the kids have grown. As the years pass, there is the additional bonus of having your recorded family history. However, great skill is required in blending interesting anecdotes with your chronicle about each family member. Be careful not to make your letter too lengthy. Remember that your goal is to produce a form letter your friends will want to read from beginning to end.

A few, personalized, handwritten sentences should always be added, either at the beginning or end of the form letter, along with your original signature. When preparing the letter, be sure to leave space for this brief handwritten comment and your signature.

Another way to send out holiday greetings to many family and friends is by using email. With the cost of holiday cards and stamps rising every few years, this will save you quite a bit and still have the benefit of keeping in touch and extending your greetings. If you can add family pictures to the email, all the better.

Each year, as the holidays begin and end, the subject of the family newsletter is discussed in newspaper columns, magazines, and at parties. Maligned by some and defended by others, the decision to send a form letter is yours alone.

* * * * * * * * * * *

BLUF

Letters, correspondence, and thank-you notes are *never* outdated

---

Chapter 5

Business and Social Cards

"A handsome card, properly presented, makes a tremendous impression."
Letitia Baldrige

There are several types of personal cards that Army spouses may, at one time or another, be interested in using. In addition to the traditional calling cards, which are seldom used today for their original purpose, there are the more frequently used address cards and business cards. Calling cards, which we will now call social cards, still have a place in Army society for a variety of purposes, for example, as address cards to give out before or after a PCS. As in the civilian world, business cards remain prevalent in the military. All of these personal cards serve the basic purpose of sharing written information about ourselves with others in a very neat, easy-to-read, easy-to-save form. Because calling cards have such a long association with our military society, traditional guidelines for their wording and use have evolved, but appear here more for your general edification and interest. Address cards and business cards, on the other hand, are far less restricted in their format, although observing certain courtesies is helpful. Following is information you will find helpful concerning these types of personal cards, as well as guidelines for ordering them.

CALLING/SOCIAL CARDS
Traditionally, calling cards, sometimes referred to as visiting cards, are small, flat cards printed with an individual's official name and title. In the case of a married woman, that usually meant "Mrs." followed by her husband's full name. These visiting cards were once used extensively for social calls, as an expected part of military society for officers and their wives. This tradition eventually expanded to include senior noncommissioned officers. However, the tradition has faded from practice in today's Army. The two remaining occasions when you might want to leave your card during a social call are: New Year's receptions held in the home, and the first time you and your servicemember are entertained in the commander's home. Although they are by no means necessary, and certainly not expected, they can be a thoughtful, old-fashioned touch on such occasions. Diplomatic assignments may require more extensive use of calling cards, but the correct protocol for each country will vary. If you are interested in finding out what is considered proper in a foreign country you are visiting or to which your servicemember is assigned, the protocol office at the American Embassy can inform you of the local guidelines.

Whether or not you want to have traditional calling cards is a personal choice, but your decision may depend in part on the views of your soldier's current commander. Many, if not most commanders today, see no need for calling cards; others see the retention of the old traditions as desirable and may express the desire for their officers and NCOs to leave calling cards when calling on them. The best way to know what a particular commander prefers is for your servicemember to check with the adjutant or aide.

If you decide as a couple to have calling cards made, you may want to order both individual cards as well as joint cards. You will find that individual cards may be used not only for social calls but for a variety of other purposes. However, remember that your servicemember's calling cards and your joint cards will have to be changed every time he or she gets promoted, and should be ordered in limited quantity. Joint cards are less practical than individual calling cards since their primary use today is for gift enclosures.

Uses Defined

As you will see from the following discussion, social cards may be used for a wide variety of purposes. Although not essential for today's Army spouse, social cards are now quite economical and can add a touch of class to your life.

Social Calls - The most well-known use of calling cards, and doubtless their original purpose, was for social calls. As with many etiquette practices, the meaning and usage of calling cards have evolved over time, and it is helpful to know their origin to appreciate how the practice has progressed and changed. Here are the traditional, rather-stilted guidelines for calling cards as Army spouses of previous generations would have known them:

- The man leaves one of his cards for every *adult male and female* of the host's household (including household guests), up to a *maximum of three cards*.
- A woman leaves one of her cards for every *adult female* of the host's household, up to a *maximum of three cards*.
- A female in the military, invited because of her military position and not as the spouse of the servicemember, uses the same formula as a gentleman.
- Joint cards are not used for social calls.
- A husband usually carries his and his wife's cards, and leaves them either on arrival or departure.
- Calling cards are left (plain, not in an envelope) on a table near the door, in a small tray if one is available there.

Business - Social cards are different from business cards in format and content, but social cards may be substituted for business cards when necessary. Avoid using business cards in place of calling cards for a social call. To use a social card as a business card, simply write in the information desired, such as address and phone number (also see "Address Cards" later in this chapter), using black ink.

Gift Enclosures - This is perhaps the most relevant purpose for traditional calling cards today. Calling cards make perfect gift enclosures because they identify the giver, provide sufficient room for a brief note, are small like a gift card, and will go with any color wrapping paper. The calling card may be "personalized" by striking through the printed name and signing your first name (in black ink). Either individual or joint calling cards may be used for gift enclosures. If you have small, matching envelopes for your calling cards, this is the perfect time to use them.

Example of calling card personalized for use as a gift enclosure:

> Mary,
> With love and
> appreciation for all
> ~~Mrs. Harold Samuel Greenberg~~
> that you've done
> Sally

Friendly Notes - This is another practical 21st-century use for calling cards: Calling cards are very handy to have on hand when you need to leave a brief note. Although it's polite to let someone know before you stop by, it's not always possible. If you do stop by without calling and find no one at home, then what? Of course, a text or instant message would suffice, but calling cards show thoughtfulness and intention. Personalize your card. Jot a brief note, and leave it where the recipient will see it when they return. It's a nice way to leave a message and, if you make a practice of carrying some calling cards in your wallet, card case, or purse, you'll never be without something to write on.

Address, Email and Phone Information - Writing your address, email, and phone number on your calling cards for friends is useful on two special occasions: (1) When welcoming a newcomer with a meal or freshly baked goodie, you can tuck in your calling card after adding your address, email, and phone number, and signing your first name. This will help the newcomer learn and remember your name; recall what you brought (especially useful when several items are delivered at once); know where to send a thank-you note, and how to call with any questions about the community. (2) The second special time to add your address to your calling card is when you are about to move. If you already know your forwarding address, give this information along with your email to your friends on your calling cards. They're neat, small, and very elegant—a splendid way to be remembered.

Example of calling card used as an address card:

> Write to me soon!
>
> Mrs. ~~James Robert Ponds~~
>   Sue
> 185 Old Post Road
> Fort Lewis, WA 98444

ADDRESS CARDS

    The increasing popularity of using the social card for address, email, and phone information led to a new type of card that gave more information. We refer to this as an "address card," since its primary purpose is to show not only the individual's name, but also address, email, and phone number. It can be a unique, personal way to give out your contact information as you meet new people in Army life.

    Address cards are an informal compromise between old-fashioned calling cards and formal or professional business cards. Your address card could include: (no title) your first name, middle name or initial, and last name in the center; your address and phone number in the lower right corner. You may also wish to include your email address and/or social media handles, perhaps on the lower left-hand corner. Whatever information you decide to include, avoid a cluttered appearance. Address cards are more casual than business cards, and they provide your contact information. This is especially useful if you have recently moved and not yet reconnected in the volunteer or job market. As with the traditional calling card, you may "personalize" these cards by striking through your name, writing a brief note on the front (or back), and signing your first name. Avoid using these address cards in lieu of calling cards at social calls, but they may be used for gift enclosures.

Example of a printed address card

>
>
>   Carmen A. Shaw
>
>   6 Texas Road
>   El Paso, TX 79902
> carmen@gmail.com   915-625-xxxx

BUSINESS CARDS

> BLUF
>
> Business cards can be used for much more than putting them in the bowl at the local deli for a chance to win a free lunch. They are a reflection of you that can be used in a multitude of ways.

Business cards are widely used today, not only by professionals, but also by self-employed Army spouses, and even teens advertising such services as lawn care and babysitting. Business cards may be ordered online. They can also be printed on a home printer using the Avery business card forms available at your local office supply store. Those are especially useful because you can print just a few at a time.

When designing your business card, remember that the business name or company trademark should be the featured information, unlike a calling card or address card that feature your name. If you are a professional, you may want business cards that give your own professional title. If you have no professional title, either use no title or use Mr. or Ms., and *your* first and last names. Additional text might include your business address, email address, websites or social media pages (where applicable), and telephone number. However, as with the address cards, use care not to "clutter" this important part of your business image. Here is an example; notice how the business name is featured.

Barbara's Brides
Wedding Coordinator
42 Cannon Drive
Fort Sill, OK 73503

Barbara Bailey   580-433-xxxx
bbbrides@gmail.com

Presenting Your Business Card

Presenting your business card gracefully, in a timely manner, and only under appropriate circumstances are important skills to master for your business image. In a strictly business setting, presenting your card or exchanging cards is usually straightforward. However, social settings present more of a challenge. The following guidelines may be useful:

- Keep a supply of business cards handy in your wallet or card case, so you will be prepared for a private presentation should the occasion arise.
- Use care in giving your card to a stranger. Wait until you are sure that the person is truly interested, or that this is someone you want to have your name and address.
- Avoid giving your card during a meal, especially in someone's home. After the meal is concluded, or as you leave, is a more appropriate time. For example, if Barbara Bailey (depicted on the business card example) attends a dinner and her dinner partner happens to mention that his daughter has just become engaged, she can certainly describe her enjoyable business as a wedding coordinator. Her dinner partner may become interested in her services and ask if she has a card, which she should give to him later or after dinner in a private moment.
- Be careful not to use military functions as an opportunity to pursue business. Army spouses need to be discreet about "talking business" at military social functions, though you may present your card if it is requested. Special care must be taken never to force your card on your servicemember's colleagues and especially subordinates.
- "Personalizing" your business card is similar to personalizing your calling card or address card. Strike through your name and write a brief note on the front or back, signing only your first name. For example, Barbara might send her card to her dinner partner, listing on the back the names and phone numbers of a few of her satisfied customers (with their permission, of course). She might add a note explaining, "Some of my best references—feel free to call them," and sign it "Barbara."

SUMMARY OF USES FOR PERSONAL CARDS AND INFORMALS

	calling cards		address cards	business cards	informals[1]	
	individual	joint			individual	joint
social calls	X					
business	X		X	X		
gift enclosures	X	X	X	X	X	X
notes and messages	X		X	X	X	
invitations					X	X

[1] Informals are discussed in Chapter 4.

Leftover-card Uses

When your servicemember gets promoted and their individual calling cards and your joint calling cards are no longer appropriate, or when your address or business cards become outdated, don't throw them away—recycle them! They have practical uses. The very best use is as place cards at the dining table; fold the cards inside-out lengthwise and, voila, you have place cards. You might score (lightly cut) the plain side along the center, lengthwise, to make them easier to fold. Also, they can be turned over and used for name tags, especially if you have plastic sleeves to slide them into. Use your own creativity to devise other uses for your leftover cards. They are such neat, sturdy cards; it's a shame to waste them.

ORDERING GUIDELINES

The following guidelines will be useful when ordering your personal cards and informals, whether online or at a brick-and-mortar stationery store:

Method of Printing

No longer must personal cards and informals be engraved. Normal printing or thermography is perfectly acceptable and much less expensive. And don't forget the possibility of using your own laser printer and Avery business card templates. However, if you decide to order engraved cards, be sure to ask for your "plate"— the copper plate, or "die," on which your name has been etched. Once you have a plate, it's less expensive to have more cards produced.

Sizes

Most printers know the appropriate sizes, which may vary slightly based on the length of the name. However, these are the norms:

military man's or military woman's calling cards—3¼" by 1⅝"
military attaché's calling cards—3½" by 2"
married non-military woman's cards—3⅛" by 2½'
joint cards—3½" by 2½"
address cards—3½" by 2"
business cards—3½" by 2"
informals—vary from 3" by 4" to 4½" by 6½

Style

White paper with black printing is best for calling cards, informals, and most business cards. However, business cards for certain types of businesses, especially those owned and operated by women, can be very effective in color. The lettering style should be rather simple, though there is a wide variety from which to choose. (A few are depicted in the examples in this chapter.) If both spouses choose to order individual calling cards, the quality of paper and lettering should match, if possible.

Traditional Wording

If you choose to add an old-fashioned touch with calling cards, you may be interested in the traditional wording guidelines. As they are from an earlier time, some of the guidelines do not reflect today's broader societal and cultural changes, and you may find yourself making your own updates where necessary. (Wording guidelines for address cards and business cards have already been discussed.)

- Never use initials. Spell everything out, even the middle name. Note: This does not apply to address cards.
- Rank should be spelled out in full, with the exception that both first and second lieutenants are simply "Lieutenant."
- On joint cards, rank may be abbreviated if the full rank and name would make the line too long.
- "Junior" may be abbreviated to "Jr." if it would otherwise make the line extremely long. When written out, use lower case j (junior) and when abbreviated, use a capital J (Jr.). Separate junior or Jr. from the surname with a comma.
- "II" or "III" should be written in Roman numerals, and separated from the surname with a comma.
- Wives used their husband's full name preceded by "Mrs."—not their own first name. (This did not apply to the very casual informals that a woman might have ordered for her own use on which she would use no title and her own given name, alone or with her last name.)

- Wives who retained their maiden name after marriage were a category not covered by most calling card guidelines. They may have used calling cards with their name and no title; the title of "Ms."; or their professional title, if they had one.
- Chaplains' cards do not show their rank.

Placement of Wording for Calling Cards

A military person places his or her full name in the center of the card with the rank and service placed on two lines in the lower right corner. (A field grade officer may follow this format or place the rank before the name.) According to tradition, a nonmilitary wife's calling card had only one line of print, centered on the card, consisting of her husband's full name preceded by her title of "Mrs." On joint cards, general officers above the rank of brigadier general may deviate from the normal format of using the full name and choose to use only their rank and surname (e.g., Lieutenant General and Mrs. Doe).

A servicemember assigned as an attaché, or who has a job involving civilian businessmen/women or foreign nationals, may need business cards giving more information than described in the preceding paragraph.

Calling Card Envelopes

Matching envelopes, only slightly larger than your calling cards, can be handy to have, especially when you want to use your calling card as a gift enclosure. However, on the rare occasions when you might want to send your calling card through the mail, the envelope must meet the minimum size the post office allows, 3½" by 5". It is acceptable to place your card in a larger envelope in order to mail it, but it's usually easier instead to use a fold-over or informal card. Remember that envelopes are not used with calling cards left during a social visit.

Care

Carry your cards in a card case or side pocket of your wallet to keep them protected. Discard any soiled cards.

* * * * * * * * * * * * *

Calling cards and business cards help others remember who we are.

Chapter 6

Telephone Manners

"Well, if I called the wrong number, why did you answer the phone?"
James Thurber

The telephone is a vital means of communication in today's busy world, and the proliferation of the smartphone makes good telephone manners more important than ever before.

THE LANDLINE—JUST SAY "HELLO"
Telephone etiquette has changed over the past several decades. Once an Army wife, and even the children in a military family, proudly and properly answered their home phone by identifying first their sponsor and then themselves: "Lieutenant Smith's quarters, Mrs. Smith speaking." By the 1970s, the tradition had been modified to a shorter, but still informative, answer: "Mrs. Smith" or "Jane Smith." Unfortunately, unscrupulous people can take advantage of this telephone information.

Today everyone, civilians and military alike, should answer a home landline with a simple "hello." This allows the caller to identify himself or herself first; then, the person receiving the call can decide whether or not to continue the conversation.

Servicemembers, too, should answer their home phones with a simple hello. This is sometimes hard because they always identify themselves as soon as they answer the phone at work. However, in this world of increased crime and terrorism, OPSEC must take precedence over polite telephone manners.

FOR THE CALLER
The name you use to identify yourself when placing a telephone call varies with the type of call you're making. For *social* telephone calls, give your first and last name—even if the person you're calling doesn't use your first name, "This is John Smith calling." For *business* calls, it is traditional to use your title and last name, "This is Mr. Smith calling." However, with the increased informality of our society, you will find that many people today use the less formal "first and last name" for both types of calls, not only to identify themselves but to address the person called as well. If you prefer not to establish such informality in your business relations, it is certainly proper to use the traditional form of identification.

When making social calls, even to a close friend, try to text first, "Do you have a few minutes to talk?" or, "Is this a good time for you to talk?" If you opt to call without checking ahead, ask them the same at the beginning of the phone call. If the answer is no, offer to call back later. This is not only courteous but will save you embarrassment if the person is busy or has another call in progress. (See "Call Waiting" below.)

Callers also need to be sensitive to the daily schedule of the person they are calling. It is not polite to call anyone either very early or very late unless it's an emergency. A good guide is not before 9 a.m. and not after 8 p.m. Also, try to avoid calling at the traditional family times: mealtimes, evenings, and weekends. Thoughtfulness toward others is the best watchword.

If you get the wrong number, apologize before you hang up.

FOR THE ANSWERER

Hopefully, the person calling you is already in your phone contacts, or will identify himself or herself as soon as you've answered "Hello." If they are not in your contacts and do not identify themselves, and they ask to speak to someone else, etiquette suggests that you not ask the caller's name. We tend to inquire out of curiosity, rather than a real need to know. However, sometimes a sense of security, or the apparent nature of the call, does necessitate asking the identity of the caller for the benefit of the one being called. Use your judgment and common sense to decide when it's appropriate to say, "May I ask who's calling?"

CHILDREN AND PHONES

Children should follow the same method of answering the phone as adults—simply say "Hello." You may find it helpful to give your children further instructions, such as:

- When the call is for someone who is at home, the proper response is, "Just a minute, please." Then put the phone down, go to the person, and tell them they have a phone call. Children should not be allowed to *shout* the person's name, at least not without first covering the mouthpiece of the phone or pressing the mute button.
- When home alone, children should say to callers that a parent is unable to come to the phone right now and will return their call later.
- When the caller asks to leave a message, always write it down. Then put the message where it will be seen.

TAKING MESSAGES

All telephone messages should be written down—not only by the children in the family, but also by the adults. Every message should include: who called, what message was left, and the time of the call. Always keep paper and pencil visible in the home for this purpose. Another important point: the family should decide on some conspicuous spot in the house where *all* messages are to be left. Select a location in full view of the front door (or whichever door the family normally uses), so that a message cannot possibly be overlooked. Consider taping them to the banister or the front-hall mirror, or getting a dry erase board. Be sure any paper notes are secured so they can't blow away. Once your family has developed this habit of writing down and posting messages for one another, no one will have to worry about not receiving a phone message. If your children have their own phones, you may consider having them call or text you with the relevant information.

INTERRUPTIONS

Frequently, when young children or others need your attention, they interrupt your telephone conversation. It's important to deal with these situations promptly, after apologizing to the person on the phone. Your child's crying may not make you uncomfortable, but it probably concerns or bothers the person to whom you are speaking.

At other times, you may need to interrupt a telephone invitation in order to check your calendar or speak with your spouse to ask if they are free to accept an invitation. That's fine to do; just remember that if you place your hand over the receiver, it only muffles the sound. Never say anything you do not want your caller to hear. If this is a concern, remember to mute the call.

ENDING PHONE CALLS

The best way to end a telephone call is to close the conversation and simply say "goodbye." However, closing the conversation can sometimes be a problem. Here are several suggestions that might help: "Sally, before I have to hang up, I'd like to…"; "John, I see by the clock that I'm due at a meeting"; "It's been great talking with you, but I'm afraid I have to run"; or "Thanks so much for your call." When all else fails, simply say, "I'm sorry, but I have to go now."

TELEPHONE TIME

Time management experts note that the telephone can be the greatest "time waster," at both the home and office. Treat the use of the telephone cautiously; it can rob you of time and money. Be especially careful of extensive phone use while living or visiting in a foreign country. Make sure to check your data plan when traveling, and update as necessary so you can stay in touch with family and friends back home as affordably and frequently as possible.

PHONE USE BY GUESTS

If a guest in your home receives or needs to make a phone call, try to provide some privacy. If you have an extension in the bedroom or a quiet part of the house, offer its use. If they are using your landline, step out of the room during the phone conversation. Since you don't know the nature of the call, presume that privacy is desired. Your guest will appreciate your courtesy. If there is an official military phone (commander's or other type) in your home, be sure that you point out which phone or line your guest is to use.

Guests placing long distance calls on a landline should always use a credit card, call collect, or offer to pay for the call.

HOW LONG TO LET THE PHONE RING

When placing a call, let the phone ring six to twelve times. Though waiting for twelve rings seems like a very long time, it actually takes only about one minute. Remember that the person you are calling might be busy and unable to answer immediately. Give him/her at least a minute to respond.

VOICEMAIL AND ANSWERING MACHINES

Voicemail and its antecedent, the answering machine, are wonderful conveniences and a time-saver for both the owner and the caller. However, just as with your telephone, good manners are important—not only for the person creating the outgoing message, but also for the person leaving a message.

Recorded Outgoing Message

When recording your outgoing message on either your voicemail or an answering machine, the essentials are these: identify yourself or location by giving your name or phone number; and ask your caller to leave their name, number, time of their call, and a short message after the beep. You might also explain that you will return their call as soon as possible and thank them for calling. Play your message back to check for accuracy, clarity, and voice tone. It's also a good idea to check your outgoing message occasionally to ensure that it says what you think it says.

The outgoing message on your home answering machine is not an appropriate place for humorous or off-handed messages. Nor should it be looked upon as an opportunity to record your children's voices or their renditions of songs or poems. People who call usually have a purpose in mind; anything other than a serious message on your part is unwelcome. Although it is often a hassle, be sure to check your voicemail from time to time so as not to miss anything important, such as a message from your servicemember's unit. The Family Readiness Group may be looking to verify your contact information so that you can receive timely unit information, or there may be a serious official command message. Often smartphones leave transcriptions so you don't even have to listen to the voicemail.

Advice for the Caller

Give your name, telephone number (with the area code, if it is different), and the time of your call. You might also state the day of your call if you have reason to believe the person you are calling might be away for several days. Smartphones will capture this information, but it is best to be safe. Try to be concise with your message, as some answering machines or voicemail boxes allow only a limited time for each incoming message. Speak distinctly and slowly, and say goodbye when you're finished so your listener will know you have concluded. Don't get discouraged and hang up when you realize you have reached a machine and not a person. Remember that the machine is there so the person you're calling won't miss your call. Using it to leave your message will save you from having to make a follow-up call. Depending on the time sensitivity of the message, you may consider sending a text after a day or so, asking whether the recipient heard your voicemail or answering machine message.

Collecting Your Calls

When you are away from home for an extended period, if you have an answering machine at home, call home and query your answering machine for messages every two or three days. Then you can respond to any that require an immediate answer. It's not a good idea to turn off your machine when you will be away if you normally use one. This can be a sure sign that no one is at home and may be an invitation to vandals or burglars. Above all, *don't* leave an outgoing message on your machine that tells everyone calling you are away and when you expect to return.

CALL WAITING

"Call waiting" is becoming a very popular telephone option. It allows all calls to ring through, rather than ending with a busy signal when the phone being called is in use. When that happens, the answerer can ask the caller to call later or terminate the ongoing call in favor of the incoming. Courtesy to both parties involved is most important so that neither feels abruptly cut off. If you're making or receiving an important call and don't want to be interrupted, you can override or ignore the call-waiting signal. You can also temporarily disconnect "call waiting" by punching in a short series of numbers; check with your phone company if you are interested in this feature.

THE BUSINESS PHONE

Whether you decide to use a separate business phone or use your home phone or smartphone as your business phone, the way you answer your business calls can be important to the success of your business. Current guides to business etiquette frequently devote an entire chapter to this subject. (See suggested reading in Chapter 38.) Normally, civilians answer with the name of the company and their name: "Fancy Catering Services, Jane Smith speaking." If you have a separate business line for your home business, this is the recommended method of answering. It sets the tone for a strictly business call. Always try to answer your business phone within two or three rings, if possible.

If you use your home phone or smartphone as your business phone, there are several options to consider for answering the telephone. During business hours, generally 9 a.m. to 5 p.m., answer just as a normal business phone is answered, "Cathy's Cakes, Cathy Carson speaking." Alternatively, you might say, "Hello, this is Cathy Carson," or simply state your phone number. After business hours, you should answer with a plain hello. Whatever business greeting you choose, the most important thing is your tone of voice. A pleasant voice lets callers know that you are a friendly person and eager to do business with them.

TEXTING/MESSAGING

Exercise consideration when texting and messaging. Many people prefer texting to calling, and as such, texting ahead to arrange a phone call may be preferable. Group texts can be useful for planning events or for keeping in touch with far-flung friend groups. Just remember different people have different sensibilities and what may be funny or pertinent to one or some of the group text members may not be appreciated by the other participants.

* * * * * * * * * * * * *

As the communication equipment we use becomes more complex, so do the guidelines for our telephone manners.

SECTION THREE: INVITATIONS

Chapter 7

Extending Invitations

"Invitations are documents of style as well as commemoration of events."
Martha Stewart

At first glance, extending invitations seems *very* easy. You simply want to let your intended guests know that you're planning an event, give them the details, and invite them to attend. However, some fairly common problems for the host or hostess can complicate this basically simple task. Whether extending formal or informal invitations, oral or written, there are important matters to consider beforehand. Let's take it step by step.

EARLY PLANNING
   When to Extend Invitations - Two to three weeks in advance is normal. This allows time for your guests to plan their activities around your event. However, four or more weeks in advance is advisable for very special or holiday events, especially when inviting foreign guests. For a major event, you may even send a "save the date" card three to four months in advance (examples provided later in the chapter).
   Confirm Date with Special Guests - When planning an event with a guest of honor, or an event that requires the presence of the senior spouse or commander, you *must* confirm the date with that special person (or couple) before inviting the other guests.
   Type of Party - Decide on the type of party you want to have, but be sure you understand the terms typically used. For example, "cocktail party," which means drinks and simple *hors d'oeuvres*, should not be confused with "cocktail buffet," which means enough drinks and food that guests will not need to eat before or after.
   Time - The type of function being planned can determine the time.
- *Coffee* usually starts in the morning at 9:30 or 10:00 a.m., or in the evening at 7:00 or 7:30 p.m.

- *Brunch* usually starts before noon.

- *Tea* is usually held from 3-5 p.m. (although large teas are often held earlier in the afternoon to prevent the guests from having to travel during peak traffic times).

- *Cocktail Party* starts around 6 p.m. (preferably with a stated ending time, e.g., "6-8 p.m.").

- *Cocktail Buffet, Buffet, or Dinner* usually starts anytime from 6:30-8 p.m.

  Location and Map - If planning to entertain somewhere other than your own home, confirm (preferably in writing) the facility's availability on your selected date *before* extending the invitations. Enclose or attach a small map for a difficult-to-find location, even if it's your own home.

  Method of Invitation - Decide how you will extend your invitations. (Examples of written invitations are provided later in this chapter.)

- *Phone Calls or Face-to-Face:* Oral invitations are often used because the hostess knows the invitation is received and hopes to get an immediate answer. If the invitation is accepted, then the hostess should follow up with a written reminder that gives the details of the event. This can be an invitation with "To remind" noted on it, or something as simple as an email or note that acknowledges acceptance and confirms the event information. (See samples of reminders at end of this chapter.)

- *Fliers:* This type of invitation is useful for large groups because it is easily and inexpensively reproduced. It also gives the hostess a great opportunity to be creative! Early dissemination and R.s.v.p.s are essential.

- *Fill-in-the-Blank Invitations:* Greeting-card companies make a bountiful assortment of these, but somehow the number in the package never seems to equal the number of invitations you need to send.

- *Handwritten Invitations:* Use flat or fold-over informals or brief notes. For formal occasions, such as large teas; write the invitations in the third person.

- *Fully or Partially Printed Invitations:* These are used only for formal-style invitations, and always written in the third person. They are normally used for large teas, weddings, and invitations extended by senior officers.

- *Electronic Invitations:* These are used for many different occasions, whether formal or informal. Recipients usually receive an email with a link directing them to their invitation that often includes animation. Most Army invitations for official events are electronic and delivered via email.

Attire - After deciding how fancy or casual you want the event to be, use the correct term on the invitations to describe your guests' expected attire; e.g., formal, informal, business suit, coat & tie, casual, or very casual. (For an explanation of all the commonly used dress terms, see Chapter 10, "Dress for the Occasion.") You may also wish to further clarify by putting in parenthesis what you would recommend as the attire: e.g., Casual (open collar); Texas casual (jeans and cowboy boots); Hawaiian luau (shorts). The only time to omit the appropriate dress term is when you are absolutely certain that the guests know what to wear to this specific type of event—usually the case for luncheons, coffees, and teas.

Response - When sending out written invitations, decide if you want responses and, if so, how you prefer to receive them. Responses (both acceptances and regrets) not only facilitate planning, but also confirm that all of your guests received their invitations. These are your options:

- *"R.s.v.p."* - When these letters are written alone, a written response is desired. R.s.v.p. stands for the French phrase, *Répondez s'il vous plait*. You may prefer to substitute the English translation, "Please reply" or "A reply is requested."

- *"R.s.v.p. (phone number or email address)"* - The guests are to respond by calling that phone number or sending a reply email.

- *"R.s.v.p. by (date)"* or *"R.s.v.p. NLT (date)"* - Although all responses should be made within 24-48 hours, "NLT" tells the guests the absolute latest time to respond. This approach is often used by the military for official functions, but seldom for private social events.

- *Omit R.s.v.p.* - This means no response is expected. However, thoughtful guests will still thank the hostess for the invitation when they see her and say whether or not they can come.

- *"Regrets"* or *"Regrets only"* - Guests are to respond only if they will not attend. This approach is not recommended, since no response may also mean that the invitation wasn't received.

When receiving written, email, or telephonic responses, prepare a guest list including the day, date and time of the event and immediately record responses. When receiving responses by phone, be sure to confirm the details of the event for the guests.

WHAT TO INCLUDE

Every invitation, from the most formal engraved invitation to an email invitation, needs to include certain basic information: (1) the type of event, (2) host's name, (3) day and date, (4) time, and (5) location. Additionally, invitations usually include: (6) how guests are to respond and (7) the proper dress for the event. Providing this information, without leaving any room for confusion or misunderstanding, is the responsibility of the host or hostess. Here are a few tips to assist you.

Ink - How invitations are written and the color of ink used are important.

- *Handwrite or professionally print, but never type,* personal invitations. Exceptions: Because of the large number of guests and the high cost of professional printing, military organizations typically use internally printed invitations from high-quality printers or electronic invitations. For the same reasons, you may occasionally see invitations to large teas prepared in a similar fashion.

- *Black ink* should be used for formal invitations, partially printed invitations, and personalized informals (no ballpoints, please). The black ink is in keeping with the formality of the invitation and the black printing. However, *colored ink* to complement the color of the invitations may be used for very informal invitations.

Day and Date - It's very important to give both the day of the week as well as the date. This is appropriate for every type of invitation, even for children's parties. Giving both pieces of information helps prevent any possible confusion and avoids embarrassing mix-ups for everyone. Be sure to check your calendar for the day and date before writing the invitations; it's too easy to make a mistake otherwise. Only invitations to events of historical value, such as changes of command and weddings, state the year.

Time - It is less confusing to give the time using the 12-hour clock, rather than the 24-hour clock that the military uses. For example, "7 o'clock" or "7 p.m." is less confusing than "1900" to most civilians. One exception is when inviting foreign guests who routinely use the 24-hour method of telling time in their country. Should you need to indicate whether an event is to be held in the afternoon or evening, the guideline is: Afternoon is before 6 p.m., evening begins at 6 p.m.

Who's Hosting - Often, fill-in-the-blank invitations fail to include a line for the name of the person giving the party. Don't let the card company's omission cause you to forget to include this important bit of information.

R.s.v.p. - Because this abbreviation represents a sentence, it is preferable to capitalize only the first word/letter, "R," rather than all the letters. This has the added advantage of not making the R.s.v.p. appear more bold and important than the body of the invitation itself. However, "RSVP" (no punctuation) is acceptable. When it's used, "R.s.v.p." appears in the lower left corner of the invitation.

Attire - Write the required attire for the occasion in the lower right corner of an invitation. The three terms most often used are formal, informal, and casual. See Chapter 10, "Dress for the Occasion," for a chart of all the options.

Map - If you wish to include a map for the convenience of your guest, slip it into the envelope behind the invitation, or it can be an attachment for an invitation that is sent via email.

Oral Invitations - Include all the basic information required for written invitations; plus, consider the following guidelines:

- *Always extend oral invitations in private.* This prevents hurting the feelings of others who are not invited, but who might overhear your invitation.

- *Never begin an invitation with a question* such as, "Are you free on Saturday?" This puts the person in the awkward situation of admitting their availability, before knowing what is about to be asked of them.

- *When an oral invitation is extended to someone at work,* the host or hostess should follow up in writing. This confirms the guest's invitation and that the invitee relayed the information to their spouse.

Guest of Honor – If there is to be a guest of honor, the other guests should be informed. On written invitations, this can be done by writing across the top "To honor...," "In honor of...," "To welcome...," "To farewell...," or "To meet...." It is appropriate to send the guest of honor an invitation as a written reminder. Do this by omitting "R.s.v.p." and writing "To Remind" or "Reminder" in either the upper left corner or lower left corner. For invitations with preprinted "R.s.v.p.," strike through this and add the reminder notation.

Senior Guests - When very senior officers are invited to a small function, it is courteous and helpful to send them a guest list. This is also appropriate for senior command sergeants major when invited to small gatherings of NCOs and their spouses. Because senior military routinely meet many people, they appreciate being able to familiarize themselves with the guest list beforehand. This can be accomplished in one of two ways (examples at the end of this chapter):

1. On the back of their invitation, write "Also invited:" followed by a listing of the other invited guests, or;
2. Several days before the event, send a separate guest list to them, in hardcopy or via email, usually through their administrative assistant. Restate the function, host, day, date, time, and location at the top.

This latter method has several advantages over the first. By sending the guest list after the responses have been received, it will reflect only those who accepted, rather than all those invited. A separate guest list also provides space for helpful information, such as spouses' names, the names they normally go by, and the guests' jobs. For example, one entry might read:

COL Margaret and Mr. John Doe (Maggie & Jay)
Commander, Division Artillery

CANCELLATION OR POSTPONEMENT

Occasionally, something unexpected may necessitate the cancellation or postponement of an event, after the invitations have been extended.

When to Cancel or Postpone - What might cause a host or hostess to cancel or postpone an event after they've already extended the invitations? Certainly a major injury, illness, death in the family, canceled wedding plans, or unexpected orders are acceptable reasons—anything that was unforeseen and is substantive enough in nature to prevent the host and hostess from carrying out their plans. However, the host and hostess should not cancel or postpone an event, after they've already extended invitations, simply because they receive an invitation for the same time—no matter how important the invitation! The only exception is for an invitation from the White House (see "An Invitation to the White House" in Chapter 9), and even that may be declined if the host and hostess have planned a family wedding for the same time.

How to Cancel or Postpone - This may be accomplished by phone, email, or in writing if time permits. However, keep the following points in mind: When *canceling* an event, courtesy requires an explanation: illness, death in the family, unexpected orders. On the other hand, *postponing* an event does not require an explanation; however, the reason for postponing may be provided. Later, when the postponed event is rescheduled, remember that the original responses do not apply; the new date may not be convenient for all of your guests.

FORMAL INVITATION GUIDELINES
Formal invitations follow specific guidelines that differ from informal invitations in several ways.

- The wording is in the third person. Any information added to a completely or partially printed invitation should be handwritten in black ink (never typed).

- Each line in the body of the invitation is centered between the left and right margins.

- Military rank is never abbreviated, always written out in full except 1st and 2nd lieutenants are simply "Lieutenant."

- The names of the host and hostess:
    - Full name is given when space allows (Brigadier General and Mrs. John Joseph Doe). An exception to this rule occurs with very senior military officials who may use their title with last name only (The USAREUR Commander in Chief and Mrs. Doe).
    - If space is a problem, the middle name is normally omitted.
    - If space is a problem, the suffix of "junior" may be abbreviated as "Jr."
    - If the military member is the wife, her first name is used as well (Colonel Mary Doe and Mr. David Doe).
    - If both are in the military and of the same rank, use one rank and both names (Lieutenant Colonels Jerry and Pat Doe).
    - If both spouses are in the military and of different ranks, the primary host/hostess is named first.
    - Chaplains use their title of "Chaplain."

- Military doctors use their military rank.
- The names of the guests:
  - On formal invitations, guests' names are written using the complete rank and last name only, with no first or middle name (Lieutenant Colonel and Mrs. Doe).
  - On the envelope, the full rank, first and last names are used (Lieutenant Colonel and Mrs. John Doe). See Chapter 8, "Addressing Envelopes," for specific categories of people.
- Abbreviations and initials are avoided whenever possible (with the common exceptions of Mr., Mrs., Dr., R.s.v.p.). However, when an individual always uses an initial for the first or middle name, it may be used.
- Spell out date and time. Half-hour times are stated as "half after seven o'clock" or "half past seven o'clock."
- Give the year only on invitations of historical significance.
- Note any honored guests by adding the information above the name of the host, usually across the top of the invitation ("In honor of Lieutenant General and Mrs. Star" or "In honor of Command Sergeant Major and Mrs. Stripes" for prominent people, and "To meet Colonel and Mrs. Eagle" or "To meet First Sergeant and Mrs. Diamond" for new arrivals and guests). The rank may be abbreviated here if necessary.

INFORMAL INVITATION GUIDELINES

The casual lifestyle of today often carries over into our entertaining. The wording guidelines may remain the same, but the type of event, whether there is a cost involved (and what the money will be used for), and the method of extending invitations and responding can vary widely. For example, today invitations are often extended as emails or evites. (See evite.com for creating an evite.) You can add designs and color to an evite to enhance the look and appeal. The responses can also be sent electronically, either to an email given on the invitation or by clicking the response button on the evite. The efficiency of both evites and emails ensures that the invitations aren't lost in the mail. And best of all, they are quicker, easier, and cheaper—a real boon for the host and invited guests.

EXAMPLES OF INVITATIONS

Printed formal invitation:

> The Soldiers and Spouses
>
> of the
>
> 3rd Battalion, 84th Field Artillery
>
> request the pleasure of your company
>
> at a Farewell Dinner
>
> honoring
>
> Lieutenant Colonel and Mrs. John James Cannon
>
> on Saturday, the thirtieth of June
>
> at seven o'clock
>
> at the Community Club
>
> Fort Sill, Oklahoma
>
> R.s.v.p. by 23 June   Army Service Uniform
> 404-987-6545   Civilian Informal
> 3-84Adjutant.mil@mail.mil

Handwritten formal invitation:

> In honor of Colonel and Mrs. John Doe
>
> Command Sergeant Major and Mrs. Jack Jones
>
> request the pleasure of your company
>
> for dinner
>
> on Saturday, the third of June
>
> at half after seven o'clock
>
> Quarters 28
>
> Fort Drum
>
> R.s.v.p. 315-234-5678   Coat & Tie
> jones@gmail.com

Informal invitations:

> Colonel Sarah Smith and Mr. Jonathan Smith
>
> cordially invite you and your guest
>
> to an Aloha Welcome Cocktail Party
>
> on Friday, the 19th of May
>
> at six o'clock
>
> Quarters One
>
> Fort U.S.A., Hawaii
>
> Please bring an hors d'oeuvre to share.   Drinks will be provided.
> R.s.v.p. xxxx@gmail.com   Aloha Casual
>   (Shorts/Sundresses)

Colonel Jane Adams and Ms. Kathy Roberts
invite you and your guest
to a Command Team Potluck Barbeque
on Friday, 20th of June
starting at six o'clock
Quarters One
Fort U.S.A., Texas

Please bring a side or dessert to share.
BBQ Chicken/Beef and Drinks will be provided.

R.s.v.p. xxxx@gmail.com

Texas Casual
(Jeans, Boots, Cowboy Hats)

Partially printed invitation:

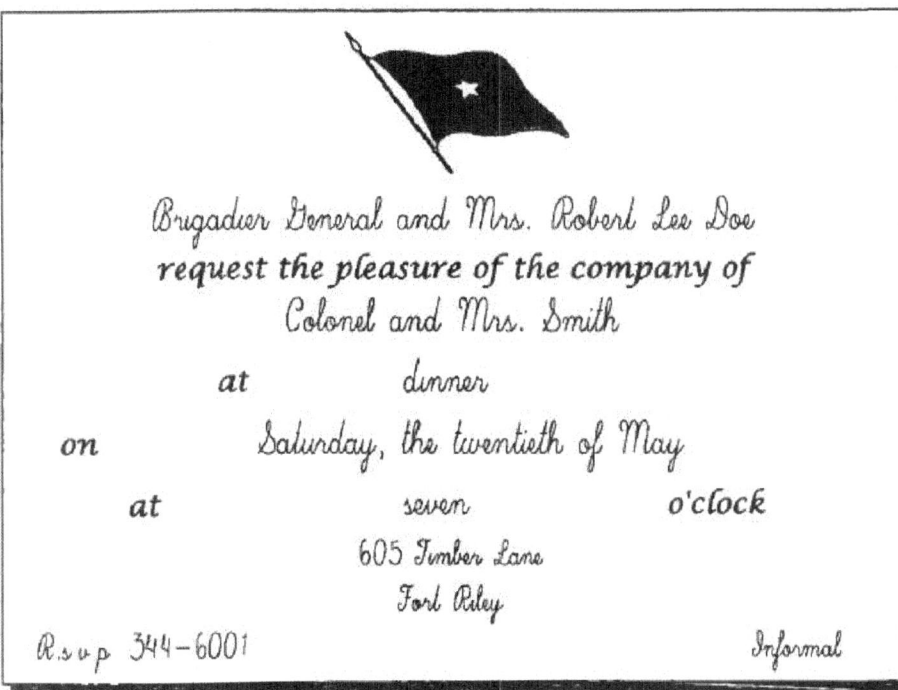

(Envelope will give guest's full unabbreviated rank, first and last name.)

Informal invitation written on a joint informal:

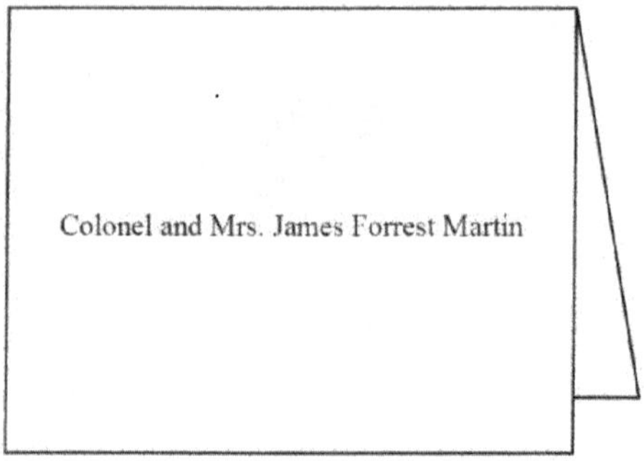

Front view

Colonel and Mrs. James Forrest Martin

Inside view

Cocktail Party
Friday August 25
6 - 8 p.m.
5605 Pine Tree Lane
Fort Lewis
R.s.v.p. 253-xxx-xxxx   *Coat & Tie*
e.g. jfmartain@gmail.com

Informal invitation written on a plain informal:
(written on the front if it's plain enough, otherwise written on the inside)

> Potluck Dinner
> Saturday, March 1
> 7:30 p.m.
> The Smith's
> 303 Old Post Road
>
> R.s.v.p. 404-633-xxxxx   Casual

Invitation written as a brief note:

> November 2
>
> Dear Janet,
>
> Please join us for lunch to bid Sally Smith farewell. The get-together is set for Friday, November 16th, 11:30, at my house (43 A Bay Shore Road).
>
> As former advisor to our board, you know how much Sally did for out club. We all hope that you will be able to join us in giving her a warm send-off.
>
> Sincerely,
> Patricia Adams
>
> R.s.v.p. 707-300-XXXX
> padams@gmail.com

"Please reserve this date" cards, printed and handwritten:

> Please reserve this date
>
> Saturday, April 14, 1993
>
> for a 25th Wedding Anniversary Celebration
>
> in honor of
>
> Colonel and Mrs. Jack Ryan
>
> Susan, Sarah, Drew Ryan

> Dear Susan,
>
> We have some exciting news! Harry will retire in May, and his parents are coming up from Georgia for the big event. They will host a dinner in his honor on May 31st. We hope you and Joe will reserve this date.
>
> Fondly,
> Betty

"Also invited" list:
(handwritten on reverse side of an invitation sent to a very senior guest)

> Also invited:
>
> C(s/M) & Mrs. Ronald Smith (Ron/Ann)
> 1st Bn, 9th Inf.
>
> C(s/M) & Mrs. Joseph Brown (Joe/Joan)
> 2nd Bn, 8th Inf.
>
> C(s/M) & Mrs. Harold Black (Harold/Mary)
> 9th Bn, 29th Inf.
>
> C(s/M) & Mrs. Robert Burns (Bob/Sally)
> 1st Bn, 99th Inf.

Separate guest list:
(usually sent via email several days before function to the senior guest's office)

Dinner Hosted by BG and Mrs. Samuel Thomas Star
Saturday, April 23
Quarters 3, Fort Benning

■■■

GUESTS:

Major General & Mrs. Mark M. Miller (Mark/Peggy)
Deputy Commander, V Corps

Brigadier General & Mrs. Henry C. Higgins (Hank/Patty)
Commander, Support Command

Dr. & Mrs. Raymond J. Richmond (Ray/Mary)
Chief of Surgery, City Hospital

INVITATION REMINDERS

If you have ever forgotten to attend a party, gone on the wrong day, arrived at the wrong time, or worse yet, had guests appear at your door on the wrong day, you will definitely understand the importance of having written details of the function. That is the purpose of a "to remind" card: to verify the day, date, time, location, and type of event.

Extending invitations by email or phone has become commonplace today. It's easy and inexpensive; there are no invitations to get lost in the mail; and the person hosting can, hopefully, get a quick or immediate response. However, the wise host doesn't stop there. For those guests who accept, follow up with an email or written reminder. "To remind" cards that are emailed or sent via US Postal service or hand delivered take exactly the same form as written or printed invitations, except that "R.s.v.p." is omitted (since the guest has already accepted) and "To Remind" is added. "To Remind" is usually written in place of "R.s.v.p." in the lower or upper left corner. Occasionally, you will see the notation, "p.m." (for *pour memoire)* or "Reminder" used instead; they serve the same purpose. All else about the invitation remains the same. You may use any type of written or printed invitation for a "To Remind" card. Invitations that have "R.s.v.p." preprinted, can be used by striking this out and writing in the reminder notation.

EXAMPLES OF INVITATION REMINDERS
Written on a commercial invitation:

IT'S PARTY TIME

WHAT 4th of July Picnic
WHEN Wednesday, July 4 – 2 p.m.
WHERE The Caldwell's, 2trs. 3

To Remind   Very Casual

Reminder written on an informal note card:
(omit the name, if printed on the front)

> Cocktail Buffet
> Saturday, November 18
> 7 o'clock
> Tom and Alice Smith
> 404 East 5th Street
> Clarksville
>
> To Remind   Informal

So the next time you extend invitations by email, phone, or in person, follow up each acceptance with an email or written reminder. You might even mention that you will send a reminder. If your guest says, "Oh, that's not necessary," do it anyway! Both you and your guests will benefit from your thoughtfulness.

****

The complete invitation is the mark of a considerate host.

Chapter 8

Addressing Envelopes

"Why do you sit there looking like an envelope without
any address on it?"
Mark Twain

Addressing envelopes correctly is as important as preparing invitations correctly. This attention to detail not only ensures that the invitations reach their intended destination, but it is also a courtesy to the guests. Using the proper form of a guest's name and title is a reflection of the host's concern for this small, but important, aspect of social etiquette.

The procedure for addressing social correspondence, such as invitations, has adapted to the recent changes in American society. Many women are now in military service or have professional civilian jobs, and some women keep their last names after marriage. There are other changes with same-gender marriages and couples living together who have never formally married. Additionally, it seems that very few individuals send letters anymore! Texts and emails often replace that form of correspondence. The guidance provided here includes consideration of these changes and their effects on addressing invitations and other social correspondence.

OFFICIAL AND SOCIAL CORRESPONDENCE
Correspondence is divided into two categories: official (business correspondence in the civilian world) and social. Official correspondence includes, among other things, invitations sent from an Army office by an individual, such as a commanding officer acting in his/her official capacity. Social correspondence consists of letters and invitations extended by an individual acting as a private citizen. Additionally, social correspondence is generally personal in nature: friendly letters, invitations, and thank-you notes. It is necessary to understand the difference between these two categories because the guidelines for addressing envelopes sometimes vary. This chapter will explain the variations.

One guideline that does remain constant is that these two types of correspondence should be sent to different addresses. Official correspondence (to include official invitations) is sent to the office address of the individual. If an official invitation is for a couple, it is sent to the office address of the individual for whom the invitation originated. Social correspondence (to include social invitations) should be sent to the home address or personal email address. While this is not always possible in foreign assignments, it should be followed whenever practical.

One guideline that is changing as a result of today's greater number of women in the military and women with civilian professional titles is the way in which social correspondence is addressed to them. Out of respect for professional spouses and the titles they have earned, it seems far more appropriate today to recognize those titles on all occasions, both social and official, extending the same courtesy to all. Accordingly, you will see that view reflected in the related examples in this chapter.

RANK AND INITIALS

For *official or business correspondence,* the Army uses the servicemember's full rank on the envelope for official correspondence. The individual's name is usually written on the envelope as: rank, first name, middle initial, last name. Typically, the envelope is addressed only to the individual in the organization with whom they are corresponding, not the spouse. If it is an invitation and both are invited, the wording in the invitation would state something like "you and your spouse or guest are invited...." However, remember that spouses do not send "official correspondence;" the examples provided below are for your understanding of the differences.

Social correspondence calls for rank to be spelled out in full as well. Exception: Both 1st and 2nd lieutenants are addressed as Lieutenant. However, when the envelope is not large enough to accommodate the full rank and name, it is permissible to abbreviate the rank. Accepted abbreviations for Army rank can be found in Chapter 36, "The United States Army." (You will notice that accepted rank abbreviations and capitalization vary with the different branches of service.) Initials for the first or the middle name are normally not used in social correspondence. Usually, the middle name is simply omitted.

The rank of a servicemember should be used on all correspondence. If the spouse holds a military rank or professional title, that should be used on the correspondence as well. When the spouse is a civilian female, it is permissible to use "Mrs." and then include either her first name or the first name of her husband, if you so desire. Most women prefer to see their own names on an envelope addressed to them personally.

EXAMPLES

Military Husband Outranks Military Wife
    Official: (1) When invited because of his official capacity
                    Colonel John W. Doe
                    (his office address)

      (2) When invited because of her official capacity
          Major Jane E. Doe
          (her office address)
Social: Colonel John Doe and Major Jane Doe
       or list on separate lines w/2nd line indented
       (home address)

Military Wife Outranks Military Husband

Official: (1) When invited because of her official capacity
          Major Jane E. Doe
          (her office address)
      (2) When invited because of his official capacity
          Captain John W. Doe
          (his office address)
Social: Major Jane Doe
       and Captain John Doe
       (home address)

Wife and Husband With Same Rank

Official: Captain John W. Doe or Captain Jane E. Doe
       (appropriate office address)
Social: The Captains John and Jane Doe
       or list on separate lines with her name first
       (home address)

Military Wife with Civilian Husband

Official: Major Jane E. Doe
       (her office address)
Social: Major Jane Doe
(military related) and Dr. John Doe
       (home address)
Social: Dr. John Doe
(civilian related) and Major Jane Doe
       (home address)

Women Who Retain Their Family Name - You will see from the following examples that addressing envelopes to couples in which a spouse has retained her last name prior to marriage follows the correspondence rules just described. Spouses who retain their family name before marriage, but who have no professional rank or title, take the title of "Ms."

Military Wife Who Retained Family Name

Official: (1) When invited because of her official capacity
Captain Jane E. Smith
(her office address)
(2) When invited because of his official capacity
Captain John W. Doe
(his office address)
Social: Captain Jane Smith and Captain John Doe
or list on separate lines (2nd line indented)
(home address)

Non-Military Wife Who Retained Family Name

Official: Captain John W. Doe
(his office address)
Social: Ms. Jane Smith and Captain John Doe*
or list on separate lines (2nd line indented)
(home address)
* It doesn't matter whose name is first.

General Officers - Although social correspondence is generally sent to everyone's home address, invitations to general officers and those intended jointly for general officers and their spouses should be sent to the office address. The reason is that every general officer has such full business and social calendars that all invitations involving the general must be coordinated with the office executive assistant. Therefore, send invitations for generals to their office and thank-you notes to their homes.

Military Doctors - All correspondence addressed to military doctors should reflect their rank rather than professional title. In personal conversation or when seeing them professionally, you may prefer to call them "Doctor," but address correspondence to them using their rank.

Chaplains - Even though chaplains are always spoken to using only their title of "Chaplain," their correspondence should be addressed with both their title of

Chaplain (may be abbreviated as Chap.) and their rank, which is usually enclosed in parenthesis.

> Chaplain (Colonel) Jane or John Doe
> or Chap. (COL) Jane or John Doe
> or Colonel Jane or John Doe
> > Brigade Chaplain [indented]

Widows - Widows are addressed using their title of "Mrs.," followed by their late-husband's first and last name. Even though her husband is no longer living, a woman retains her husband's complete name, until she dies or remarries. So Mrs. John Doe is the traditional form for a widow.

Divorcées - Women who are divorced may retain and use the title of "Mrs." and their former husband's last name, but not his given names (so long as the divorcées have not legally taken back their family name). Thus, the difference between the way correspondence is addressed to a divorcée and to a widow appears in the first name. A divorcée who continues to use the title of "Mrs." may be addressed using either her first name or maiden/family name followed by her married name, as in Mrs. Jane Smith or Mrs. Johnson Smith. Since she can no longer be "Mrs. John Smith," she inserts her given name or maiden/family name so everyone would be clear that she is not married anymore.

"Ms. Jane Doe" is the contemporary form that does not suggest a marital status. More young women use this form professionally and anytime they think their marital status is not pertinent to the communication or conversation. Simply "Jane Doe" is an informal form that is perfectly correct too. However, the best answer is to find out the preference of the person you are addressing.

Promotables - When an officer in the Army is on a promotion list, his or her *official correspondence and official invitations* are addressed using the "present" rank, with no indication of promotion selection. However, it is appropriate as a courtesy in addressing his or her *social correspondence* to add "(P)" after the current rank, as in Captain (P) John Doe, for Captain Doe who is on the promotion list for major. The officer should never use the (P) with his or her own name or signature.

Other services do not always follow the same etiquette with regard to those who have been selected for promotion as the Army. For example, when a military member in the Air Force is on a promotion list, it is appropriate as a courtesy to address his or her social correspondence using the *new* rank, followed by "(Sel)," as in Major (Sel) John Doe, for Captain Doe who is on the promotion list.

Retired Military and Other Officials - Social correspondence envelopes for retired military may continue to be addressed using their rank, with no mention of their retired status. Only for official correspondence (including official invitations) is the word "Retired" added after the name. When used, this indication of retirement follows only the retiree's name, not the spouse's name, as in:

Official: Colonel John J. Doe, USA Retired
and Mrs. Doe
Social: Colonel and Mrs. John Doe

Presidents and vice presidents do not retain these titles after they leave office. However, governors, senators, ambassadors, justices of the Supreme Court, and some judges after they retire may be addressed, as a courtesy, by the title held when they retired, as in:

Social: The Honorable
Margaret Livingston
Salutation: Dear Judge Livingston

Diplomats and Dignitaries - For those very special situations when you need to know how to address correspondence to high-ranking diplomatic and civil dignitaries, the title "The Honorable" is most often used, formally written on the line above the name. To save space or for other practical reasons, it may be on the same line with the name. However, it is not used in the salutation of a letter, or in speaking to the person. If you refer in writing to someone with this title, "The" is not capitalized unless it is the first word in the sentence.

Invitations, whether official or social, are addressed using the social format shown on the following pages.

American Ambassador*
* In Latin American countries, it is preferable to use the full title of The Ambassador of the United States of America.

Official: (at post)	The Honorable John (or Jane) Doe American Ambassador
(away from post)	The Honorable John (or Jane) Doe American Ambassador to (country)

Social: The Honorable
(at post) American Ambassador
 and Mrs. (or Ms./Mr.) Doe

(away from post) The Honorable
 The American Ambassador to (country)
 and Mrs. (or Ms./Mr.) Doe

Salutation Dear Mr. (or Madam) Ambassador

U.S. Senator
 Official: The Honorable
 (D.C. office) John (or Jane) Doe
  United States Senate

 Social: The Honorable
  John (or Jane) Doe
   and Mrs. (or Ms./Mr.) Doe

 Salutation: Dear Senator Doe

Governor
 Official: The Honorable
  John (or Jane) Doe
  Governor of (state)

 Social: The Honorable
  Governor of (state)
   and Mrs. (or Ms./Mr.) Doe

 Salutation: Dear Governor Doe

U S. Representative
    Official: The Honorable
    (D.C. office)   John (or Jane) Doe
                            U.S. House of Representatives

    Social: The Honorable
                            John (or Jane) Doe
                                and Mrs. (or Ms./Mr.) Doe

    Salutation: Dear Representative Doe

Secretary of the Army (Navy, Air Force)
    Official: The Honorable
                            John (or Jane) Doe
                            Secretary of the Army

    Social: The Honorable
                            Secretary of the Army
                                and Mrs. (or Ms./Mr.) Doe

    Salutation: Dear Mr. (or Madam) Secretary

Chief of Staff of the Army
    Official: General John (or Jane) Doe
                            Chief of Staff of the Army

    Social: General John (or Jane) Doe
                            Chief of Staff of the Army
                                and Mrs. (or Ms./Mr.) Doe

    Salutation: Dear General Doe

American Consul General
    Official: Mr. John (or Mrs. Jane) Doe
                            American Consul General

    Social: The American Consul General
                                and Mrs. (or Ms./Mr.) John Doe

    Salutation: Dear Mr. (or Mrs.) Doe

Mayor

Official:	The Honorable John Doe Mayor of (city)
Social:	The Honorable Mayor of (city) and Mrs. (or Ms./Mr.) Doe
Salutation:	Dear Mayor Doe

FOREIGN TITLES AND NAMES

When living overseas, it is courteous to use local forms of address for your foreign guests. However, addressing envelopes correctly to persons of other nations can be a challenge. For example, German men and women prefer to be addressed separately and with all of their titles.

Herr Doktor Ernst Schmidt
Frau Professor Irmgard Schmidt

Many Spanish surnames are composed of both the father's and the mother's family name. It is never incorrect to use both, but tradition or the person's own preference sometimes dictates the use of only one (e.g., Maria Falla y Ortega or Senora Maria Ortega). Forms of address in Italy frequently include the retention of the married woman's maiden name. The best way to ensure that you are using the correct form is to check with your local protocol office, consulate, or embassy.

RETURN ADDRESS

Normally, a return address is written in the upper left corner on the front of the envelope. (For examples, see Chapter 4, "Correspondence.")

By necessity, the space available for writing your return address is small. For that reason, the use of rank abbreviation is acceptable, even though the rank of the addressee is usually written out in full. For social correspondence, the middle name or initial is not necessary. The return address might read "COL and Mrs. John Doe." If the social correspondence is from the wife only, it could read "Mrs. John Doe" or perhaps "Mary Doe."

HAND DELIVERY

You may often prefer the convenience and speed of hand delivering social correspondence, such as thank-you notes and invitations. For a casual note, simply put the name on the envelope that you normally use when speaking to the person (e.g., Sally Carter), tuck the flap inside (not sealed), and write "By hand" in the lower left comer. For less casual correspondence and all invitations, use the correct written form of address on the envelope (e.g., Mrs. Robert Carter). "By hand" communications are not sealed, as they may be frequently delivered by others, and to seal the envelope would indicate a lack of trust. Confidential correspondence is never sent in this manner.

HOLIDAY INVITATIONS

Because people receive so many cards and letters during the holiday season, it is a good idea to do something to call attention to your holiday invitations. This can be accomplished by writing "Invitation" on the lower left comer on the envelope. This alerts your guests to the fact that this piece of mail requires immediate attention and response.

```
CSM and Mrs. Tom Reilly
230 Tank Trail Road
Fort Knox, KY 88888


                    CSM and Mrs. Fred Fremont
                    665 Howitzer Road
                    Fort Sill, OK 99999


Invitation
```

Note: It is correct to use the official state abbreviation on both formal and informal correspondence.

* * * * * * * * * * * *

Attention to detail while addressing correspondence is always appreciated by the recipient, and the mail carrier.

Chapter 9

Responding to Invitations

"All invitations are flattering, for they
express friendliness and offer hospitality."
Lillian Eichler

So you've just received an invitation—great! Whether you accept or regret is your decision, but *your social responsibility is to reply immediately*. The manner in which you respond is determined by the invitation itself.

RESPONSE INDICATORS
Invitations can be either written or oral. Written invitations may be printed and mailed to the recipients or, as is growing in popularity, they may be typed using an electronic invitation program and emailed to the recipients. Look on the lower left corner of any printed (virtual or hard copy) invitation to learn how the response should be made. There are a variety of options and each has a different meaning. The various possibilities are:

Blank Lower Left Corner
When no mention of a response appears on the invitation, then none is expected. You may attend if you are free, or not attend if you have a previous commitment or would prefer not to; it's that simple. However, the next time you talk to the person who invited you or see that person in a private setting, it is polite to thank him/her for the invitation, and indicate whether or not you will be able to attend.

R.s.v.p.
Although these letters stand for the French phrase, *Répondez s'il vous plait,* which means "Respond if you please," they really convey a much stronger meaning. R.s.v.p. means that a response is expected! When written alone, as shown above, this indicates that the host or hostess wants a concrete response.

R.s.v.p. (Phone Number/Email address)
This implies that the host or hostess wants the response made by telephone or email message. If a cellular phone number is provided, a text response may also be appropriate. In our busy lives, this is the most expedient method of obtaining responses for written informal invitations.

R.s.v.p. by (Date), or R.s.v.p. NLT (Date)
This gives the very latest date that a response can be accepted, very commonly used for baby showers and weddings. However, the courteous response is *always* made within 24-48 hours of receiving an invitation.

Regrets, or Regrets Only
This indicates respond only if you cannot accept the invitation. It is a reverse approach to most response requests; in other words, the host or hostess hopes to learn which guests will attend by hearing only from those who will not attend. Even though the use of "Regrets" is not recommended, if that is the request, follow those instructions. If a phone number/email address is given, you may call, text or email to regret. Although not very common, if no phone number or email address is given, a mailed written regret is expected if you cannot attend.

ORAL/EMAIL RESPONSE

When an oral response to an invitation is appropriate, it's better to make it over the telephone, rather than in person. Or, you could text or email as soon as possible after receiving the invitation. There are several important reasons for this. First, the host or hostess usually has a list of invited guests nearby; if you call, text or email, your response can be immediately recorded. On the other hand, if you give your response to the person who invited you when you see him or her in the commissary or at a luncheon, they may forget your answer, or that you even mentioned the invitation. It's not fair to put the burden of remembering on the one who extended the invitation; it's the invited guest's responsibility. Another reason for phoning, texting or emailing responses is to keep them private, so others cannot overhear your conversation and get their feelings hurt if they weren't invited.

WRITTEN RESPONSE

Very few invitations today require a written response. Once in a while, you might receive a wedding invitation that has no response card enclosed; in that case, a written response is appropriate. For guidance on writing a response to a formal invitation, such as a wedding invitation, see the examples given in Chapter 27, "Military Weddings."

RESPONSE TIME

When responding to an invitation, whether you do so in writing, text, email or phone, your response must be prompt—*within 24 to 48 hours!* If you wait any longer, several unfortunate things may happen. First, the individual hosting may worry that you didn't receive the invitation. Then, he or she may begin to feel that you aren't pleased to have been invited. Finally, your manners may be in question for not answering the invitation. None of these is desirable, nor designed to make a good impression. To start things off right, answer every invitation in a timely manner.

FAILURE TO RESPOND
More has been written, said, and thought about this deplorable state of affairs than any other etiquette topic. Failure to respond to an invitation that requests a response creates a real dilemma for those doing the inviting. Did you not receive their invitation or did you simply forget to respond? Maybe you just don't want to accept. Should they call you? If so, how soon? What should they say? How can they word future invitations so that the guests will respond? These questions have plagued us through the decades.

Guests need to remember that an invitation is issued from a desire to offer hospitality and to please. Those who do the inviting are offering their time, energy, and money to entertain their invited guests. Therefore, to receive an invitation that requests a response and *not* respond is extremely rude! There is just no nice way to describe it, and no excuse is acceptable.

Responding promptly to invitations is one of the basic courtesies of society, both military and civilian. There are some rules of etiquette that may be bent to meet your needs, but this is not one of them! You are not obligated to *accept* every invitation you receive, but you are obligated to *reply* when one is requested.

ORAL INVITATIONS AND RESPONSES
Oral invitations differ from written informal invitations in that you are expected to respond immediately. That's why many people prefer this method of extending invitations for small gatherings. Hopefully, if you accept the invitation, you will receive a written reminder.

When you receive an oral invitation, accept immediately and with enthusiasm if you will be free and want to attend. Be sure to record all of the particulars you will need to know: host, day, date, time, place, dress, and type of party. If you are not free to accept, briefly explain why. Even saying you are "not free" is more polite than brusquely saying you "can't come."

Often, you can't give an immediate response. Perhaps you need to check with your spouse to confirm that both of you are free, or arrange for a babysitter first. Whatever the reason, assure the person inviting you that you will contact him or her promptly with your answer—promptly means in the next day or two. A longer delay may require another phone call to you, which would be embarrassing for both of you.

RESPONSE COMPLICATIONS
Can't Meet the Time Limit - When you are unable to give a definite reply to an invitation within forty-eight hours, you are obliged to regret. If it is an informal invitation, you should give an explanation for your regret. This may result in your host offering that you take a little more time. For example, if you can't find a babysitter or don't know yet when you'll return from a trip, the host might offer that

you keep trying to find a sitter or wait until your vacation plans firm up. This is most likely to happen if the host doesn't need an exact guest list right away—usually the case for a large, informal party. However, the number of guests for a small seated dinner is far more critical and, if you take more time to give your response, that limits the time the host has to invite someone else in your place. In such a situation, don't be surprised if your prompt regret is accepted without offering an extension of time. The important point is that the 24-48 hour response time doesn't change, regardless of your personal situation. If you can't give a prompt acceptance, then you should regret. The individual hosting can accept your regret or give you longer to respond—but that's not your prerogative.

When Only One Is Free - If you and your spouse receive an invitation, but only one of you is free to accept, consider the type of event before deciding whether one may accept or both must regret.

- *Any social event in the home* (except commander's New Year's reception)—The polite course of action is to regret both, explaining that one of you cannot come. Those hosting have the prerogative to accept the regret or invite the available one to attend alone.

- *Commander's New Year's Reception*—The military member is expected to attend, even if his or her spouse cannot. The only acceptable reasons to regret are illness or being out of town. On the other hand, the spouse should not attend the commander's New Year's reception without the military sponsor.

- *Most official functions* (such as parades and receptions)—All invited guests are free to accept and attend, with or without their spouses. However, with an event using Official Representational Funds (ORF), typically for divisional units and up, personal guests are paid for by the host. So don't be offended if your spouse is deployed and you are not invited to certain high-level military functions.

When You Have Houseguests – If you receive a personal invitation for a time when you expect to have houseguests, you should not ask to include them. Either accept the invitation and leave your guests on their own, or regret and explain the reason. If you have already accepted an invitation and then find that you are going to have unexpected houseguests, either tell your guests you have a previous commitment, or explain to the person who invited you why you must change your acceptance to a regret. He or she may or may not choose to extend the invitation to your guests as well. For official invitations to parades and other ceremonies, notify the adjutant or protocol office responsible for the invitations. This will usually result in your houseguests being included.

KEEPING TRACK OF YOUR INVITATIONS
As your calendar fills up, it's sometimes difficult to keep track of every commitment. You will have invitations, doctor's appointments, even email invitations, which you should print out. It's important to develop a logical system that works for you, and follow it!

The most common system in use is that of taping invitations to the refrigerator door, kitchen cabinet, bulletin board, or wherever there's room to put them. The flaw in this system is the *random order* that results. Instead, try organizing your invitations and appointments in *chronological order* in a single column, with the earliest-dated item on top. This calls your attention to the next commitment and provides ready access to its details. Develop the habit of checking your calendar at the beginning of each week for the week's events, and every morning for the day's events. Checking your calendar regularly should eliminate the embarrassing situation of forgetting an invitation.

It helps to note on each invitation the date you responded. This provides a ready check that you haven't inadvertently taped the invitation to the refrigerator without writing or phoning your acceptance. Also, take the invitation with you when you go to the event as a handy reference for address, date, and time. When the event is over and a thank-you note is in order, put the invitation in a conspicuous place, such as your kitchen counter, to remind you to write that note.

TWO INVITATIONS FOR THE SAME TIME
If two invitations for the same time arrive simultaneously, decide which you prefer to attend, accept that one, and regret the other. However, should you receive the second invitation *after* having accepted the first, you may not go back and regret the first, in order to accept the second. The second invitation might sound more interesting or be from someone very special, like your spouse's commander; nevertheless, you are not free to accept if you have already accepted another invitation. Only an emergency or illness should cause you to regret an invitation once you have accepted it.

WHEN NO RESPONSE IS RECEIVED
What should you do if you have asked for responses, but some of your invitations are not answered? This is a common problem because so many people are careless about responding. Here's the recommended course of action:

1. Wait about one week for the invitations to reach their destinations and the guests to respond. Exception: If the invitations said "R.s.v.p. by (date)" or "NLT (date)," you must wait until that deadline before checking on the guests who fail to respond.
2. Then contact the guests who have not responded and politely ask,
    a. Did they receive their invitation?
    b. Can they attend?

Invitations do sometimes get lost or delayed in the U.S. mail and military distribution. (There must be a giant basket in the sky for all of the invitations that have gone astray; perhaps they go to the same place as the socks that disappear in the dryer!) And, with emailed invitations, items often end up in the "Junk" folder instead of the Inbox. Consequently, the first concern is whether or not the invitation has been received. A gentle approach to the inquiry might be, "John and I are planning a dinner party for the 19th, and we sent you and Bob an invitation last week. I'm calling to see if you have received it yet, because we haven't heard from you."

Do not presume that those who don't respond to invitations are not coming. This could be a serious miscalculation. Additionally, this should not be turned into a "teaching" situation by turning away at the door those guests who didn't respond, or calling the offending guests onto the carpet in front of the boss's desk for an etiquette lesson. The subtleties of etiquette should not be taught with an iron fist.

EDUCATE POLITELY

Everyone can help to set the example of responding promptly to invitations by always replying within 24 to 48 hours. Unit commanders, command sergeants major, and coffee-group leaders can be especially influential in this. Their visibility and leadership roles put them in a unique position not only to set a good example, but also to provide opportunities for others to learn the importance of this social responsibility and other military manners. However, the way in which this education process is handled is very important to its success. The following are ideas for educating politely that others have used with good success:

- *Use R.s.v.p. on unit-coffee invitations.* Even if you've had poor results with this in the past, don't give up and abandon this courtesy or resort to a standing reservation list. Ask the spouse who is hosting the event to call those who don't respond. It's important for the party planner to know how many to expect, and good for the young spouses—and some of the older ones as well—to get into the habit of responding to invitations.
- *Have brief protocol discussions with the coffee group.* Periodically explain R.s.v.p. and the rationale for responding within 24 to 48 hours. Remind them that, when it's their turn to host an event, they will appreciate everyone's prompt response. Some coffee-group leaders set aside five minutes during the business meeting of every coffee to discuss a different point of etiquette.
- *Prepare and send out an information letter* prior to an upcoming major event, such as a unit dining-out, that light-heartedly describes the appropriate dress, receiving-line procedure, posting of the colors, toasts, and other special events of the evening.

AN INVITATION TO THE WHITE HOUSE

If you are ever fortunate enough to receive an invitation to the White House, you will be pleased to know guidance is available to you for every aspect of the protocol involved. The White House Social Office coordinates all invitations to the president's home, and stands ready to assist you and answer any questions you may have. That's especially important since the social style changes from one administration to another, and no enduring guide can be written that will always apply.

Response - An invitation to the White House should be answered immediately. As with other invitations, you respond in the manner indicated on the invitation. Often, an invitation to the White House includes an insert card with a telephone number and a request for you to respond by phone to the White House social secretary. An "admit" card, noting which entrance to use, will also be included.

White House invitations are rarely declined. Traditionally, there are only four reasons for a regret: a family wedding, a family death, illness, or travel. If you should find yourself in the unfortunate circumstance of having to decline a White House invitation, explain the reason when you regret. Though such an explanation is considered optional for most formal invitations, it is not for a White House invitation.

Dress - The proper dress for the occasion will be stated on the invitation. However, invited guests may request additional clarification from the Social Office. Though current fashions always prevail, the natural tendency is for White House guests to dress up. Use care not to overdress; a good guide is simplicity, both in dress design and color. There's a practical reason for that advice. Photographs are normally taken of the president and first lady with their guests at social events, and plain attire produces the best results—it doesn't detract from the faces in the photo. Another recommendation is not to carry a purse, especially for a White House reception where you want your hands free for greeting others and holding refreshments. Though a photo ID is required, you may always slip it into a pocket or alternatively, carry a small bag with a shoulder strap. Military servicemembers are given specific uniform guidance by the White House Military Office.

Arrival - Gate entrance information and times are usually included with the invitation, and should be carefully followed. Plan to arrive early, especially for a reception where there may be several hundred guests. Keep in mind that nearby parking is limited. Whether invited for a reception, state dinner, or luncheon, all guests go through a metal detector similar to those used in airport security.

Receiving Line - Once inside the White House, guests wait for the president and first lady and any honored guests to enter and form the receiving line. The traditional guideline for White House receiving lines, that of a man going first, is changing in response to the ever-increasing number of women taking key roles in the professional world. Today, whoever is more likely the reason for the invitation goes first; i.e., the guest who has been invited because of his/her official capacity, precedes his/her spouse or date through the line. As with other receiving lines, only brief greetings are exchanged. If no receiving line is planned, the president and first lady will enter and walk around shaking hands, briefly greeting their guests. The president is addressed as "Mr. President," and his wife as "Mrs. (last name)." The title "First Lady" is an unofficial one and not used when addressing her, but is often used in conjunction with "Mrs. (last name)" when introducing her.

Etiquette Pointers - For state dinners, everyone waits to be seated until the president, first lady, and honored guests are seated. The menu card and personal place card may be taken as souvenirs. Guests do not leave until the president and first lady have made their exit. After any White House event you've attended (other than a stag affair), it is most acceptable to send a note of thanks to the first lady. Guests do not take hostess gifts to the White House.

Photographs - Guests inside the White House usually refrain from taking pictures. A White House photographer is present on social occasions to photograph the president and his guests, and these photos will be sent to the guests as a memory of the event. However, for events held outside on the White House grounds, guests are welcome to bring their video and/or still cameras and take as many photographs as they wish.

* * * * * * * * * * * * *

Another way to translate R.s.v.p. is, "Real snappy—very pronto!"

SECTION FOUR: SOCIAL GRACES

Chapter 10

Dress for the Occasion

"Fashions fade, style is eternal."
Yves Saint Laurent

If you have ever discovered too late that you wore the wrong clothes to a party, you *know* the importance of learning the meaning of the dress terms commonly used in Army social circles. For those of you who have found yourselves in this embarrassing predicament, it may be some small consolation to know that you certainly aren't the first, nor will you be the last, to commit such a faux pas. Unless you know what dress terms mean, it's easy to make such a mistake. The first rule to remember is this: If you receive an invitation and are not certain of the dress requirement, never hesitate to ask the hostess/host for more specific information. The second rule is: If you discover that you've worn the wrong type of clothes to a function, laugh it off and enjoy the party.

DRESS TERMS

BLUF

"Understanding the difference in dress terms can help you be the belle of the ball instead of wearing bell-bottoms to the ball."
MAJ Cassandra Perkins

Actually, there are only three basic dress terms: formal, informal and casual. The problems are that informal and casual are often confused for one another, and occasionally a few different dress terms are thrown in which may add to the guests' uncertainty. These factors are further complicated by the type of event, climate, location, and the fact that civilians may interpret the dress terms differently. For example, casual for Hawaiians and much of the Pacific means "Island Casual"— open-collared, colorful print shirt for men; muumuus, sarongs, or walking shorts for the women. Casual for Europeans may mean coat and tie for men and "Sunday dress" for ladies. It's important to learn the standard meaning of dress terms used in Army circles today, but also to consider any other factors that might affect their meaning before deciding what to wear.

FORMAL
This term is the least often seen, but the best understood.

Military formal function—The invitation may say "Formal," "Black Tie," or a specific uniform or choice of uniforms. Regardless of the uniform worn, the bow tie is standard for formal functions. Officers are expected to wear either the Army Mess uniform or Army Service Uniform, also referred to as Army Blues. For NCOs and enlisted personnel, Army Service Uniform (ASU) are preferable. Civilian guests will wear tuxedos.

Civilian formal function—The invitation may say "Formal," "Black Tie," "Tuxedo," or "Smoking" (European), but they all mean the same thing. Military men may wear either a tuxedo or an appropriately formal military uniform. (For an explanation of the *most* formal category, "White Tie," see the "Dress Terms Chart" section later in this chapter.) Although formal dress styles vary with the times and from one location to another, for women, whether military or civilian, a long dress is always appropriate for formal functions. Additionally, a very fancy short dress, cocktail dress, and evening trousers with dressy top are sometimes in fashion. If you're new in an area and wondering what to wear to a formal function, it's a good idea to ask a friend what is normally worn in that location. It's also important to consider the type of event before deciding what to wear; for example, a full-skirted ball gown would be lovely at a formal ball, but a bit too much for a formal reception. Note: Long, dressy dresses and long skirts are not normally worn before five o'clock in the evening, unless you are the hostess and the party is in your own home.

INFORMAL
This is the least-understood term of dress. "Informal" really describes a *dressy* occasion, but the term is often confused with "casual." Think of "informal" as having the same root word as "formal"; it's just one step removed from that. Sometimes, an invitation will state, "Civilian Informal," meaning civilian-style informal clothes for the military, not uniforms. Other more descriptive terms often used for informal functions are, "Business Suit" and "Coat and Tie."

Military informal function—The invitation will state the uniform, usually Army Blues with four-in-hand tie (long tie). The four-in-hand tie is always worn with Army Blues at functions held in the afternoon and those that begin before Retreat.

Civilian informal function—When the invitation says "Informal," men are expected to wear a suit—a dark (subdued) business suit with tie. The phrase "Coat and Tie" is a shade less demanding because this implies that men may wear a suit *or* sports jacket—but always *with a tie* for informal functions.

For the ladies, a civilian informal function requires a dressy dress; just how dressy depends on: the type of function, time of day, hostess's preference, and local customs. An informal dinner-party invitation usually means that the ladies will wear dressy clothes, jewelry, and heels; whereas at an informal daytime reception, luncheon, or seminar, they will wear more subdued dresses or suits. When in doubt, never hesitate to ask your hostess what she plans to wear.

CASUAL

This dress term has the least dressy meaning, but offers the broadest range of possibilities. For extremely casual events, a more descriptive term than "Casual" may be used. Another option is for the host or hostess to use "Casual" on their invitations and tell their guests what he or she will be wearing, as a way of clarifying how they might dress.

Military casual function—The invitation will state the uniform, probably "Class A" or "Duty Uniform."

Civilian casual function—The important point for men to remember is that a casual function means *no tie*. They may or may not wear a sports jacket, depending on the casualness of the event. When in doubt, they can wear a jacket and plan to take it off if the host isn't wearing one. Other terms occasionally used for events that are more casual are "Open Collar," "Very Casual," "Sporty," or "Jeans;" these terms imply that no jacket is expected.

Because casual functions can range from dinner parties to backyard barbecues, the range of dress is extensive. Guests need to consider the type of party and, if still in doubt, ask the hostess what she plans to wear. An invitation to a "casual" function usually means a lady wears a simple skirt, dress, or nice slacks, with simple jewelry and either low-heeled or flat shoes. Unit coffees are usually considered "casual." For even more casual events, very casual slacks, jeans, or shorts may be appropriate.

Another area of casual dress is that of sports-related invitations. Sports attire is definitely more "relaxed" today. All-white is no longer expected on the tennis courts, and there are few rules for golfers' dress other than comfort. The one possible exception to sports dress "requirements" is riding. If you accept an invitation that will involve riding and are a novice, you need to know what is considered proper attire. Traditional riding attire means trousers (preferably jodhpurs), boots, riding coat, gloves, and a hard hat (necessary for safety). For less traditional riding, you will at least need slacks and boots or shoes with a flat heel. Usually, if a hard hat is needed, your host will have one you can borrow. Riding western saddle is more informal: jeans, boots, and hat.

DRESS TERM CHART

The following chart shows the commonly used dress terms, along with the corresponding appropriate dress for military, and civilian men and women, along with tie requirements. The terms are listed on the chart in descending order of formality.

It should be noted that, historically, "Formal" referred to "White Tie" (tails), and "Semiformal" referred to "Black Tie" (tuxedo or dinner jacket). However, because so few attend these very formal affairs, the accepted meaning of these labels has changed. The term semiformal has disappeared from common usage, and what was semiformal has become formal. This leaves no term in common usage for the very formal or white-tie attire.

Occasionally, you may see "Semiformal" incorrectly used in place of "Informal" on an invitation. However, it is not a common dress term in use today and should be avoided to prevent confusion. In the event you receive an invitation that designates the dress as "Semiformal," the intent for civilian clothes is probably for the men to wear suits and the ladies to wear cocktail dresses. Check with the sponsoring unit or hostess to be certain, especially if you think a military uniform is desired.

Army Service Uniform (ASU) and Army Combat Uniform (ACU), along with Army Greens, are a few newer terms.

Commonly Used Terms	Army Uniform	Dress for Civilian Men	Dress for Civilian Ladies
Formal (White Tie)[1]	Blue or White Evening Mess	Tuxedo, bow tie is standard	Cocktail or Evening Formal Gown
Semiformal (Black Tie)	Blue or White Mess or ASU with bow tie	Dark Dinner Jacket or Tuxedo	Long or short evening dress
Uniform Informal	ASU with four-in-hand tie; or Army Greens	Business Attire	Cocktail Dress
Duty Uniform	ACU (or local policy); or Army Greens	Business Attire	Business Attire; afternoon dress or suit
Civilian Informal	See columns to right for civilian attire	Business Casual, Business suit or sports coat and tie	Dress, business suit or pantsuit
Casual	See columns to right for civilian attire	Business Casual, Slacks, Open collar shirt, jacket or sweater	Casual dress or blouse and simple skirt, slacks with jacket
Very Casual	See columns to right for civilian attire	Open collar shirt and slacks	Slacks with blouse

[1] "White Tie" is the most formal style of dress, but is seldom seen in Army circles. Civilian "White Tie" means "tails"—a long black tailcoat with matching trousers, white wing-collared shirt, white bow tie, and white pique waistcoat. Military "White Tie" for Army officers is the Army Blue Evening Mess uniform and White Evening Mess, which differs in certain details from the Army Mess uniform. See *Army Officer's Guide* for more information.

WEARING SPECIAL AWARDS

Occasionally, Army spouses are honored with special awards for their long-term unit and community efforts. An example of such an award is The Artillery Order of Molly Pitcher, which includes a certificate and a medallion. This medallion may be worn by the proud recipient whenever they please. Certainly, it is appropriate to wear such an award at a branch formal function, such as the Artillery's annual Saint Barbara's Day Ball. However, its wearing need not be limited to this once-a-year function. Any formal or informal occasion when the recipient cares to display their award and thinks it complements their outfit is a good time to wear it.

TRAVEL

Travel is a substantial part of the American way of life, and this is especially true for those in the military. It helps to know how to dress and pack for commercial travel. For extended overseas air travel, keep your hand luggage to a minimum and always wear comfortable shoes for the often-lengthy walks through airports. Dark clothes, even in the summer, travel better because they won't show soil as quickly. When weather permits, select cotton-blend knits for both comfort and wrinkle-free qualities. Slacks for women are accepted in most countries today, and certainly are more practical for extended travel than skirts or dresses. A light-weight jacket or sweater is always advisable, even during warm weather; airplanes are cool and blankets seem to be a rapidly disappearing item. Select clothing with lots of pockets. Use a large shoulder bag that closes securely. Alternatively, you might consider replacing your purse with a waist pack, backpack, or messenger bag both for security and for leaving your hands free—a special boon when traveling with small children.

SOCK COLOR FOR CIVILIAN CLOTHES

Most soldiers are so accustomed to wearing black socks with their black uniform shoes that they automatically select those same black socks when wearing black shoes, regardless of the color of their civilian trousers. However, socks should match or complement the trousers (and a belt should match the shoes). That's why civilian fashion designers have created so many different colors of socks! There are occasions when one might break this rule by wearing vivid socks to make a fashion statement, but the basic rule of matching cloth-to-cloth and leather-to-leather is the standard. You'll see that it makes you look well put together.

TYING A BOW TIE

With the increased popularity of wing-collared tux shirts and the accompanying hook-on bow ties, the use of proper bow ties is decreasing. However, since wing-collared tux shirts are only worn with tuxedos and the most formal Army Evening Mess uniforms, there are still plenty of opportunities for a man to wear a proper bow tie.

Few men ever learn to tie their own bow ties, perhaps because it's easier to tie one on someone else than to tie one on yourself. Learning to tie a bow tie isn't hard, and the results look so much better than a clip-on. Somehow, clip-on ties always look just a little too exact and flat—maybe even a little artificial. Learning to tie a bow tie is as easy as tying a bow and only takes a little practice.

Before you begin, here are a few tips learned from experience:

- The best places to shop for proper bow ties are tie shops and tux-rental shops.
- When selecting a bow tie to be worn with a military uniform, select a black one of moderate width and conservative material.
- The sculpture-shape or contour tie is easier to work with than the straight bow tie.
- Practice a few times on a friend or the bedpost before demonstrating your newly acquired skill.

After you've purchased the bow tie, it's time to practice, following the easy steps on the next pages. Remember that tying a bow tie is just like tying a bow. When you have perfected your technique and are actually tying your or your spouse's bow tie, begin with the shirt collar unbuttoned and turned up. Lengthen the tie about one inch by moving the slide and put the tie around the collar, then turn the collar down and button it.

1. Start with the left end extending 1½ inches below the right.

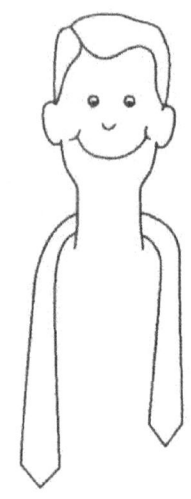

2. Cross longer end over shorter end, and pass up through the loop. Pull both ends to tighten.

3. Form front loop of bow by folding over shorter end (hanging) and placing it across collar points.

4. Hold this front loop with thumb and forefinger of left hand. Drop long end down over the front.

5. Bring long end around front loop of tie and poke it through knot behind. Adjust ends of bow and tighten in front by pulling on both sides of neck band. Take up slack at slide that's under the collar.

* * * * * * * * * * * * *

Understanding the difference between "informal" and "casual" is the key to understanding dress terms.

Chapter 11

Guests' Responsibilities

"The ornament of a house is the friends who frequent it."
Ralph Waldo Emerson

A great deal has been written about hosts' and hostesses' duties; very little mention is ever made about guests. Yet, guests have responsibilities as well. Receiving an invitation involves more than just going to someone's party, partaking of their food and drink, and talking with friends. Whether you are invited to share an evening, an overnight, a weekend, or a more extended stay with friends, you have responsibilities as a guest. When a host or hostess sends you an invitation, they expect you to assume certain basic courtesies toward them and the event they're planning. The following discussion describes the courtesies expected of both the party guest and the houseguest on receiving an invitation, attending the event, and expressing appreciation afterward.

PARTY GUESTS

It has probably never been easier to plan and give a party than it is today with our relaxed lifestyle. Parties are more fluid, harmonious, and easy going. That's not to say there's no work involved. Your hosts will go to great effort to ensure success; but you, the guest, are a vital ingredient. A great party can be more than the sum of its parts when each guest does his or her share.

Response - Respond promptly to an invitation, preferably within 24 to 48 hours. Understand the meanings of the various response indicators (for example, "R.s.v.p." and "Regrets Only") and respond accordingly. For more details, see Chapter 9, "Responding to Invitations." At the time of your response, note any dress term indicated on your invitation. If you are uncertain about what style of dress is expected, now is the time to consult with your host or hostess and request clarification.

If either you or your spouse has an extreme allergy to a certain food, by all means, inform the host or hostess. They would rather learn of that before planning the menu than at the event after the food has already been prepared.

Who Attends - Only those invited should attend. If you receive an invitation for a time when you will have houseguests, either plan to go without them, or regret the invitation and explain your reason to the host or hostess. They may offer to include them, or they may simply accept your regrets. If duty will prevent your spouse from attending, but you are free (or vice versa), again you should regret, giving your reason. If the party isn't a seated dinner, the hostess probably will invite you to come anyway. Don't plan to take babies or children, unless they are specifically invited or the invitation states that the function is for families.

Type of Party - Know the type of function you are attending. For example, if it's a cocktail party, you may need to eat dinner before or after, because only simple hors d'oeuvres will be served. On the other hand, a cocktail buffet promises more substantial food.

Proper Dress and Social Cards - Dress appropriately. Understand what the dress terms used on invitations mean (see Chapter 10, "Dress for the Occasion"); but, if still uncertain, consult with your host or hostess. Consider whether or not you want to take cards; appropriate times would be the first time you are invited to the commander's home, or a commander's New Year's reception in their home.

Arrival Time - Arrive on time—not even five minutes early, or more than fifteen minutes late. If necessary, kill time for a few minutes to avoid arriving early. If you will be unavoidably detained beyond fifteen minutes, call or text the host or hostess either a few minutes before the party begins or a few minutes after—but not right at the start of the party when they are greeting the other guests at the door. It's a good idea to take the invitation with you to have the address, invitation time, and phone number handy.

Hostess Gift - Taking a hostess gift when you have dinner in someone's home is becoming a common practice. The gift need not—indeed, should not—be expensive; an expenditure of no more than several dollars is the norm. Typical gifts are flowers (perhaps from your own garden), wine, candy, hostess soap, note cards, and home-made goodies. Wrap the gift, and be sure to attach or include a card with your name (a good use for a social card), because you may not have the opportunity to give the gift to the host or hostess personally.

When considering what hostess gift to take, remember that flowers, unless prearranged in an appropriate vase or bowl, will have to be given special attention. If your host or hostess does not have a helper, they will have to leave the other guests in order to put your flowers in water and arrange them attractively. While cut flowers in most foreign countries are plentiful, inexpensive, and a traditional hostess gift, in America you might consider giving your hostess another type of gift which will not take them away from their guests.

When a couple takes a gift, traditionally the wife usually carries it and presents it to the hostess (except for wine which is normally carried by the husband and presented to the host). The gift is handed to the host or hostess once inside the door, just after you greet them and shake hands. However, if both the host and hostess are busy elsewhere, leave the gift (except cut flowers) on a table near the front door, or ask whoever answers the door to give it to the hostess. If you take wine or candy, the host may or may not choose to offer it to the guests, as they already have the food and drink planned for the evening—and besides, your gift is for them personally.

A host or hostess will deal with any hostess gifts they receive differently, depending on what they are. Cut flowers will be taken away and arranged in a vase; then, along with any potted plants received, the flowers will be placed around the entertaining area where the other guests can enjoy them. Other gifts will be put aside to be opened later. Host or hostess gifts are never opened in front of guests, in order not to make those who didn't bring a gift feel uncomfortable. The hostess or host may, however, find a moment sometime during the evening to open their gifts quickly, out of sight of the guests, so that they can thank the gift givers as they leave. The thoughtful hostess/host will acknowledge the gifts they receive and, for very special gifts, write a brief note of thanks later.

The proper etiquette concerning hostess gifts varies with different countries. This is especially true with regard to flowers. Therefore, when selecting flowers to take to a hostess in a foreign country, consider their nation's customs. In some countries, such as Germany, red roses are reserved for sweethearts, and a flowering plant is considered inappropriate as a hostess gift. In other countries, such as Italy, chrysanthemums and asters are used only for funerals. See "Host Nation Courtesy and Customs" in Chapter 34 for more information on hostess-gift etiquette in different countries.

As You Arrive - Greet the host, hostess, and any guests of honor before you start socializing. Normally, the host and hostess meet their guests at the door, and greetings are exchanged there. However, at very large parties someone else may open the door and take your coat. In that event, remember to seek out the host, hostess, and any guests of honor before you get a drink or visit with friends.

Circulate and Talk - Guests are expected to help make a party successful by engaging in pleasant conversation and mingling with the other guests. Interesting, meaningful topics of conversation are encouraged, but the courteous guest avoids getting involved in discussions of controversial subjects that may create unpleasantness. If there's a guest of honor, take time during the event to talk with him or her. If you sit down, don't stay in one place all evening; circulate and talk with as many guests as you can.

Offers of Help - Unless the event is very casual, do not offer to help the host/hostess or go into the kitchen, either before or after the meal. If they need help, they will ask. When you go into the kitchen, it distracts the host/hostess from their last-minute preparation, and lets you oversee the food preparation (which a host or hostess may not enjoy). If you offer to help and the host or hostess should accept, any accident that might occur (such as dropping the salad bowl or getting a spot on your dress), would embarrass you both and complicate the party timeline. The modern, organized host or hostess (see Chapter 15, "Party Preparations") has carefully planned the party, so relax and enjoy the party, even while the host or hostess is working—just as they will do when you return the invitation.

Accidents - If you cause or are near when an accident occurs (e.g., spilled food or beverage, broken china, cigarette burn), take whatever immediate action is necessary and then be sure the host or hostess is informed. If something has spilled, they have access to the cleaning supplies and can take action before the stain is set. If something has been broken or burned, the damage isn't so easily undone; nevertheless, there may be cleaning up that should be done. The guest who caused the accident needs to face the host and/or hostess, apologize, and offer to replace or repair the damaged item. Accidents are sometimes unavoidable, but trying to cover up what has happened only makes matters worse.

Buffet Manners - Buffet dinners are becoming increasingly popular. It is a more casual way to dine than seated dinners; nevertheless, certain rules apply.
- Let the senior person and any guest of honor serve themselves first. The host and/or hostess should indicate when the food is ready and ask the appropriate person to start the line.
- If lap trays are available, it's better to take one *after* you fill your plate. In that way, you hold your plate directly next to or slightly over the serving dish, and nothing can spill as it might if your plate is centered on a tray.
- The host or hostess may indicate in which room they would like you to sit. Otherwise, sit wherever you like, but preferably not next to your spouse.
- Find a level, secure place on which to set your beverage. Hopefully, the host will have positioned the seats so that each one is near a table or hard surface. Avoid trying to balance your glass on your lap tray while you eat.
- Attempt to sit so that your lap forms a level, stable platform for your plate or tray.
- You may begin eating as soon as you've served yourself and been seated.
- Going back for seconds is a compliment to the hostess, but be sure everyone has gone through the line before you return.
- When you are finished eating, leave your napkin in your lap until you are ready to get up or someone takes your plate or tray. At that time, gather up your napkin and either hand it along with your plate, or lay it on your tray to the left of your plate.
- Taking your plate or tray to the kitchen is *not* a help to the host, especially if they are still eating. Wait until they begin removing the plates or trays, then one or two may offer their assistance. It's easier for the host/hostess to cope with a few plates or trays at a time than to have them all arrive in the kitchen at once.
- If you do assist the host or hostess in collecting plates or trays, do not stack one on top of another; carry one at a time or one in each hand, but no more. Those who help carry the plates or trays should not remain in the kitchen to assist in the cleanup.

Seated-Dinner Manners - There is much more etiquette concerning seated dinners than buffets. Read Chapter 12, "Table Manners," and Chapter 17, "Seated

Dinners," for more in-depth information; however, following are the most important points to be remembered:
- Traditionally, a gentleman should seat the lady on his right.
- Ladies sit down immediately. The gentlemen remain standing until the hostess is seated.
- Leave your napkin on the table until the hostess takes hers. Usually, if a blessing is said, it's done before the napkins are taken.
- Watch the host and hostess for your cue to begin eating. At a large dinner party, your host may ask that everyone begin once they are served. Otherwise, no one should begin eating the first course or dessert until the hostess does. Other courses may be eaten as soon as several guests have been served.
- If you are uncertain about anything—e.g., which utensil to use or how a particular food should be eaten—follow the lead of your hosts. The general rule concerning eating utensils is that you begin with the ones furthest from the plate. Any fork or spoon that has been placed above your plate is for dessert.
- Don't feel obliged to drink alcoholic beverages, even if they are poured for you—water should always be available. If you do drink alcoholic beverages, guard against drinking too much.
- Talk with the people seated on both sides of you during dinner.
- When finished eating, place your knife and fork together in the middle of your plate. This signals that you are finished, and prevents the flatware from falling off the plate when it is removed.
- If you must leave the table temporarily during the meal, place the napkin on your chair seat. That way, your used napkin will not spoil the appearance of the table. At the end of the meal, wait for the hostess to place her napkin beside her plate, a signal that everyone should prepare to leave the table. Then remove your napkin from your lap and place it in soft folds to the left of your plate.

When to Leave - When attending official functions and formal dinner parties, it is a courtesy not to leave before the senior person or guest of honor. If you must leave earlier, make your apologies to them first. Once the senior person or couple leaves, their departure usually signals the end of the party. If the invitation gave a stop time, as well as a start time, abide by it.

Spoken Thanks - Thank the host and hostess as you leave. Even at large unit functions, someone worked hard to arrange the party; never leave without expressing your appreciation for those efforts.

Written Thanks - Write a thank-you note promptly, normally within 48 hours, for any occasion when others have spent time, effort, and money to entertain you. (How to write thank-you notes is discussed at length in Chapter 13, "Expressing Thanks.")

Reciprocate - Accepting an invitation is like accepting a gift; you should return the favor. (In most cases, the only occasion that guests need not repay is the commander's New Year's reception; attendance for that event is expected of all military members, unless they're ill or out of town, and no social obligation is incurred.) High rank or position of the host does not lessen the guests' indebtedness. However, that should not cause concern, because it's never necessary to repay in kind; a simple cook-out or dessert-and-coffee invitation can repay a formal dinner. Also, there is no established time in which to reciprocate (even as much as six months later is acceptable). It helps to keep a list of those you need to entertain so you remember your obligations. Note: An invitation extended (whether accepted or not) fulfills your obligation to reciprocate. There are a few spouse functions that do not require reciprocation: luncheons, coffees, teas, pot-lucks, and baby showers.

HOUSEGUESTS

The mobile lifestyle of many Army families means you may frequently be invited to stay with friends, or invite a houseguest to your home. An invitation to visit friends is always welcome, but may be doubly so during a PCS. Not only does it afford the traveler an opportunity to spend time with old friends, but also a respite to catch up on laundry, ironing, and sleep, and perhaps save a little money. Your first responsibility as a houseguest (after you have accepted) is to let your hosts know when you will arrive; how long you can stay; and any special requests you may have, such as doing your laundry or visiting your favorite shop. If possible, mutually agree on a tentative schedule before you arrive. Once you arrive, make every possible effort to fit into your host's household routine. It helps for everyone to understand everyone else's agenda.

Luggage - If your visit comes during an extended travel period and you have a loaded car, try to pack a small bag that is easily accessible and adequate for your brief visit in route. When accepting an invitation that includes attending a particular function or participating in a special event, your host should let you know the appropriate dress and the anticipated weather.

Helping Out - As a houseguest, you have a duty to help make your visit as effortless as possible for your hosts. Whether your visit is for one night, a weekend, or longer, act as you would in your own home—make your bed, store away any bedding or bedrolls, keep your bedroom and bath picked up, etc. The longer you stay, the more you will want to make yourself useful with the household chores. When you depart, ask your hosts if they would prefer for you to take the sheets off the bed and replace them with clean ones (if there is time, offer to launder them), or simply to make the bed. Try to be helpful without upsetting the household routine.

If your hosts have a limited income, you might consider offering to help out with the added expense of your visit. If they decline your monetary offer, don't insist. Whether your offer is accepted or not, your thoughtfulness will be appreciated.

Visiting with Children - Visiting friends with your children can be a pleasant experience for all, particularly when your hosts also have children of approximately the same age and interests. However, the reverse can just as easily be true. To make the most of such occasions requires extra planning by both the hosts and the guests.

When planning such a visit, discuss your children's needs with your hosts in advance. Will they need special equipment—a crib, high chair? Are there events in your children's daily routines that must be considered when planning other activities? How will your children be supervised and what activities will be available for them? Everyone needs to know in advance what to expect. If your hosts do not have children or have older children no longer at home, they may need to do some "childproofing" of their home before your arrival. It would also help if they have an unlimited supply of patience. In the event it becomes obvious as you plan your visit that your family's needs are not compatible with your host's, consider staying instead at the post guest house or a nearby motel. You can still enjoy seeing one another during the day. Having children doesn't mean you must give up visiting your friends; it just means that the children's needs have to be considered and more planning done.

Thank-you Gifts - When you are a guest in someone's home, you will want to give your host/hostess a thank-you gift. It's usually better to select something in advance, make something and bring it with you, or perhaps purchase a gift while you're there, than it is to send something after your visit. A gift sent later can appear to be an afterthought and, if your host and hostess are living in certain overseas areas, it's even possible that they would have to pay customs duty on it. Your thank-you gift need not be expensive to be appreciated; selecting or making something that is especially appropriate for the host/hostess or their home says you truly care about them and want to show your appreciation. If funds are a consideration, remember the simple formula for an inexpensive but meaningful gift—*make it, bake it, or grow it.*

There are other possibilities to consider. If your hosts have children, it's nice to bring them gifts in lieu of a gift for the parents. Examples might be toys or stuffed animals for young children, or the latest board game for young adults. Another very meaningful way to say thank you includes the gift of your time. Perhaps you could offer to take their children on an excursion—to the zoo, museum, or a movie. If you are a skilled cook, you might prepare your special recipe (bring any unusual ingredients with you); or simply offer to take everyone out for a meal. Your thoughtfulness toward your hosts and their family tells them that you genuinely appreciate their hospitality.

* * * * * * * * * * * *

Being a good guest means more than just having a good time.

Chapter 12

Table Manners

"It is you who sets the example to your children for table manners."
Letitia Baldrige

Table manners, like all social behavior, are an important aspect of how we relate to others. Our eating behavior, loosely defined as our table manners, should reflect the same thoughtfulness and courtesy toward others that guide our social interaction.

GOOD TABLE MANNERS

Learning good table manners is something best done as a child, under the watchful eyes of caring parents. However, it's never too late to learn, and those who feel that their education in this area has been neglected may find this chapter especially helpful. When in doubt, one can always fall back on the old advice to watch your host and hostess and follow their lead. However, that has been known on occasion to lead to rather embarrassing results. Charlotte Ford, in the 1988 edition of her book *Etiquette,* tells a marvelous story about some Vermonters who went to dine with Calvin Coolidge at the White House. Because they wanted to be proper, they decided to watch the president and do exactly as he did. Things went wonderfully well until the coffee was served. The president very carefully poured coffee into his saucer, then added cream and sugar. They quickly followed his actions. Imagine their horror when he bent down and put the saucer on the floor for the cat!

Table manners traditionally revolve around what goes on at the table while eating. The basics are pretty simple. Sit with good posture, hands in your lap when not eating and perhaps forearms on the table between courses. Don't take your napkin until the hostess takes hers. Remember that serving dishes are passed around the table in a counterclockwise direction; food that is served to you will be served from your left. Begin eating when the hostess starts to eat, except at a buffet when you may begin to eat as soon as you have served yourself and are seated. If you don't know which eating utensil to use, see what your host or hostess is using (see below for more details). If you must remove something from your mouth, the rule is to remove it the same way it went in— if you used your fork, take it out with your fork; if you used your fingers, take it out with your fingers. The one exception is for fish bones; for safety, these are always removed from your mouth with your fingers. Talk with the guests who are sitting on either side of you. If you must leave the table before the end of the meal, lay your napkin in your chair; when you leave the table at the end of the meal, lay the gently gathered napkin by your plate.

That's about it; if you follow these basics, you will be comfortable even at the most formal dinner.

WHICH FORK TO USE

Ever since forks became widely used, which was about two hundred years ago, this has been the most frequently asked etiquette question. Everyone's fear has been of making a mistake in selecting the right table implement, or knowing how to use an unfamiliar one. The most practical answer is that it shouldn't greatly matter. Emily Post once remarked, "No rule of etiquette is of less importance than which fork we use"; this remains true today. Common courtesy demands that we relegate the great fork question to the realm of trivia, and overlook any transgressions.

That isn't to say that you need not attempt to learn what the different eating utensils are used for, the order in which they are typically used, and how to use them correctly. Even though we acknowledge that using the incorrect piece of flatware is no major calamity, it's equally true that knowing which one to use and how to use it will help you relax and enjoy even the most formal dinner. This knowledge will also help you set your own table correctly, and train your children in the proper use of these implements. (See Chapter 17, "Seated Dinners," for a place setting diagram.)

Place Setting

No Required Place Setting - There is not a required place setting that is always used, because only flatware that will be needed during the course of a meal should be placed on the table. The number and variety of flatware in any place setting can vary, based upon: the menu, order of courses, and style of the meal (formal or informal). However, there are limits. Tradition says that there should be no more than *three* of any one type of flatware used in a place setting for even the most formal dining. In other words, no more than three forks, three knives, or three spoons should be used. If more than three are needed, selected ones may be omitted from the place setting (such as the salad fork and knife, and the dessert fork and spoon) and brought in with the appropriate course. The exceptions to that rule are that the cocktail fork may become the fourth fork, and the butter knife may become the fourth knife used in a place setting.

Standard Placement - There is a standard placement for the different types of flatware used in a place setting. Forks are traditionally placed to the left of the plate (except the seafood fork that is always placed to the extreme right, and the dessert fork that may be placed above the plate). Knives are placed to the immediate right of the plate (except the butter knife that is placed across the top or side of the bread plate). Spoons follow the knives to the right of the plate (except the dessert spoon that may be placed above the plate).

The guiding principle in the placement of flatware is this: Eating utensils are placed on either side of the plate in such an order that, as they are used, the place setting remains intact in appearance. In other words, the utensils that will be used first are placed farthest from the plate; as each is used, the remaining utensils still appear neatly arranged next to the plate. Therefore, assuming a place setting has been correctly arranged and that you use every utensil provided for each course, you always use the outside piece of flatware. If a particular course requires two implements, select the outside fork on the left and the outside knife on the right. By the time you get to the dessert course, there should be no more flatware on either side of your plate. For formal dinners, the dessert fork and/or spoon will be brought in with the dessert; for informal dinners, these utensils may have been placed above your plate and you simply reach up and move them down (fork to the left, spoon to the right). Spoons for coffee will be served with the coffee. Serving utensils may be used for serving any type of food that seems appropriate. For example, a cold-meat fork may be used for serving hot meat, or anything else you care to serve with it. Don't be limited by the names of the serving utensils.

AMERICAN AND EUROPEAN STYLES

Because many Americans have lived or traveled in Europe, you will now find a large number adopting the European style of eating. In fact, the European style is actually easier, quieter, and more efficient—especially for left-handed people—and most of the world uses this manner of eating. The origin of the American custom of transferring the fork is unknown, although eating with the fork in the tines-up position seems more in keeping with the design of a fork. Certainly, in today's world, both American and European styles are acceptable.

American Style - The American style of eating means that all food is eaten primarily from the fork held in the right hand, tines up. In order to cut, the fork is transferred to the left hand, and the knife is picked up and held with the right hand. When cutting is finished, the knife is placed across the right, upper edge of the plate, sharp side of the blade toward the center of the plate; and the fork is transferred to the right hand to begin eating again. When finished eating, the fork and then the knife (if both have been used) are placed in the middle of the plate, fork with tines up and knife with sharp side of the blade facing in, handles resting on the side of the plate. (This placement is important because it signals that the person has finished eating, and it precludes the knife falling off the plate when it is removed.) Another feature of American-style eating is that the left hand generally remains in the lap, out of sight, except when it is needed for cutting or eating.

European Style - To eat using the European style, hold the fork in the left hand (tines pointed down, except for slippery food), and the knife in the right. Cut in the customary manner. When pausing during the meal, the knife and fork may be placed on the plate in an inverted "v" position, with the fork tines down. When finished eating, the fork and knife are placed side by side in the middle of the plate, as with the American style of eating. Because both hands are used in the European style and because of custom, both hands remain above the table at all times. In fact, in Europe it is considered bad manners to put your hand in your lap. When not using one or both of your hands for eating, simply rest the arm just above the wrist on the table edge.

SPECIAL CHALLENGES

Table manners consist of more than simply knowing which eating utensils to use. They also involve knowing how to eat the different types of food, hold your glass, use your napkin, and a myriad of other special challenges that often confront you at the dinner table.

Etiquette for Special Foods

Artichokes - These unusual vegetables are a quite popular finger food, especially with children who enjoy taking them apart. Peel leaves off one at a time with your fingers, beginning with the outer-most leaves. If a sauce is provided, dip the fleshy end in, and eat the edible portion of the leaf by pulling the soft end between your front teeth. Leaves are neatly discarded on the side of the plate or your bread plate. When you reach the choke (thistle-like center), hold the artichoke with a fork or fingers, cut away the "feathers" by sliding the tip of the knife underneath, then cut and eat the remaining "heart" with your fork. If the artichoke is filled with a sauce, the choke will have already been removed.

Asparagus - Most people eat asparagus with a knife and fork; however, small, fresh asparagus may be eaten with your fingers by picking them up by the stalk end. Europeans have a passion for asparagus in season, especially the Germans who regard their white asparagus as a culinary treat.

Bacon – Very crisp bacon may be eaten with your fingers. Otherwise, it should be eaten with a fork.

Chicken - Chicken should be eaten with a knife and fork. The only exception is fried chicken served at a very casual function, such as a picnic or family meal.

Rolls, Muffins, Biscuits, and Toast - These finger foods require different techniques. Rolls and French bread are eaten by breaking off a small piece. Everything else is cut—but in different directions. Muffins are cut in half vertically, and biscuits are cut horizontally. Bread or toast is cut in half. To butter these items, there are a few more points of etiquette to remember: Biscuits and toast are the only breads completely buttered while they are hot; everything else is buttered in small pieces just before being eaten. To apply butter, hold the piece of bread on the bread plate, not up in the air.

Salad - Salad is usually prepared so that the ingredients are small enough to be eaten without cutting. If something does need cutting, first try using the side of your fork; then, if necessary, use your knife and fork.

Seafood - When fish is served whole, begin by cutting off the head. Then hold the fish with your fork while making a cut down the top of the back, remove the top fillet, and lift the backbone off of the bottom fillet. A separate plate is usually provided for the head and bones. Use your fingers to remove any bones that get into your mouth. Jumbo shrimp served as shrimp cocktail in a compote are too large to be eaten in one bite. Such shrimp, served with the tails left on, may be held by the tail to eat; without tails, they are eaten with a cocktail fork. Either way, after taking a bite, it is perfectly permissible to redip the shrimp before eating the remainder. It is not a good idea to try to cut the shrimp first. Lobster and crab are meant to be a feast, so don a bib if you like, take up the "crackers," and enjoy. The entire soft-shell crab is to be eaten.

Snails - Snails served in the shells require special utensils. They are served on a hot metal or ceramic plate with an indentation for each shell. Snail tongs and small picks, or seafood forks, are usually provided. The snail tong is held in the left hand, squeezed together to open the end for the hot snail shell, and held to steady the shell on the plate while the snail is removed with the pick or small fork. Snails are eaten whole. When the shells have cooled a bit, they may be tilted and drained for the delicious garlic butter remaining, either into the mouth or onto the plate where it may be soaked up with bread.

Soft-boiled Egg - Soft-boiled eggs are served in eggcups, small end up. To begin, hold the egg in place with one hand and slice off the end of the shell horizontally with your knife blade about half an inch from the top; cut all the way across and remove the top. The egg inside of the top piece may be eaten with your spoon before the shell cap is placed on the side of your plate. Add salt and pepper, if you like, to the remaining egg, and eat from the shell with your spoon.

Soup - The way soup is eaten depends on the type of soup and the container in which it is served. If the soup contains large pieces of the ingredients or floating garnish, these should be eaten before the liquid. Soup is spooned away from you and then sipped from the edge of the spoon. In order to spoon up the last few mouthfuls, the soup plate may be tipped slightly away from you. When soup is served in a bowl with handles or in a cup, it may be drunk by picking up the bowl or cup, or part may be eaten with a spoon and the remainder drunk. During pauses while eating soup and when finished, the soup spoon is normally placed on the place plate beneath the soup plate or bowl. The only exception is for a soup plate or bowl that is so large a spoon cannot rest securely there.

Spaghetti - If you are given a place spoon (tablespoon), hold it in one hand, and rest the side of the spoon against the plate; with the other hand, twist a small amount of pasta onto the fork by turning the tines around in the bowl of the spoon. With practice, this same technique can be accomplished without the use of the place spoon. Alternatively, spaghetti may be cut with the fork.

Etiquette at the Table

Stemmed Glasses - Large, stemmed glasses are properly held by gripping the base of the bowl with the thumb and fingertips, not by gripping the entire bowl. An exception is made when chilled wine is served in a stemmed glass; then, the glass should be held only by the stem, in order not to warm the wine with your fingers.

Mugs - When coffee or other hot beverage is served in a mug *without* a saucer, care must be taken with the spoon to prevent soiling the table. If you have a plate, put your spoon there. Otherwise, be sure to remove the excess liquid from the spoon onto the edge of the mug, after you stir, before placing the spoon on the table.

Napkins - Traditional napkin placement before the meal is either on the plate or, if the first course is in place, to the left of the forks. However, intricate napkin folds and unusual placements are very popular, so don't be surprised to find the napkin in such a non-traditional place as your empty wine glass. Napkin rings are also very fashionable and collectible. Guests should not take their napkins from the table until the host or hostess has done so, nor return them to the table at the end of the meal until the host or hostess does. Small luncheon napkins should be unfolded completely and placed across the lap. Larger dinner napkins are left partially folded (usually with the folded edge placed away from you) so that you can wipe your fingertips on the inside and not get stains on your clothing. If you must leave the table during dinner, put your napkin in your chair so that its crumbled and perhaps soiled appearance will not be visible. When dinner is over and you are about to leave the table, gather the napkin in soft folds and lay it to the left of your plate. You may or may not return it to a napkin ring; it's your choice.

Salt and Pepper - Only add salt and pepper to your food *after* you have tasted it; to do otherwise is an insult to the hostess, who presumably has ensured that everything is seasoned to perfection. If the salt and pepper are passed before you have had an opportunity to taste everything, you may put some on the edge of your plate for later dipping.

Smoking - When dining in someone's home, the absence of ashtrays is a clear indication that this is a nonsmoking home; if you feel that you must smoke, step outside to do so. There may be a cigar bar after dinner, but the cigars are usually smoked outside as well.

Spills - If you spill something in someone's home and it's a serious stain, such as red wine, be sure that the host or hostess is made aware of it right away. They will want to take steps to correct the problem while it can still be done easily. For example, a red wine stain can effectively be treated with salt (for spills on the tablecloth) or club soda (for spills on the carpet), if applied promptly. In the case of a badly soiled tablecloth, you should also offer later, in private, to have the linen cleaned. When you have a real "disaster" while dining, apologize but don't dwell on the accident. To do so only calls more attention to it. Your hosts and friends will forgive you; anyone who doesn't isn't a real friend.

* * * * * * * * * * * *

Common sense and courtesy to others are the
essential guides to good table manners.

Chapter 13

Expressing Thanks

"From my heart, thank you."
Ginger Perkins

We've all been taught to say thank you, but often this is not enough. This is not to say that a thank-you note is always necessary. On some occasions a phone call, another word of thanks extended in person, a returned kindness, or even a small gift may be more appropriate. The keys to developing good manners in expressing your thanks are these: Don't take other people's kindness for granted, be prompt in expressing your additional thanks, and practice!

When Is It Appropriate? - Knowing when additional thanks are appropriate is easy. Almost every kindness, except the smallest gesture, requires some form of appreciation beyond the initial thanks. For example, when you are entertained by someone—a dinner or a party (even in someone else's honor or a promotion party)—saying "thank you" as you walk out the door is not enough. No doubt the host has gone to a lot of trouble and expense. The efforts that went into preparing for the event, and the enjoyment you've derived from those efforts, warrant an additional expression of appreciation. Another time when an additional thank-you is needed is when someone has gone out of their way to do something special for you. For example, a neighbor drives you to a hospital appointment when you don't have a car, or you need volunteers and they willingly offer to help—simply to say "thank you" once for such kindness may appear to be ungrateful.

How Should You Express Your Thanks? - Knowing how to express additional appreciation is equally easy. Do what is necessary to ensure that your appreciation is perceived to be genuine. For most kindnesses a thank-you note is commonly used, usually proper, but not always necessary. In some cases, a phone call or text may be more appropriate, especially when it's to someone you call routinely—but do so within a day or two, and say your thanks at the beginning of the conversation. (One exception: After a big party a note is always better, even if the host or hostess is a very good friend—so that he or she isn't inundated with thank you phone calls.) There are also many situations when expressing your appreciation person-to-person is suitable—for example, when you see your hostess or host before you have had a chance to write or call to say thank you (but never say thanks in front of others who weren't at the party), also when you receive a small gift or kindness. Returning a kindness is another appropriate way to express your thanks for small neighborly gestures of friendship, such as sharing one of your culinary specialties with a neighbor who has shared theirs with you. Then there are those rare occasions when a kindness is so great that you feel words and simple repayment are not enough. On those special occasions: for an individual, you can select a small gift to accompany your note of thanks; for a group, you can do something out of the ordinary, like have an appreciation tea. In other words, your appreciation needs to be appropriate for the kindness.

WRITTEN THANKS

> BLUF
>
> Expressing gratitude lowers stress hormones.

For those occasions when it is appropriate to write a note of thanks, you will want to be able to express your feelings in just a few sentences. Writing a sincere thank-you note is an acquired skill that takes careful thought and a bit of practice. There are basic guidelines. They concern who normally writes a note, whom to address it to, and what is included. However, your promptness and expression of appreciation are what really matter, not how well you follow the guidelines. When you are unsure if a thank-you note is necessary, the best guidance is: When in doubt, write a note.

Written thanks can take several forms. The typical thank-you note is written on a small note card. However, you may also use a small sheet of stationery; an informal (if personalized, use your personal informal, not joint); your personal social card (not joint); or a commercial thank-you card with a personal, handwritten note added.

BASIC GUIDELINES

Regardless of which partner serves as the family's social secretary—sending invitations, responding to invitations, and writing the thank-you notes—either spouse may write a note of thanks.

- Traditionally, thank-you notes for social events are addressed and written only to the hostess, even when the host was involved in the event. You, as the guest, express your appreciation to the hostess. Of course, you also mention your spouse's appreciation and extend special thanks to the host.

- Notes written for a gift or kindness are addressed to the individual or group responsible.
- Thanks seem more sincere when something special about the event or gift is specifically mentioned. Notes that only say, "Thanks for inviting us—we had a great time," and "Thanks for the great gift—it's just what I've always wanted," simply don't say enough.
- The person who writes the note signs only his or her name to the note. Joint signatures are considered proper only on greeting cards, postcards, and gift enclosures.
- Thank-you notes may be mailed or hand delivered (see "Hand Delivery" in Chapter 8).

HOW SOON TO WRITE

The courteous thank-you note is written the day after the event or receipt of a gift—at most, within forty-eight hours. Such promptness has several important benefits: (1) Your response seems more spontaneous and, therefore more sincere. (2) You can more easily recall details of the occasion, or your feelings about the gift, to help you write a meaningful note. (3) Your thank-you note for a gift (e.g., farewell, welcome, birthday, anniversary) assures the sender that the gift was received, which is especially important for gifts sent by mail.

Whether or not a hostess writes a thank-you note for the hostess gifts she receives depends on the gift. A small, inexpensive gift is best acknowledged at the time it is received, as the guest is departing, or the next time the hostess sees or talks with the guest. However, a very special hostess gift (perhaps a handmade item, a gift selected especially for the host/hostess, or an obviously expensive gift) should be acknowledged with a written note. The host/hostess need not write quite so promptly as the guest should. They courteously give the guest time to write first, then follow up with a very brief note. However, if the guest hasn't written in four or five days, the host/hostess shouldn't wait any longer to send their thanks.

Despite good intentions, we all occasionally slip up and delay writing. What then? Do we just forget it or write a note anyway—even a week or two late? You can either write your note with a brief apology, or call. A phone call at this point is always acceptable; the host/hostess isn't being inundated with calls from the other guests and, with a simple apology, you can sincerely thank them for the enjoyable time you had.

There are a few occasions when circumstances explain a delay in writing thank-you notes. For example, a mother-to-be who goes to the hospital to deliver shortly after receiving gifts at a baby shower is not expected to take her stationery with her to the hospital. For such a busy lady, a few weeks' delay in writing those thank-you notes is certainly understandable and requires no apology. The same is true for other unusual circumstances. Let common sense be your guide, while remembering that the sooner the thank-you notes are written, the better.

The spouses of command sergeants major, unit commanders, and general officers should remember that their manners help set the standard for those around them. When they write thank-you notes promptly, others notice and are encouraged to do the same.

Similarly, parents set the standard for their children, and should teach them to write thank-you notes as soon as they can write. Their thoughtfulness will be very much appreciated, even if their penmanship is lacking, and the habit of writing thank-you notes will become an enduring practice in their lives.

THANK-YOU NOTE FORMULA

A typical thank-you note follows the basic format shown below. Notes written for gifts or kindnesses can follow this same format by substituting the appropriate information—name the gift or kindness, highlight why you like it or how you will use it, and sign off with an appropriate closing phrase.

Dear (host/hostess's name),

(Soldier's name) and I would like to thank you and (host/hostess's name) for (day of event). The (name the event) was (few complimentary words describing the event). (Short sentence expressing appreciation for being invited). (A descriptive sentence or two praising the delicious meal, the beautiful home, or the gracious hospitality). (A sentence that compliments the host.)

(A "thanks-again" sentence. Final sentence to express your hope of seeing them soon or a final compliment).

Sincerely,

(writer's signature)

(day note is written)

Your choice of words and the enthusiasm you express in describing your enjoyment of the occasion, gift, or kindness will personalize and add life to this stiff format. Be certain that your thanks are specific enough that the note couldn't have been written before the event or to any other person. Avoid overuse of flowery adjectives, but don't omit them completely and end up with a stiff note. Writing a rough draft can be very helpful because you are trying to convey a great deal in a few well-chosen words. If you feel really uncomfortable writing a note of any length, a few words jotted on your social card—such as "Mary, Saturday evening was great! Thanks for including us."—can suffice.

However you extend your thanks, try to do so within a few days of the event. When it comes to saying thank you, promptness equates to sincerity.

Dear Janet,

Thank you and Mary for hostessing the December coffee. It was a big success!

This is such a busy time of year for everyone, and to have your tree up and the house all decorated for us was really special - and beautiful! It certainly put everyone in the mood for our cookie exchange.

Please share my appreciation with Mary. Her gingerbread house centerpiece was the prettiest I've ever seen.

Happy holidays,
Margaret

Thursday, 12th

Harry—This is my favorite champagne. Hope you like it too!

~~Mr. Thomas Adam Cook~~

Thomas

Notes on calling cards:

> Mary, it was such a delight to see you again. Thanks so much for ~~Mrs. Joe Daniel Farmer~~ including me.
>   Jamie

> Beth, we had a wonderful time. Many thanks.
>   ~~Mrs. Harold Grant Banker~~
>   Love,
>   Carol

* * * * * * * * * * * *

Expressing prompt and appropriate thanks is another way of showing thoughtfulness toward others.

SECTION FIVE: ENTERTAINING

Chapter 14

Entertaining with Ease

"The secret of successful entertaining is not to take on a style that is burdensome and laborious, but to find one's personal style
– whatever it may be."
Marjabelle Young Stewart

BLUF

When entertaining, be yourself,
be prepared, and be spontaneous.

We entertain for a variety of reasons. The most important ones are to please our friends, both old and new, to bring people together, and to enjoy their company. Entertaining in the Army differs little from the civilian world. However, as an Army spouse, you will probably entertain, and be entertained, more than your civilian counterpart. There are a multitude of ways to entertain and many resources—books, magazines, internet—available to help a host. This chapter, hopefully, will provide you with another resource. Nowhere is it written that as an Army spouse you must entertain, but remember that opening your home and providing a welcoming environment, regardless of what is provided, is a huge step in building friendships and relationships.

EASY STEPS TO BECOMING A SUCCESSFUL HOST/HOSTESS
Enjoy people and well-prepared food.

The first requirement for becoming a successful host/hostess is simply to enjoy being with people. But, you have to relax in order to do that. Remember that entertaining can be fun when you don't worry and get uptight about it. Even when you do "official entertaining," it's important to relax and be yourself. You'll discover in the process that you can enjoy your guests much more, and your enjoyment will be contagious.

In addition to enjoying people, a successful host/hostess cares about what is being offered to the guests. That's not to say the menu has to be expensive or formal—

quality, freshness, a pleasing blend of tastes and colors, and care in preparation are the essential ingredients. If you're not an experienced cook, simply concentrate on mastering a few recipes that are appropriate for entertaining. Bookstores and libraries have cookbooks to help even the most unskilled cook plan and prepare menus and entertain. There are many websites that have sample menus and ideas for entertaining anywhere from 4 to 20 people, from simple to extravagant, and all on a limited budget. Finally, don't be afraid to ask for help or advice from more-experienced spouses.

Decide whom you want to entertain.

Guest List - Your guest list is important. Whether it's an informal supper, buffet, or cocktail party for fifty, you will want to invite a harmonious group of guests. It's certainly appropriate to include on your guest list those people who have entertained you, but that shouldn't be the only criterion. Consider how your guests will get along and enjoy one another's company. While it's certainly important for guests to make every effort to mingle and talk with the other guests, it's equally important for the hosts to invite a pleasing mix so that the guests have interests enough in common to stimulate conversation.

The Norm - As an Army spouse, you will find that you tend to entertain in order to foster camaraderie and to help establish the friendships that support you through this vagabond life. Those you entertain will probably be a mix of your soldier's military colleagues and their spouses, along with friends, civilian guests, and your own colleagues.

Army spouses typically entertain the following groups:
- your soldier's immediate supervisors (to welcome or farewell them, or repay their hospitality)
- your soldier's unit contemporaries (often to celebrate some successfully accomplished major task at work, or their return from an extended field exercise or deployment)
- your soldier's subordinates, one level down (to get to know them better and give them an opportunity to know you)
- unit coffee group (when it's your turn)
- those who have entertained you (to reciprocate, but not necessarily in the same style)
- neighbors (to have fun and get to know them better)
- your own acquaintances—business colleagues, bridge group, sporting companions, church group
- last-minute guests whom your soldier invites on the spur of the moment (Don't worry or apologize for not having the house in perfect order or not providing a fancy meal; they truly can take "potluck.")

This list may look intimidating, but don't let it frighten you. It's only presented as an overview of typical Army entertaining. No one entertains all of these groups in a short period of time. Additionally, as you move around, you will find that various locations and assignments call for different amounts of entertaining. Your personal family situation also affects how much you entertain. Some periods in your life may be relatively free, while others too full with family, job, or outside interests for you to entertain any more than the barest minimum. In sum, do what feels right for you and your soldier.

Entertain in a manner that suits your budget and your style.

Plan entertaining that's easy on your budget, and don't let the manner in which others entertain influence you. Never be tempted to spend so much on entertaining that it affects your budget for the rest of the month. Remember that fancy food, crystal, and silver are not important for successful entertaining; people can enjoy a very casual get-together with simple food as much as a formal party with an elaborate menu. It's the simple act of getting together that's most important.

Remember that no one expects to be repaid in kind for their hospitality. If you and your soldier are invited to a formal dinner at the boss's home and the best china, silver, and crystal are used, this doesn't mean you must reciprocate in the same way. A gourmet dinner can be repaid with something as simple as a backyard cookout where hot dogs are served on paper plates. Simply select a menu suitable for the type of get-together you are planning—one that is eye-appealing, tasty, and well prepared.

Entertain in a style that is pleasing to you. If you prefer casual get-togethers, then that's what you should have. You will have more success when you feel comfortable. Also, plan a party that you can accommodate with ease. If you don't have all the fancy accessories for entertaining formally or enough place settings for the number you want to invite, then plan something casual using your everyday things, or substitute plastic or paper products. Pretty, coordinated paper plates, cups, and tablecloths are available today, as well as attractive plastic ware that is inexpensive and reusable. By simply coordinating the things you use, a very pretty appearance can be created. Certainly, never apologize for the things that you don't have. Your guests will not notice if you don't call it to their attention.

Follow the etiquette guidelines that pertain to invitations.

- Start early. If you're inviting a special person or hosting a unit coffee, check with that special person or the coffee-group leader *before* setting the date.
- Send out invitations at least two to three weeks in advance.
- If you extend invitations by email, follow up with an email reminder.
- If you use written invitations and indicate a response is expected, don't hesitate to call those who don't respond in a reasonable time. They may not have received the invitation.
- Understand the terms frequently used for parties and styles of dress:

cocktail party—drinks and simple *hors d'oeuvres*

cocktail buffet, heavy hors d'oeuvres—stand up and eat, with substantial enough food so that guests don't have to eat before or after

buffet dinner—food on table or buffet, guests sit elsewhere to eat

dinner—guests sit together at table(s) to eat

formal, informal, casual dress (See definitions in Chapter 10, "Dress for the Occasion.")

- Keep a careful record of the guests' acceptances and regrets.

(For more detailed information about invitations, see Chapter 7, "Extending Invitations.")

Select the menu with care.

- Plan your menu well in advance (usually with your soldier's input). Consider your budget; your cooking capabilities (don't experiment on guests); your kitchen capacity (stove, oven, refrigerator, freezer); your serving pieces and cooking pot sizes; guests' food allergies or religious restrictions; and pleasing food combinations (colors, shapes, tastes, temperatures, textures).
- Plan a menu that permits as much pre-party preparation as possible (some for freezer, some for refrigerator). Leave as little preparation as possible for the day of the party and for after the guests arrive.
- It's nice to feature your own personal specialties, regional cooking, or dishes you've learned to prepare from your travels and life overseas. If you are featuring the cuisine of a particular region (e.g., Hawaiian, Tex-Mex, Cajun, German), consider giving your guests the option of dressing accordingly. Theme parties, such as '80s or Mardi Gras, are another possibility.
- Once you've found a proven menu or two, they can be "recycled" for the different groups you entertain.
- Decide on the drinks you will offer—a full bar, or wine and beer as the only alcoholic drinks. Always have water available. Do you want to feature a special drink? Or do you prefer to serve no alcohol?
- If your budget permits, another option is to have all or part of the meal catered. Most military communities have at least one Army spouse who has a gourmet-dessert or cake-decorating business. It's often fun to have a really spectacular dessert created by someone else, leaving you more time to concentrate on the entrée. (Your post newspaper and spouse club newsletter are often good sources for such information.)
- Prepare more food than you think you'll need. It's better to have extra than not enough.

Plan and prepare the physical layout.

- Decide on the following: method of service, traffic pattern, size of dining table, centerpiece, serving pieces and their placement on the table (experiment to

make sure everything fits and balances), location for serving dessert and beverages, seating plan and place cards.
- In preparing your home for the guests, there are a number of things to consider. Look at every area that will be seen or used from the standpoint of your guests and their comfort. Ensure that the front yard, sidewalk, and porch are clean and tidy. Make room in the closet for coats and provide sufficient hangers, or prepare a bedroom where coats can be laid on the bed. Provide enough seating, coasters, napkins, and a table within reach of every chair (if guests will be eating on their laps). See that the bathroom has fresh soap, towels, and plenty of toilet paper. Provide an outside area if there are any tobacco users.
- Look at the various possibilities your home offers for expanding your entertainment areas. What about using the patio, basement, garage or carport?

Make party planning a prerequisite.
(See Chapter 15, "Party Preparations.")

- Plan to allow yourself extra time on the day of the party. If you plan well and follow your plan, you should have time to rest during the afternoon before the party. This rest is important so that you won't be tired when your guests arrive. This cushion of time also provides for anything unexpected—your never-fail recipe fails, the dog gets sick and has to be taken to the vet, or a million other possibilities. The allotted time, whether used for rest or the unexpected, will help you be composed and unhurried. If you are comfortable and relaxed, your guests will be too.
- Discuss with your soldier the help you will need from him or her during the preparation and after the guests arrive. Discuss where everyone will sit, and when to stop serving drinks before dinner.
- So that you are not distracted, plan how your young children and pets will be fed and taken care of immediately prior to and during the party.

Don't forget the last-minute details—

- Be dressed one hour early.
- Take food from refrigerator and/or freezer that needs to come to room temperature or defrost. That includes the butter.
- Turn on the porch light and appropriate house lighting. Light any candles you want to use. Turn on any music you have planned.
- Don't hurry! If you catch yourself racing, purposely slow down your movements. You will retain your composure and stay calm only if you don't allow yourself to rush.

Remember your host and hostess party manners.

- Preferably both you and your soldier greet your guests at the door. Shake their hands as you welcome them, remember to smile and make eye contact, and take their coats (or give directions as to where they should be put).
- Don't go to the door with a drink in your hand.
- Introduce newcomers to those already there, or to a small group.
- Place your guest book (if you plan to use one) in an easily accessible place. Have several pens available.
- Talk to each of your guests sometime during the evening, even at a large party.
- American etiquette says don't open hostess gifts in front of the other guests. If wine or candy is brought, don't feel as though you have to serve the wine or offer the candy.
- Be prepared for flowers by having vases and clippers handy. Quickly put flowers in water and place in the rooms the guests will be using.
- Be alert to guests' needs. Is anyone being left out of the conversation? Do any glasses need to be refilled? (Of course, never serve so much alcohol that a guest becomes tipsy.) Do the ashtrays need to be emptied?
- Stay with your guests as much as possible, both before and after dinner. Plan a menu that requires minimal last-minute kitchen time. Any kitchen cleanup and dishwashing should not begin until the last guest has departed.
- See the guests to the door when they indicate they must leave. Help them with their coats and shake their hands, thanking them for coming and for any hostess gift they brought. Don't close the door right away, but remain at the open door until they have walked or driven away.

Practice good manners in service, as well as dining.

- At seated dinners, follow the rules of protocol in seating your guests. (For more information, see Chapter 18, "Seating Arrangements.")
- Use place cards for eight or more. (Old calling cards folded inside-out can be used very nicely for place cards.)
- Consider foreign guests' linguistic abilities. If they don't speak English well, be certain to seat them next to someone who speaks their language.
- Review the seating arrangement with your soldier so that you both can help guests find their places.
- For formal dinners, assign dinner partners. (For more details, see Chapter 17, "Seated Dinners.")
- The hostess should never serve herself first, or be served first. However, because at a seated dinner guests should wait for the hostess to "lift her fork" before they begin eating, she shouldn't delay once everyone has been served.
- At seated dinners, dishes are passed or served counterclockwise (to the right) around the table.

- When food or drink is served to guests seated at a table by someone standing behind them, the general rule is: Serve from the guest's left and remove items from the guest's right.
- If second servings are appropriate, the host or hostess should take seconds to encourage the guests to do so.
- When dessert is to be served at the table, everything from the previous courses (including the salt and pepper) should be removed from the table first.
- Once a seated dinner is over, suggest that everyone move to the living room rather than remain at the table for a long time. Be careful, however, not to interrupt a lively and entertaining discussion that all your guests are obviously enjoying. Often this mood is difficult to recapture.
- When dessert and/or coffee are served away from the table, traditionally, ladies are served before the men (beginning with the most senior).
- Serving coffee and dessert in the living room or another area is a convenient way to give your guests an opportunity to mix and talk with someone other than their dinner partner. This may also allow the table to be cleared (if you have a helper) and provide a break without signaling that the party is over.

Keep good records.

- What did you serve for dinner the last time you hosted some of these same guests? What did you wear? Records will help you add variety and avoid repetition.
- Keeping complete records of your parties can eliminate many frustrations, such as how much to prepare and what problems you've encountered with certain recipes in the past.
- An entertainment record (see Chapter 15, "Party Preparations") provides a wonderful history of past parties, and facilitates menu planning and estimating expenses.

DON'T OVERLOOK SPONTANEITY

This entire chapter has dealt with how to plan and organize your entertaining. However, be careful not to miss celebrating special moments because you're afraid everything won't be perfectly organized. Having time to plan does make entertaining easier, but that isn't always the most important thing to consider. The spontaneous gathering of friends and family to celebrate a special moment will be remembered long after you've forgotten a perfectly organized dinner party. Consider such occasions as these—a promotion list, command assignment, victorious sports event, successful completion of an exercise, or an unexpected honor bestowed on one of your family members. Such special moments in life should not pass by unnoticed for want of time to plan and prepare.

There are a variety of ways to make up for the lack of planning time, but the key is to keep the menu as simple as possible. It's easier to prepare large portions of just a few items than small portions of many. Guests understand that a spur-of-the-moment get-together means there was little time to prepare. They will appreciate any type of event and enjoy the spontaneity of the celebration. For example, a backyard cookout or potluck get-together can be quickly organized. In the United States, take-out restaurants or home-delivery can make short-notice dining just a phone call away. If it's a family event being celebrated, don't forget to enlist the children's help; they are usually anxious to be involved and appreciate being needed. Be flexible and enjoy the moment!

* * * * * * * * * * * * *

Thorough planning and preparation before your guests arrive
will ensure that you have a good time at your own party.

Chapter 15

Party Preparations

"Opening your home is opening your heart."
Ginger Perkins

There are two ways to plan and prepare for a party: the hard way and the easy way. The hard way allows you to do nothing until the last minute and then, in a flurry of activity, try to accomplish everything—cleaning the house, setting up for the party, and doing all the shopping and cooking—all in a day or two. By the time the guests arrive, the host who follows this method is usually too exhausted to enjoy the party. Then there is the easy way to plan and prepare for a party, which is what this chapter is about.

Whatever event you are planning, large or small—coffee group, FRG event, buffet dinner, cocktail party, formal seated dinner party—prior planning is the key to managing a successful event. The guidelines in this chapter may help you, and an internet search for "party planning" will provide many different websites, apps, and professional services you might use. It doesn't matter which one you follow; they all offer help that will keep things in perspective and hopefully easier for you.

Your planning usually starts with a group and then a date. If you host the event in your home, plan on at least 3 weeks from concept to completion. Once the group and date have been set, decide if there will be a theme, i.e., Fall Fest, Academy Awards party, Winter Wonderland, Appetizers and Cocktails. Of course, you don't have to use a theme but, if you do, it will be fun and you can plan the invitation and menu around the theme. Most importantly, how you entertain in your home should be based on your style.

If you are planning a function focused on your spouse's unit, keep in mind that entertaining in the Army is about getting to know each other, developing friendships, and having fun. You might invite those who are new to the unit as well as members who are leaving. Army families are often in transition, and partying together helps develop camaraderie. You might consider events where folks are teamed up to play trivia games, outside team sports, or icebreaker activities. The most memorable events tend to be ones where people are engaged in common activities and goals. It will help the new members meet the group on a more casual footing.

PRE-PARTY SCHEDULE

Having thought through the basic details of who, what, when, and where, it helps to make a pre-party schedule for the weeks prior to the event, and include all the tasks that need to be accomplished. As you divide up the work, try to balance the load so that you don't have too much to do on any one day.

The sample pre-party schedule that follows is written in general terms for a party you might have in your home; however, it can easily be adapted for a function held at another location. Develop your own pre-party schedule and list the specifics for this event, such as the dishes to be prepared on each day. Try to squeeze in a cushion of time for any unexpected problem. Notice the reminder for you and your spouse to be dressed one hour early. This ensures that you and your spouse are ready when the first guests knock on the door or ring the bell, even if they're early. Also, by dressing early, you don't have to rush in fear that you won't be ready on time. Rushing is the worst thing a host can do. When you rush, you often become flustered and it shows.

Three weeks prior:
- Send the invitations—mail, email, or phone.
- Plan menu and beverages.
- Decide on service style (buffet, sit-down dinner, preserved-plates).
- Decide on table size and location.
- Decide on rooms you will use for the party.
- Think about decorations/accessories.
- Determine if you need help, ask and delegate tasks.
- Do you want to take pictures?

Two weeks prior:
- Clean party serving dishes - crystal, silver, trays.
- Decide if purchasing plastic/paper plates, glasses or using your dishes.
- First round of grocery shopping and cooking dishes that can be frozen or stored for two weeks.
- Select music and make a playlist.
- If the event is "adult only" and you have children, consider contracting a sitter.
- If children are included in the event, plan age-appropriate activities.

One week prior:
- Clean the house—this may require an overall cleaning.
- Determine where tables and chairs will be set; you may need to move furniture to improve flow.
- If children are included, where will their area be?
- Prepare party decoration/accessories—linens, tablecloth, centerpiece, table decorations.

- Make place cards—for buffet or seated dinner.
- If it is a special dinner, you may consider typing the menu. All can sign the menu as the evening's memento.
- Check with helpers/tasks.
- Decide on your and spouse's wardrobe.
- Make a checklist for the day of the event: when foods should be thawed, reheated, brought to room temperature.
- Inventory dishes.
- Stock the bar.
- Determine if name tags are needed.

Four to three days prior:
- Let your neighbors know when your event is.
- Put out theme decorations, candles.
- Finish grocery shopping—prepare foods that can be refrigerated up to 2 days prior to the event.
- Designate and clean bathroom for guests to use.
- Put together a "spill box"—cleaning materials required for spills (wine/alcohol/food spillage).
- Put together a "first aid" kit with Band-Aids, gauze, bug spray and sewing kit for emergency mending.
- Specify a place for purses/coats/winter gear.

One day prior:
- Purchase any fresh flowers that you will use.
- Set up tables and chairs.
- Finish as much cooking as possible.
- Check with helpers/tasks—let them know their timeline and how they can help.
- Make extra ice cubes.
- Set table/dishes/serving trays.
- Set out name tags, if needed.
- Make your party checklist (described below).

Party Day:
- Morning—Do last-minute cooking.
  Set table/dishes/serving trays.
  Arrange centerpiece.
- Afternoon—Make last check of front of house, guest bathroom, and coat closet with hangers.
  Take freezer/refrigerator items out as appropriate—review your checklist.
  Get beverages readied.

- Evening—Ice beverages as appropriate.
  If children are to be included, check on their area and entertainment.
  If using a sitter, have arrive 1 hour early.
  Be dressed 1 hour before guests arrive.
  Review with spouse and helpers when to stop serving drinks before dinner; review seating arrangement with spouse.
  Don't forget lighting and music.

BLUF

Preparation is the key to success.

PARTY CHECKLIST

It helps to create a party checklist. Start by listing the time in half-hour increments, beginning several hours before the party, and extend it until dinner is to be served. (Dinner is usually served about one hour after the guests arrive.) Note on the schedule the guests' arrival time and when you plan to eat. Next, fill in *everything* that must be done prior to your inviting the guests to the table.

A party checklist provides more advantages than just balancing the workload. First, it provides a handy checklist that precludes your forgetting or overlooking any necessary details. Second, as you follow the schedule, once the tasks for each time slot are accomplished, you know you can relax until the next half-hour. This is especially important once your guests begin to arrive.

Don't forget to plan for cleaning up. It doesn't have to be listed on your schedule, but it needs to be considered. If you can afford a helper for the evening, then much or all of the clean up will be taken care of for you. If you do not have a helper, you and your spouse will have to clean up after the guests leave—and you'll find that it's much easier to clean up the night of the party, even though it means staying up late. A little planning ahead can help you. For example, leave as much counter space as possible clear to accommodate the dirty dishes and glasses. Follow the advice in Chapter 19, "Entertaining Tips," on how to soak your silver. Make sure, if you have a dishwasher, that it's empty and ready to load. If your kitchen is removed from the entertaining area and you know the guests will not hear the dishwasher, you can quickly load and start your first wash as you prepare dessert. However, do not start washing dishes by hand before your guests have departed.

Here is a sample party checklist; guests arrive at 7:30 and dinner's at 8:30.

	Party Checklist
5:00	Take meat and rice casserole out of refrigerator. Put white wine in refrigerator to chill. Set out guest book with pen. Set out name tags.
5:30	Fix supper for children.
6:00	Get dressed.
6:30	Take appetizer out of refrigerator. Put out crackers & nuts. Set up bar: ice, beer, wines, specialty cocktails or champagne.
7:00	Preheat oven—325°. Put vegetables into steamer. Put appetizer in living room. Check lighting—living room, porch. Start music playlist for the evening. Light candles—living room & bathroom. Open red wine for dinner.
7:30	Put meat & casserole (covered) into oven for 45 min. (Guests arrive.)
8:00	Plug in coffee pot. Cook vegetables. Put ice in water pitcher. Salad—from refrigerator to table. Heat sauce & rolls.
8:30	Light candles on table. Don't forget the butter. Take frozen pie from freezer. (Serve dinner.)

ENTERTAINMENT RECORD

Keeping a record of your entertaining provides you with another useful party-planning tool. It helps you remember whom you've entertained, what you served, and what you learned from that particular menu. Your entertainment record need not be fancy. A 3-ring binder or notebook will suffice—anything that provides space enough to record the necessary details of each event. However, nice commercial entertainment logs are available, if you want to create a more lasting remembrance.

You will find that an entertainment record is handy to use from the time you send out the invitations until the party is over. It contains your guest list, so it can be used to record your guests' responses. This is especially important if someone else in the family should take any of these calls; they also will have the time and date of the event readily available in your entertainment record to verify this information for your guests. Also, your menu is recorded in your entertainment record, so it will be helpful as you prepare your shopping list, pre-party schedule, and party checklist. In addition, this is where you diagram your seating arrangement, so keep it handy as you put the place cards on the table. Then, put your entertainment record in the kitchen so that, just before inviting the guests to the table, you can read through the menu one last time to verify that nothing has been forgotten. Afterward, don't forget to jot down any comments you have with regard to cooking temperatures, cooking times, or quantity prepared for the next time you prepare these dishes; this becomes sort of an "after action report" where you can note things that worked and those that need improvement. You might also want to jot down what you wore that evening (so you won't wear the same thing the next time you entertain these same people), and what hostess gifts the guests brought (to aid in writing thank-you notes).

Other ideas that you might find useful include: a guest sign-in book that serves as a wonderful reminder of the event; a typed menu card that the guests sign; a scrapbook with a copy of the invitation, menu, and a page for the group to sign; a camera to take pictures during the party for the scrapbook. You might even save your party schedule and party checklist in a 3-ringed notebook for future party planning. Any of these can provide a great memory and record of the event. Have a great party!

Entertainment Record

Type of Entertainment:

Date:

Time:

Dress:

Guest List   Accept   Regret

Menu   Beverages

Seating Chart

Decorations

Comments

$$* * * * * * * * * * * * *$$

"The hostess must be like the duck—calm and unruffled on the surface, and paddling like hell underneath."
Anonymous

Chapter 16

Buffet Dinners

"Sharing a meal with friends is one of the most
enjoyable social encounters."
Julee Ross and Sheila Lukins,
The Silver Palate

Buffet dinners are the most popular and useful form of dinner entertaining today. They allow the host to invite more guests and serve them with less effort. Unlike a seated dinner, the guest list for a buffet dinner need not be limited by the table size. The host has only to prepare the house and a delicious meal, and to set out the buffet. There's no need to decide on a seating arrangement, set out individual place settings, or worry about serving and clearing the table under the watchful eyes of guests. Buffets are ideal for large gatherings and offer a convenient way to serve a Sunday brunch, luncheon, or the most casual of summer suppers. It's little wonder that buffet (pronounced buff-ay) dinners are a very popular form of entertainment.

One type of buffet dinner is a cross between a buffet and a seated dinner. The food is placed on the sideboard buffet for the guests to serve themselves, but the guests sit at a dining table to eat. Though the invitation may say either "dinner" or "buffet," for clarity this type of dinner will be referred to in this chapter as semi-buffet. This form of entertaining doesn't offer quite the ease of preparation as a regular buffet since the table must be set. However, it does offer the ease of having guests serve themselves. It is less formal than a regular dinner, but more formal than a regular buffet. Since regular buffets are far more popular, the discussion in this chapter centers around them, unless semi-buffets are specifically mentioned.

Despite their simplicity, buffet dinners nevertheless require considerable planning and careful execution in order to be successful. The following tips should help the host plan an enjoyable, smooth-flowing buffet dinner.

MENU SELECTION

Seating arrangements are important to your menu selection, so decide where the guests will sit during the meal before planning what you will serve. If guests must eat in their laps, then plan to prepare food that has no bones and requires little cutting. For example, it's all right to serve boned chicken breasts for a buffet, but not steak.

BUFFET TABLE LOCATION
Consider the options available for where to place the food: the dining table, the sideboard buffet, possibly a picnic table on the patio, or a combination of these. Then decide which option best suits the occasion, the number of invited guests, and even the weather. If the dining table will be used, should it remain centered in the room, or moved to one side? On the patio, is the table better placed against the wall, or away from the wall so that guests can move around it?

If the food is to be served on the dining table, remove the chairs from around the table. This allows the guests to move freely around the table to serve themselves. Additionally, the chairs will probably be useful in the living room for the guests when it is time to eat.

ROOM SETUP
Wherever guests will be sitting to eat, be sure there are tables or other flat, level surfaces nearby for their drink glasses—even if they will be using lap trays. Though great for holding plates, trays are not very satisfactory for balancing glasses.

TRAFFIC PATTERN
Consider the best traffic-flow pattern before deciding how to arrange the dishes on the serving table. Where do you want the guests to begin and end when they are serving their plates? When deciding, consider the flow of other guests as they enter and leave the room. Movement around a table that is not next to a wall can be either clockwise or counterclockwise. Movement along a table that is next to the wall can be from left or right. There are no rules about the direction of traffic; do whatever works best for the space you have available.

TABLE ORGANIZATION
Arrange the items on the serving table in the order in which they will be needed: plates first, food in a logical order (if the entrée is to be served over rice, place the rice before the entrée), eating utensils and napkins last.

Try to balance the appearance of the serving dishes by placing the largest ones on the ends of the table, if that permits a logical progression.

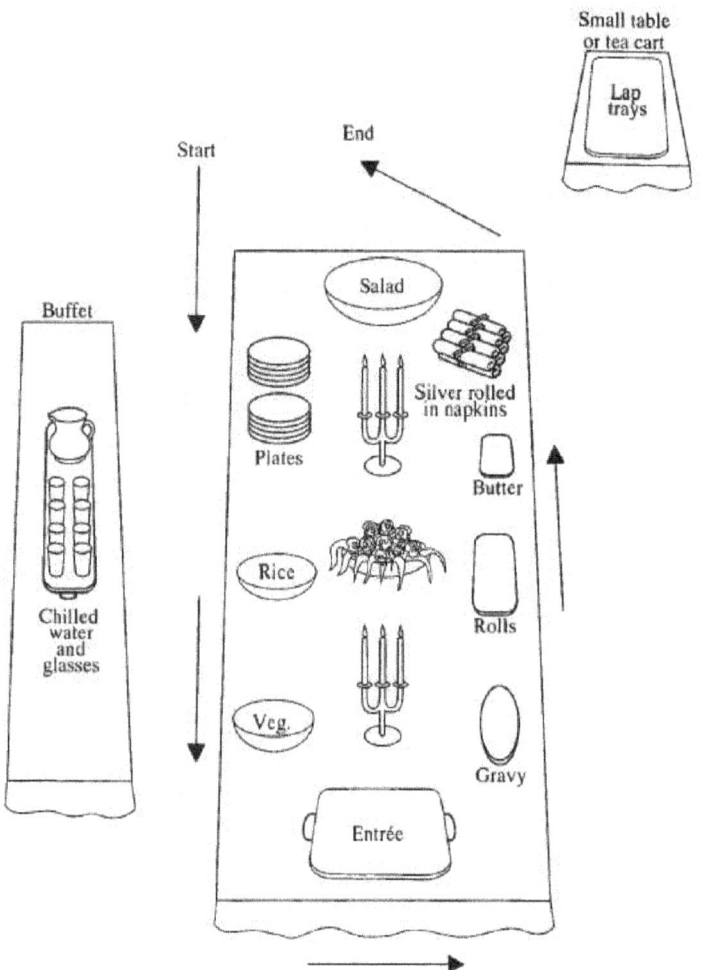

LAP TRAYS

If lap trays are to be used, place them at the end of the buffet line so that guests take them *after serving their plate, not before*. Without a tray, guests can serve their plate by holding it immediately adjacent to or slightly over the serving dishes. If they've already taken a tray before they serve their plate, guests must serve themselves across the tray. Thus, the serving spoon travels farther—providing more chance for a spill!

SILVERWARE AND NAPKINS

Provide only the eating utensils that will be needed. Place them last on the table, or on the lap trays if they are being used. (Guests do not need to carry their silverware and napkin as they serve their plate—it only complicates handling the plate.) If you plan to place the eating utensils on the table, it is easier for guests to pick them up when they have been rolled in a napkin. (A nice touch is to tie a pretty ribbon around each bundle to hold it together and add color.) Alternatively, for a casual buffet, the silverware can be placed on the table in pretty containers, such as baskets or large mugs, with the napkins arranged nearby. On the other hand, if you plan to place the eating utensils on individual lap trays and the trays are then stacked, everything will balance better when each set of silverware is placed flat on a napkin and on alternating ends of the trays. This allows the trays to be stacked on one another, and still remain level. Dessert spoons or forks and coffee spoons are not served until later, to avoid guests having to keep them after their trays are collected.

CENTERPIECE

Ordinarily, there is a centerpiece on the serving table to add height, beauty, and interest. Since no one will be sitting at this table during a buffet dinner, the centerpiece should be tall enough that it's not lost among the serving dishes. For semi-buffet dinners, when the food is placed on the sideboard buffet and the guests sit at the dining table to eat, the centerpiece on the dining table must be sufficiently low for the guests to see across it.

CANDLES

For evening buffets, candles may be used on the serving table, but *light* the candles before inviting the guests to begin serving themselves. Clip the wicks ahead of time to ¼-½ inch to ensure a low flame and less smoke when the candles are lighted.

APPETIZERS

If the host plans to serve an appetizer(s), it may be available when the first guests arrive or offered later—depending primarily on the type of food being served. Generally, those appetizers that are best eaten as soon as they come from the oven, and those that are presented individually (e.g., soup cups), are not served until most of the guests have arrived and had time to get their drinks and chat a while. Other types of appetizers may be prepared in advance and, prior to the start of the party, placed around the area where the guests will be during the "cocktail hour."

Any appetizers offered should complement the meal to come and be served in small enough quantity so as not to dull the guests' appetite. All remnants of the appetizers and the empty glasses should be removed once dinner is served—with the possible exception of nuts, which may remain throughout the party.

BEVERAGES

The choice of beverages to be served depends on the meal and the personal preferences of the hosts. However, it's always important to have non-alcoholic beverages (e.g., sodas, mineral water, juice) available for the cocktail hour, and chilled water available during dinner. Water is usually not served to everyone at a buffet, but placed on the sideboard buffet or a side table for those guests who care to take it. Wine is usually the beverage served to guests for dinner. However, since a glass is difficult to balance while walking with a tray of food, it's better not to serve the dinner wine to guests until after they are seated with their trays.

MULTIPLE ROOMS

If the guests will be seated in more than one room, conversation can be stimulated by separating couples, and arranging for an interesting conversation group in each room. The hosts should also sit in different rooms to avoid giving the impression that one room is "preferred." Guests can be told where they are to sit by giving each guest a card that designates the room where they are to sit; for smaller gatherings, the host can suggest a room to each guest as they come from the buffet table. For special occasions, seating cards can be tied to small party favors, such as candy canes, ornaments used to decorate a small tree, or small individually wrapped gifts placed in a pretty basket.

HOST MAY SERVE

It is helpful for one of the hosts to stay near the food as the guests are serving themselves. That is for several reasons: to explain any unusual dishes, to take care of any problems that might occur, and to be aware when a dish needs replenishing. A good way to do this without feeling awkward is to serve the entrée, or the dish on the opposite end of the table from the entrée—possibly a salad, fruit dish, or meat accompaniment. During this time, the other host can pick up empty glasses and any appetizer remnants in the living room, suggest which rooms the guests should sit in, and serve wine to those who are already seated.

If major spills occur, the host should take care of any necessary cleanup quickly; however, no thorough cleanup should be done until after the guests depart, unless delay would cause permanent damage. (See "Treating Spills" in Chapter 19.) For example, an unsightly spill on the tablecloth can be obscured with a clean napkin. The quicker a spill can be dealt with, the better everyone feels. All things being equal, your guests' comfort is always the prime concern.

MOVING TO THE BUFFET

When the meal is ready, there are several ways to invite the guests to begin:
- For a buffet served in the home, one of the hosts approaches the senior guest present and says that dinner is ready; invites that person to begin the line, and asks the other guests to follow. The host should speak loudly enough for most of the guests to hear. The host then walks with the senior guest to the table, describes the menu or any unusual dishes, points out where the line begins, and removes any serving dish covers. As the guests serve themselves, the hosts attend to their other duties and go through the line last. Guests begin eating as soon as they are seated.
- For very large groups, like unit get-togethers, it may be advisable for couples to go through the line together, even though they may sit apart to eat. In this case, the person serving as host invites the senior couple to begin. Other couples will naturally follow. Guests begin eating as soon as they get their food and are seated.
- For a small semi-buffet dinner (guests sit at the dining table), the host invites the senior couple to serve themselves first from the sideboard buffet, with the others to follow. In this case, the host may prefer not to serve any of the food but, instead, spend time

directing the guests to their places. Remember to use place cards for eight or more. The host couple serves themselves and are seated last. Normally, in this situation, the guests don't begin eating until everyone is seated.
- For a large semi-buffet dinner with guests assigned to several tables, the host may ask the guests to be seated first and then go to the buffet area by tables, or two-by-two with dinner partners. For this type of seating arrangement, the host couple should be seated at separate tables, and go to the buffet with the others at their table. With such a large group, the host should be sure to invite guests to begin eating as soon as they have their food.

SECOND SERVINGS

The host couple should encourage guests to return to the buffet table for seconds by doing so themselves. Even if it means taking small portions the first time, they should do so; replenishing their own plates lets the guests know it's all right for them to do the same.

At a regular buffet dinner (not semi-buffet), the hosts should take the time, before serving their plates, to replenish the serving dishes. By doing this, the serving dishes are refilled and ready for those who return for second servings, even if the hosts haven't yet had time to return themselves. If there are covers for the serving dishes, they should be put on to keep the food warm after everyone has been served.

ASK FOR NO HELP

As the guests who have been eating from trays or their laps finish their meal, the host should discourage them from taking their trays or plates to the kitchen. The host can ask anyone who starts to do so to wait and allow the host to take their tray (plate) as soon as the host is finished eating. In this way, the kitchen will not be cluttered with trays stacked everywhere and the cleanup can begin by working with one or two trays at a time. For a large crowd, the host may ask one or two other guests to bring things to the kitchen, while the host stacks. After all the trays or plates have been collected, the host should send the helpers back to the party, while the host organizes the kitchen and prepares to serve dessert and coffee.

CLEARING THE TABLE

For a buffet at which the main course and, later, the dessert are placed on the dining table for the guests to serve themselves, the table needs to be cleared between courses. *Everything* should be removed except the centerpiece and candles. Then the dessert can be placed at one end of the table and coffee at the other. Both clearing the table between courses and setting out the dessert and coffee should be done as quickly and quietly as possible. A tea cart or large tray can be of great help, and reduce the number of trips to and from the kitchen.

For a semi-buffet dinner at which the guests serve themselves from the sideboard and sit at the dining table, the hostess must decide where dessert will be served. If it is to be served at the table, the dinner plates need to be removed, and the table and sideboard cleared between the courses. This is considerable work under the watchful eyes of the guests, and best accomplished with the aid of a helper. Without help, it's better to serve dessert at a semi-buffet dinner in another area, and leave the clearing of the table and sideboard until after the guests depart.

DESSERT AND COFFEE

At a buffet dinner, dessert and coffee are either served on the cleared buffet table, or carried to the living room (or some other suitable area) on a tray or tea cart. Tradition says that the senior lady has the honor of being served first or, if the guests are to serve themselves from the table, she would be invited to serve herself first. However, in today's Army, the senior spouse may be a male. In that case, he should be accorded the same courtesy. He may well defer to the ladies, in which case things would proceed as usual. Milk and sugar and, perhaps, artificial sweetener, should be offered, even when refilling cups.

* * * * * * * * * * * *

Buffet dinners provide a more relaxed setting, but require
adequate planning and setup.

Chapter 17

Seated Dinners

BLUF

"Don't be intimidated. Have a plan, sequence preparation and service, and ask for help."
Angel Mangum

Seated dinners are special! Whether as simple as an old-fashioned, family-style dinner where great bowls and platters of steaming-hot food are passed around the table, or as formal as a meal served by the household help, an invitation to dinner always promises a memorable meal. Sitting around a pretty table with friends and interesting dinner guests, enjoying the delicious dinner prepared especially for you, engaging in pleasant and interesting conversations, relaxing in the comfort of someone's home—these are what make every seated dinner very special.

Planning, preparing, and hostessing a seated dinner requires not only extra effort, but also special knowledge. Very formal dinners are especially challenging because of the specific etiquette associated with them—though no one should let that fact deter them, because learning what's required isn't difficult. Whatever type of dinner you're planning, from the simplest family-style to the most formal, the following guidelines will be helpful.

SEATING PLAN

Always decide on a seating plan, even for a dinner for four. Follow the order of precedence, based on rank and position, and consider your guests' special situations, such as linguistic ability. (See Chapter 18, "Seating Arrangements.")

PLACE CARDS

Place cards are essential when seating eight or more. They are helpful both to the guests and to the host and hostess, who should assist the guests in finding their seats. Place cards need not be fancy or held in place-card holders, though there are beautiful varieties of both available. Outdated (previous rank) calling cards, folded lengthwise (inside out), make fine place cards; they don't cost anything, need no holders, and provide space for the name to be written on both sides.

The names to use on place cards vary with the formality of the occasion. For formal occasions, use "conversational rank" and last names; e.g., Lieutenant Colonel Ware would be written Colonel Ware. For small informal occasions, use only first names on the place cards, when everyone knows one another. For those informal occasions when first-names-only on the place cards might seem too familiar, but formal place cards seem too stiff, a compromise between the two methods would be to abbreviate the rank for the military member, and use the spouse's first and last names without a title. This would be especially appropriate for a large, informal gathering at which the guests are not all close friends.

Write or print the names on place cards large and dark enough to be read from a distance. For groups, a computer with a large calligraphy-style font can make short work of this task, and the results can look most elegant. It is very helpful for the names to be written or printed on both sides of the place cards. This aids the guests in finding their places from either side of the table, and shows everyone's name to the others at the table. In all the examples below, the names on the back side of the place cards would be exactly the same as on the front.

Examples of formal place cards
Handwritten:

Colonel Ware   *Mrs. Ware*

Computer printed with calligraphy-style fonts:

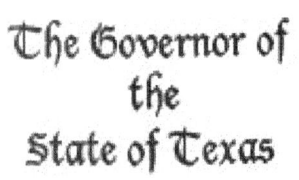 

Examples of informal place cards

For large groups:

For small groups when everyone knows one another:

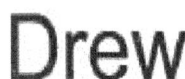

COURSE ORDER

Though most of us never serve seven-course dinners, it's helpful to know the order in which the various courses are presented at a formal dinner and what wine is usually served with each. For less formal dinners with fewer courses, the salad is frequently served before, or with, the entrée.

Number	Course	Wine
1	Appetizer	White
2	Soup	Sherry
3	Fish	White
4	Entrée with vegetables	Red*
5	Salad	Red
6	Dessert	Champagne
7	Fruit	Champagne

* White wine may go better with some entrées. The general guideline is: red wine with red meat, white wine with white meat.

METHODS OF SERVICE

The hostess may choose from a variety of service methods, or a combination of methods. One requires additional help; the others do not. Each method is described below.

- **Household Helper** - Each course is served to the guests by a helper. For a seated dinner, food is presented to the guests from their left side, and their plates are usually removed from their right side—though, if it's more convenient, plates may be removed from both sides. Beverages are served, and glasses removed, from the right.

- **Pre-served** - Any appetizer and/or soup (in a cup and saucer or mug) may be served to the guests before they come to the table. The main course is pre-served on each plate, and these are placed on the table before the guests enter the dining room. This style of service is especially useful for food that looks best when arranged on the plate, for example whole artichokes, individual gelatin molds, or Eggs Benedict. Also, the new American cuisine often features smaller portions presented in artistic arrangements that requires the plates to be pre-served. Place pre-served plates at each place with the entrée at the "4 o'clock" position.

  Plates that you have prepared before your guests arrive give you an opportunity to be creative and ensure a perfect presentation. Here are some examples. Coat the rim of a salad plate with margarine and dust with dried parsley, then place vegetable aspic in the center and garnish with fresh dill. Place a poached pear on a bed of strawberry purée to which you have added a spoon of sour cream and drawn a knife through the cream to create a design. Dust a glass dessert plate with powdered sugar, garnish with mint and/or slivers of fresh fruit to dress up even the simplest slice of cake.

- **Host Carves** - The host carves at his place or at the buffet, and serves the entrée onto each dinner plate. Then, one of two methods is used to serve the other dishes: (1) the host passes the dinner plates to the guests, after which the serving dishes are passed counterclockwise around the table for the guests to help themselves; (2) the hostess stands beside the host, takes the plates as they are served with the entrée, serves the other dishes from the buffet, and passes the filled dinner plates to the guests.

- **Tea Cart** - The hostess serves everything from her seat. The various serving dishes are placed on a tea cart and/or the table in front of her, with the plates stacked together at her place. Each plate, once filled, is passed down the table to the guest indicated by the hostess. She usually begins with the lady #1 seated on the host's right (left, as the hostess faces him); finishes that side of the table; then, beginning with the host, finishes serving those seated on her right. She serves herself last. The following diagram depicts the usual order for plates that are served and passed by the hostess from her seat:

Order of service from a teacart with hostess serving all the plates:

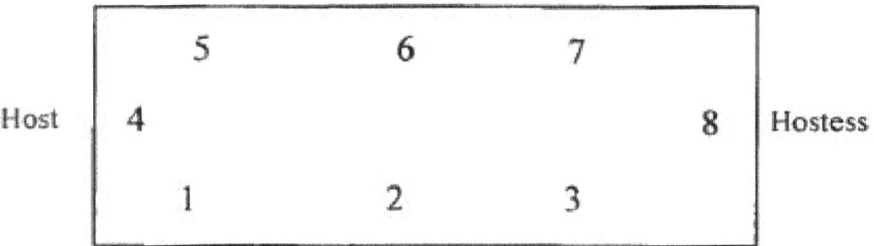

- **Semi-buffet** - Guests serve their plates from the sideboard buffet, then sit at the dining table to eat. (See Chapter 16, "Buffet Dinners.")
- **Family Style** - All the serving dishes, including the entrée, are passed around the table (counterclockwise) for the guests to serve themselves. This is the most casual type of service.

PLACE SETTING

A complete place setting is properly set as shown in the following diagram. Use only what your menu requires. If fewer glasses are needed, move the outer ones in. The water glass always goes at the tip of the first knife. Iced teaspoons, when used, are placed to the right of the soup spoon. No more than three of any one implement should be included in a place setting, with a few exceptions; the seafood fork can be included with three other forks, and the butter knife can be used with three other knives. Dessert silver may be placed above the dinner plate or passed with the dessert. Coffee spoons are passed with the coffee. Individual salt and pepper shakers can be placed near the place card, or a few can be placed around the table within reach of several guests. For individual shakers, the salt shaker is on the right, the pepper on the left. This was the international standard for the visually impaired but has become the tradition. The glass and silverware intended for each course should be removed from the table before the next course is served.

1. Fish fork
2. Meat fork
3. Salad fork*
4. Butter knife
5. Bread plate
6. Place card
7. Dessert fork & spoon
8. Meat knife
9. Fish knife
10. Soup spoon
11. Cocktail fork
12. Water glass
13. Red wine glass
14. White wine glass
15. Champagne glass
16. Napkin

* If salad is served before the entrée, move the salad fork (#3) to the left of the meat fork (#2).

PLACE PLATES/CHARGERS AND SERVICE PLATES

Place Plates/Chargers - Place plates/chargers are highly decorative, ornamental plates, their principal function being to enhance the appearance of the dining table. Their presence also ensures that guests have a plate before them from the time they sit down until dessert is served. These place plates remain at each place setting, with the appetizer and the soup bowls placed directly on them—even though the appetizer may be served on its own smaller service plate. When the soup bowls are removed, the place plates are removed at the same time, but immediately replaced by the plates for the next course. Obviously, this traditional use of place plates requires a helper to go behind each guest removing and replacing plates at the appropriate time, and it requires a large number of plates. For these reasons, place plates are seldom seen in Army entertaining.

If you have pretty place plates/chargers and would like to use them, but don't have a helper, don't hesitate to adapt the formal usage of place plates to suit your method of service. Just be sure to remove the place plates before dessert is served, and preferably before the entrée. Don't worry about always having a plate in front of your guests; that is an old-fashioned rule that only applied to "very formal" dinners.

Service Plates - Service plates are different from place places/chargers. Service plates are used for any type of seated dinner when individual containers of food require a base. In other words, service plates are like saucers. They are used under such dishes as stemware (for serving dishes like seafood cocktail and dessert), small dishes (such as ramekins and *pots de creme*), and soup bowls. These service plates are useful in several ways: to provide a solid base for the dish, give the server something to hold without touching the dish of food, keep any dribbles off the tablecloth, and provide a place for resting the eating utensil.

TABLE LINENS

Table linen is the generic term applied to whatever covering you choose for your table. Whether you select your grandmother's fragile Brussels lace tablecloth, an elegant Army-Navy tablecloth, a colorful designer sheet, a remnant of fabric edged with pinking shears, or even plain butcher paper, they may all be referred to as "table linens." No doubt that is because linen used to be the fabric of choice for the dining table. It is one of the oldest known fibers to have been woven into cloth, and its natural stiffness adds a bit of crisp formality to the fabric. Since the twentieth century, Irish linen has been prized as the best quality available.

Military wives have traditionally taken great pride in their beautifully embroidered linen and lace Army Navy or Air Force pattern tablecloths. Carefully laundered, snowy white, and faultlessly ironed, these table linens add charm and elegance to any table setting. (See "Ironing Linens" in Chapter 19.) However, the care and preservation of these linens are time-consuming, and the linen revolution caused by polyester and easy-care, tumble-dry, no-iron blends was welcomed by all. Not only are the new fabrics easier to care for but, they also offer a greater variety of colors and patterns. Stripes, floral patterns, and bold colors have replaced the traditional white and ecru on many of our casual and informal tables. Not surprisingly though, traditional linen tablecloths are still preferred for more formal dining. Certainly, when you want to set a truly elegant table, nothing can surpass the heirloom linens.

Table linens are used for a variety of reasons: to protect the surface of a beautiful table; to hide the surface of one that's not in the best of condition after too many moves; and, in general, to improve the appearance of the table. When choosing your table covering, whether it's a traditional linen-and-lace tablecloth, one of the more modern man-made fabrics, or artfully decorated butcher paper, your aim is to create a pleasing effect that will be inviting to your guests.

NAPKINS

Napkins are a meal-time necessity. Beyond that, they can add to the appearance of your table and the mood you want to create. Except for the most formal dinners, napkins may contrast in color to the tablecloth, be folded into intricate shapes, be placed in a variety of locations, and even have decorative or practical items inserted in them. How the napkins are folded can be influenced by where they will be placed.

The basic napkin fold results in a rectangular shape that is placed either in the center of the place setting or, in the event the plates are pre-served, to the left of the fork(s). (See napkin placement in the place setting pictured earlier in this chapter.) This rectangular shape is obtained by folding the napkins once in each direction and then folding the resulting square into: thirds for large dinner napkins; or halves for smaller, luncheon-size napkins. When placed to the left of the fork(s), the open edges of the napkins may be placed toward either the left or the right, so long as they are consistent. This simple rectangular shape is the most appropriate for very formal settings. However, for any other type of table setting, the napkins may be folded in a variety of fancy and imaginative shapes that add interest to the table.

The Victorians perfected the method of "crimping" napkins into intricate shapes. However, by the turn of the century, the "new manners" considered this practice to be "showy" and in very poor taste. The only fold considered acceptable, beyond the basic, was to turn the corner under and "thrust" the dinner roll between the folds. Fortunately, we have come full circle, and a fancifully folded napkin is one of the easiest and least-expensive ways to decorate your table. You can create a variety of shapes for different effects and occasions. Interesting books on the subject of napkin folding can be found in large bookstores, libraries, and online.

CENTERPIECE

Every dining table needs a centerpiece that is appropriate for the type of meal being served and the occasion. However, it need not be a floral arrangement. Though fresh flowers, beautifully arranged by your favorite florist, would be lovely and enjoyed by all, they can be terribly expensive. You can save money by creating your own floral centerpiece; try using a small store-bought bouquet, flowers and greenery clipped from a potted flowering plant, or blooms from your own garden. Also, consider creating an interesting centerpiece using small live plants in pretty containers, fruits, vegetables, green foliage, dried or silk flowers, or accent items gathered from around your home—shells, figurines, or other small collectibles. A small group of such items can be combined and attractively arranged on the dining table to create a stunning centerpiece. Strive to add interest and beauty, but not to overwhelm the table. Keeping your center decorations simple helps to make your table look welcoming. Watch the magazines for pictures of interesting centerpieces to get ideas.

The guidelines for centerpieces are simple. For a seated dinner, the centerpiece should be low enough to enable the seated guests to look across and see one another, but large enough to add beauty and interest to the table. For a buffet, cocktail buffet, reception, tea, or any other type of stand-up affair, the centerpiece should be taller. It needs to be tall enough that it's not lost among the chafing dishes or other large serving dishes on the table, but not so tall as to create an unbalanced effect. A centerpiece should always enhance your color scheme, linens, and the table setting you have chosen.

CANDLES

Candles now come in a wide variety of colors and can be used to great advantage to enhance the color scheme you have selected. However, for formal dinners, white or ecru candles are still considered the most appropriate color. Light the candles placed on the dining table just before the guests enter the dining room, and extinguish them when all the guests have left the table. The wicks should be clipped in advance to ¼-½ inches so that they burn with a smaller flame and less smoke.

TABLE APPEARANCE

The first thing the guests see when they walk into the dining room is the dining table. Its appearance sets the tone for the meal to come. In addition to the centerpiece being attractive and everything sparkling clean, the table should appear balanced. The following guidelines will help achieve the best effect.

- The tablecloth usually has an overhang of between 12-18 inches, even all around. It should be neatly pressed; a center crease where it was folded is acceptable, but any other fold creases should be ironed out. (See "Ironing Linens" in Chapter 19.)
- When candles are used, their positioning adds to the balanced appearance of the table. Place them halfway between the centerpiece and the end of the table, or halfway between the centerpiece and the top of the place setting at each end of the table. Try to place them so that they will not obscure the guests' view of one another.
- Silverware, napkins (if placed directly on the table), and plates are all positioned one inch from the table's edge.
- Napkins can be put in a variety of places—on the plates; in the center of the place setting when the plates are not there; on the table to the left of the forks (not under the forks); and for informal and casual dinners, even at the top of the plate or puffing out of a glass.
- Place settings are usually lined up with those directly across the table, unless there is an odd number of guests.
- Any serving dishes, salt and pepper, and butter, placed on the table in advance should be logically positioned and contribute to the overall, balanced appearance of the table.

CHILLED WATER

Chilled water should always be served, regardless of what other beverages you have selected for the menu. Chilled water is water in which ice has recently melted, so add ice to a pitcher of water shortly before you are ready to fill the water glasses. This is done, rather than using water and ice cubes, because chilled water causes glasses to "sweat" less. Glasses should only be filled two-thirds to three-fourths full. That way, they won't spill if the table is slightly bumped as the guests are being seated.

POURING BEVERAGES

- **With a Helper** - Chilled water is poured in the water glasses before the guests sit down. The helper pours the wine or other beverage with the appropriate courses, and also refills the glasses as needed.
- **Without a Helper** - Chilled water is poured into the water glasses before the guests sit down. If only one wine is to be served, it may also be pre-poured. If more than one wine is to be served, this may be done by the host at the appropriate time during the meal, if there are just a few guests. For larger groups, the opened wine bottles or carafes are placed on the table or passed around; the host pours for the ladies around him and then himself, and asks the other gentlemen to pour for those around them. In all cases, glasses are replenished by asking the gentlemen around the table to pour, rather than the host or hostess getting up frequently to walk around the table and refill glasses.

DINNER PARTNERS

For formal dinners, each guest is assigned a dinner partner. A gentleman's dinner partner is usually the lady who will sit to his right. They go into the dining room together; he seats her, then remains standing by his chair until the hostess has been seated. The one exception to dinner partners sitting side-by-side is the gentleman who will sit to the hostess's left; his dinner partner is the lady who sits to the right of the male guest of honor (since this latter gentleman seats the hostess).

Diagram of dinner-partner couples:

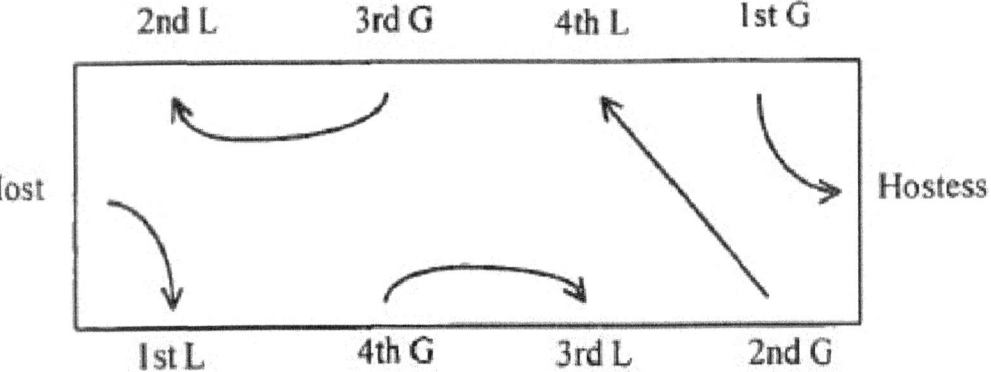

The gentleman guests are given the name of their dinner partner by either having the information: (1) announced by the host before everyone proceeds to the dining room, or (2) written down and passed to them. The written form is prepared on small cards (escort cards), and may be with or without envelopes. With envelopes, a gentleman's name is written on the envelope, and his dinner guest's name on the card inside. Without envelopes, a gentleman's name is written on the front of a small folded card with his dinner guest's name inside. In both cases, it is helpful to provide an indication of where the two will sit at the table (see following examples). Escort cards are arranged alphabetically (based on the gentlemen's names) on a small tray which is either placed on a table in the front hall so the gentlemen can take their card as they arrive, or carried around to the gentlemen sometime before dinner (by the host or household help).

Escort card with envelope; seating location indicated by hash marks:

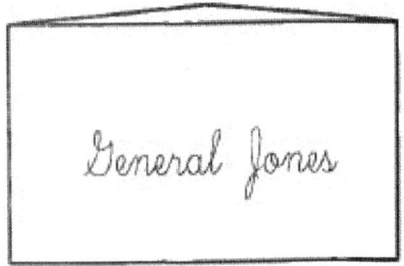

Escort card without envelope; seating indication for multiple tables:

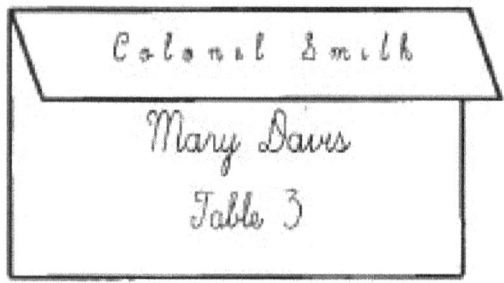

ENTERING THE DINING ROOM
The order in which the guests enter the dining room depends on the formality of the dinner and the method of service the hostess has selected. For formal dinners with assigned dinner partners, the host and lady #1 enter first, the other dinner-partner couples follow, and the hostess and gentleman #1 enter last. For buffet dinners, lady #1 is invited to go first. For other types of service and less formal meals, there is no special order.

HOSTESS'S MANNERS
The hostess should never serve herself first, nor be served first. However, unless she has encouraged the guests to begin eating, they will wait for her to begin eating, or "lift her fork," as they used to say. Therefore, she should not delay the meal by becoming distracted in conversation.

PORTIONS
When putting food on a plate for someone else, avoid serving overly generous portions. It is better for the guests to have second servings than for them to be embarrassed about leaving uneaten food.

DESSERT AT THE TABLE
If dessert is served at the dining table, everything from the previous courses (including salt and pepper) should be removed from the table first. If there is a household helper, he or she removes these items. Otherwise, the guests may be asked to pass their plates and the other things near them to the hostess, who quietly stacks them on a tea cart at her side and then rolls it slowly to the kitchen. There the tea cart can be unloaded and reloaded with the dessert, coffee, and appropriate china and silverware. It is not a good idea to use this latter method of clearing the table more than once during a meal; therefore, multi-course dinners served entirely at the dining table should not be attempted without additional help.

Coffee and liqueurs may be served after dinner at the dining table, or away from the table. The latter choice offers a welcome opportunity for the guests to stand and move around a bit after a lengthy dinner, and to talk with some of the other dinner guests.

LAST-MINUTE CANCELLATION
If a dinner guest must regret at the last minute and you would like to keep the same number of guests, it is acceptable to invite a good friend to "fill in" on short notice. Explain your predicament, that you've just had a last-minute cancellation for your dinner party, and ask if your friend would be so kind as to fill in. To anyone other than a close friend, this late invitation would be offensive, but a good friend will understand.

LATE GUESTS
If a guest is late arriving and dinner is ready, wait about fifteen minutes—no longer than twenty—past the time dinner was planned, then begin without him. If the latecomer arrives during dinner, he starts with whatever course is being served at that time. If dessert has already been served, the hostess might prepare the late guest a dinner plate, if it is convenient for her to do so.

* * * * * * * * * * * *

Planning a seated dinner is like orchestrating a concert;
all of the parts come together to create a grand finale.

Chapter 18

Seating Arrangements

"A place for everyone, and everyone in their place."
Anonymous

Seating guests according to protocol is the hallmark of an organized host. It's important for almost every type of dining occasion—formal, informal, and even semi-buffet. That is because the proper seating arrangement reflects the respect a host has for guests' positions, and guests feel more comfortable when they're shown where to sit. As you prepare for your guests, it only takes a few minutes to plan the seating arrangement, and it is a very important part of being a good host.

SEATING PLAN
Planning the seating arrangement for a dinner party is easy to do because it follows specific guidelines based on protocol and the order of precedence. Armed with these guidelines, which are described in this chapter, decide on your seating plan in advance. It helps to draw a diagram of the table and indicate where each guest is to sit. Before the guests arrive, the host and hostess should familiarize themselves with the seating arrangement so they can help their guests find their places.

The Basics
The specific seating guidelines that appear on the following pages are based upon these basic principles; (1) do not seat two gentlemen or two ladies side by side, (2) do not seat a husband and wife side by side, and (3) do not seat a lady at the end of a head table. There are, of course, circumstances that occasionally require exceptions (see "Other Considerations" later in this chapter).

Place Cards
Once you've decided on the seating plan, consider whether or not to use place cards. It's a good idea to use them for all but the smallest groups, but especially when seating eight or more. These name cards serve several useful purposes—they show the guests where they are to sit; they help the host and hostess remember where everyone is to sit, useful if they are directing the guests to their places; finally, they serve to keep everyone's name readily visible for those who might have forgotten another guest's name. It's helpful if the names can be written on both sides of the place cards so they can be read from either side of the table. (For more information about place cards, see Chapter 17, "Seated Dinners.")

PLACES OF HONOR

> **BLUF**
>
> The Guests of Honor are always seated to the right of the host and hostess!

The most important point in the protocol of seating arrangements is that the seat on the right (from the standpoint of the one seated) is always the place of greatest honor. Therefore, the seat to the *right* of the host, and the seat to the *right* of the hostess are the most important seats for guests at a dinner table. The next most-honored places are to the host and hostess's left. This means that spouse #1 should be seated to the host's right and soldier #1 to the hostess's right. Then, spouse #2 is seated to the host's left and soldier #2 to the hostess's left. For more guests, the progression continues in this same manner, alternating sexes.

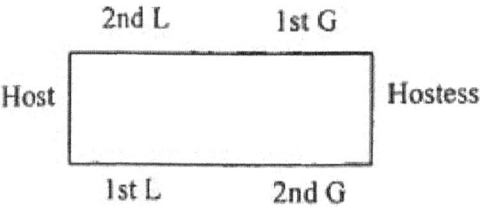

For an all-spouse affair, the seating progresses in a similar manner, except that guest #1 would take the host's seat.

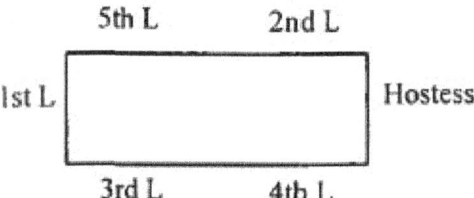

MULTIPLES OF FOUR

An exception to the normal progression of assigning seats must occur when the total number at the table is a multiple of four (4, 8, 12), in order for the seating arrangement to remain a man-woman-man-woman progression. Most people don't worry about this when only four are present, using this first plan:

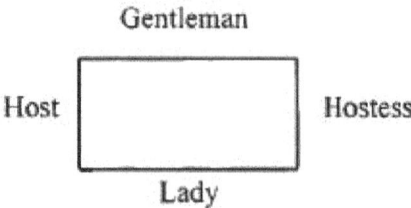

However, with eight or twelve, the hostess usually gives her seat at one end of the table to gentleman #1, while placing herself so that he remains on her *right*. The following diagram depicts this seating arrangement:

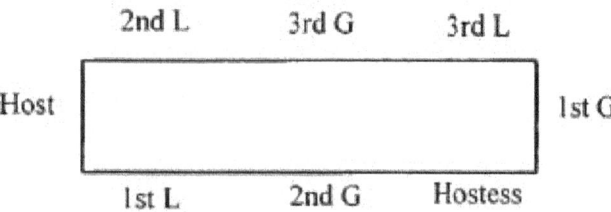

Two alternate arrangements that might be used when couple #3 is a married couple (in order that they not sit side by side) are as follows:

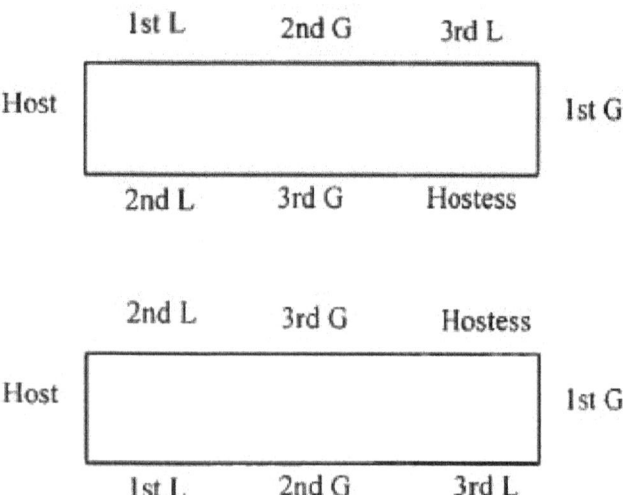

Seating for twelve, when all married couples are seated apart:

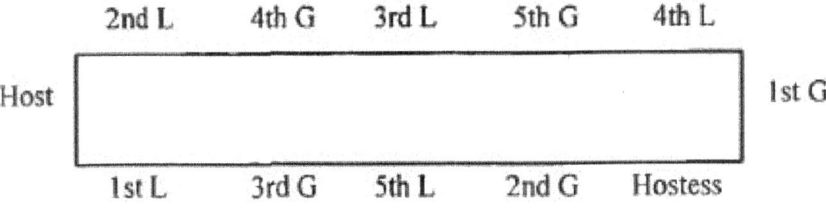

Seating for twelve, when couple #5 is unmarried and seated together:

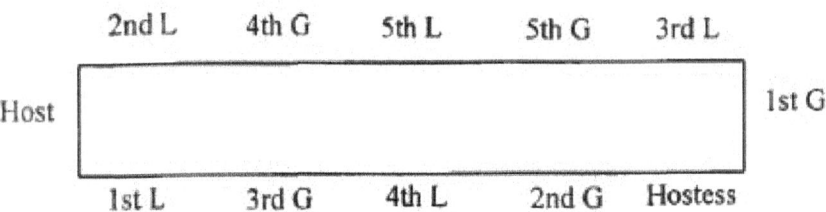

OTHER CONSIDERATIONS

Occasionally, circumstances may require deviating from the accepted seating format. For example, the hostess may need to sit to the right (rather than the left) of gentleman #1, in order to reach the kitchen or serving cart more easily. A left-handed guest may be moved to a corner position for his comfort in eating. A foreigner may be placed beside someone who speaks his language. As always, common sense should prevail.

ROUND-TABLE SEATING

Seating plans for round tables follow the same basic guidelines as for rectangular tables.

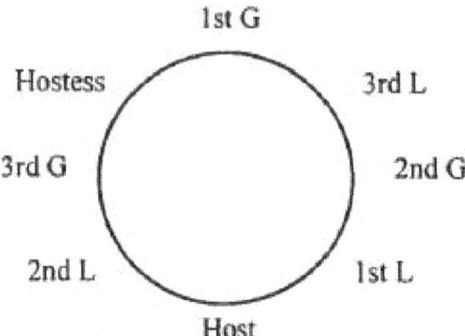

Alternatively:

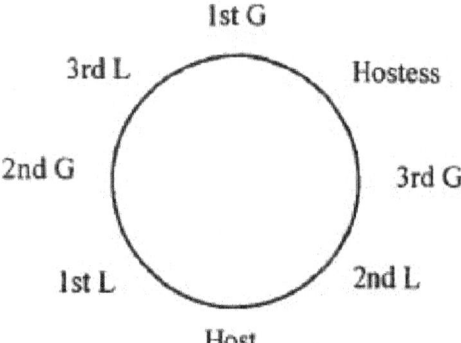

HEAD-TABLE SEATING

Round Tables - Head-table seating at round tables follows the same guideline as general seating at round tables, as shown above and on the preceding page.

Long Tables - Head-table seating at long tables follows the same progression as for other seating arrangements, except that people are seated on only one side of the table and the most important people sit nearest the middle. Men and women are alternated, except for the tradition of seating male guests at each end of the head table. Couples usually do not sit together, with the possible exception of the hosts. Two different examples follow:

Example #1—head-table diagram with hosts seated together

| 3rd G | 2nd L | 1st G | Hostess | Host | 1st L | 3rd L | 2nd G |

Example #2—head-table diagram with hosts separated by guest #1 (Lower tables show the progression of seating, in relation to the head table.)

| 5th G | 3rd L | 2nd G | Hostess | 1st G | Host | 1st L | 3rd G | 2nd L | 4th G |

Senior end

Junior end

Senior end

Junior end

ORDER OF PRECEDENCE

> BLUF
>
> Proper seating is based on rank and protocol.

The hosts determine who is guest #1, guest #2, etc. based upon each guest's job or ranking in relation to one another. This is known as the order of precedence—who precedes or comes before whom. In the military, the order of precedence is easy to determine by each military member's rank. Normally, the spouse is accorded the same precedence as the military sponsor. When two or more guests hold the same rank, their dates of rank and job assignments are considered to determine their relative precedence. At a spouses' club function, the position of the spouse on that organization's board, rather than their soldier's rank, determines the seating. However, guest lists are not always made up entirely of military members and their spouses. Then how is the seating determined?

An unofficial order of precedence is established by the White House and the Department of State that lists everyone from the President of the United States to noncommissioned officers. This order can vary with administrations. Extracts from that list, which are of most interest to Army spouses, are provided on the following pages in descending order:

Extracts from the Unofficial Order of Precedence

Governor of a state (in his own state)

Ambassadors of foreign countries

U.S. Senators

Governors (not in their own state)

Members of the House of Representatives

Secretary of the Army

Chairman, Joint Chiefs of Staff

Vice Chairman, Joint Chiefs of Staff

Chief of Staff of the Army

U.S. Ambassadors

Under Secretary of the Army

Commanders-in-Chief of Unified & Specified Commands (4-star)

Generals (4-star)

Assistant Secretaries of the Army

Lieutenant Generals (3-star)

Former U.S. Ambassadors

Civilian Aides to the Secretary of the Army

SES Civilians, DV4

Major Generals (2-star)

SES Civilians, DV5

Brigadier Generals (1-star)

Consul Generals (in their own district)

SES Civilians, DV6

Colonels/Captains (Navy) & GS-15 Civilians, DV7

Consuls

GS-14 Civilians

Lieutenant Colonels/Commanders & GS-13 Civilians

GS-12 Civilians

Majors/Lieutenant Commanders (Navy)

GS-11 Civilians

Captains/Lieutenants (Navy)

GS-10 Civilians

Vice Consuls

First Lieutenants/Lieutenants (JG) & GS-9 Civilians

GS-8 Civilians

Second Lieutenants/Ensigns & GS-7 Civilians

Warrant Officers

Sergeants Major

Noncommissioned Officers, in order of rank

Additional Considerations:
- All promotion selectees are placed at the top of their rank group.
- All retired military are placed immediately after the active duty of their retired rank.

- State and local government officials are placed according to the purpose of the event and the level of the guests. When in their own state or city, these officials take a higher order of precedence than when visiting another location.
- For those other civilian guests who do not easily fit into one of the above descriptions, the host must decide on their relative position in relation to their other guests. Consideration should be given to the guests' various ages, current positions in the community, and any past positions of importance. It is, of course, always important to seat your guests in such a way as to ensure a lively flow of conversation and mutual interests.

GUEST OF HONOR

A guest of honor does not automatically become "guest #1" in the seating plan. The senior-ranking person in the order of precedence still retains the position of guest #1 and should be seated accordingly. If the host would like to give the most honored seating position to a guest of honor who is not the senior person, the host must privately ask the senior-ranking person if he or she would mind relinquishing it to the guest of honor.

* * * * * * * * * * * *

Planning the seating may be one of the last things you do in your dinner-party preparation, but it is one of the most important.

Chapter 19

Entertaining Tips

"Self-confidence is the first requisite to great undertakings."
Samuel Johnson

There are many household tips that successful hosts have learned over the years to help them entertain more easily. These tips are presented here in the hope that they will do the same for you. Many have application in your family life and will enhance the daily appearance of your home—even when you're not entertaining.

ENTERTAINING EQUIPMENT
The entertaining equipment absolutely necessary for an Army spouse to have is minimal—the basic plates, glasses, and eating utensils—and they need not be expensive. However, added accessories and equipment make entertaining easier and add variety to the appearance of your table. As you begin to acquire items for entertaining, consider the following:

- All china used during a meal doesn't need to match; however, for a seated dinner, the plates used for each course should be alike. For a buffet, the plates need not match. Serving pieces need not be alike or necessarily match the dinner plates and silverware, but they should all blend together in appearance.

- Watch for good buys in flatware, tableware, and serving pieces (garage sales, markets, thrift shops) that you can use for different types of entertaining. Simple flatware, plates, punch cups, bowls, and stemware can be very useful, inexpensive, and easy to maintain. They will pay for themselves quickly, as compared to the expense of purchasing paper or plastic products over the years.

- Table linens are not necessary if you have a nice, well-polished table. You can serve on the bare table or use placemats. However, if you care to purchase or make table linens, remember that they need not be expensive or fancy in order to be attractive. Something as simple as a new sheet can be used on a rectangular table at a picnic, or it can be cut and hemmed to make an elegant, floor-length, round tablecloth that you might use on a round patio table. Napkins can be made from a matching or complementing sheet. A plain white sheet, dressed up with ribbons, flowers, or colored napkins, has graced many a table.

- Before you buy ready-made table linens, consider the care they require. Today, many table linens are made from no-iron fabrics, and machine-made lace tablecloths often do not require even touch-up ironing. While these may not be as formal as an ironed, real linen tablecloth, they can be real time-savers and money-savers for the busy Army spouse.

- If you own a good dining table and expect to use a tablecloth on it, spend the extra money and buy custom table pads. These folding pads are used under a tablecloth to cover the table top and protect it from spills and heat. It's a good idea to buy pads the same color as the wood of your table so that they won't be seen when used under lace or cut-work tablecloths.

- The following items will be useful for any hostess doing considerable entertaining:

    electric mixer
    microwave oven
    blender and/or food processor
    large coffee urn
    tea cart
    pitcher
    large salad bowl
    assorted serving trays, platters, bowls
    Pyrex casserole dishes with attractive holders
    chafing dish or food warmer w/candle
    bamboo lap trays
    punch bowl with cups

IRONING LINENS

The best way to iron table linens at home is to do so immediately after they come from the washing machine—not from the dryer. The moisture content of linen just as it comes from a "normal" cycle of the washer is perfect for ironing. If you are concerned about the tablecloth's dragging on the floor while you iron, put a sheet down first. Another option is to iron next to a large table and let the cloth slide out over the table as it is ironed. After ironing a tablecloth flat, you may want to fold it in half lengthwise, forming a crease down the middle. This facilitates hanging it over a large hanger or folding it for storage. Creases caused by folding, other than the long center one, should be ironed out before the tablecloth is used.

Linen has a natural stiffness when ironed while damp—starch is not necessary. Try it and see for yourself. Certainly, table linens that are to be stored should not be starched, as starch will cause yellowing with time.

Because ironing a tablecloth is so time-consuming, a tablecloth that has been used once and is still clean or has only one or two small spots does not need to be totally washed. Simply remove the spots with a damp cloth, dry, and put the tablecloth away to be used on another occasion.

If you are thinking about sending your table linens to a laundry, consider the following points. American laundries do not, on the whole, wash and iron table linens very well, but many overseas laundries do. For example, German laundries use large mangles to press the linens, and normally do such a good job that the tablecloths come back looking almost brand new. If you are on an overseas assignment, don't overlook the possibility of using a local laundry for your table linens. Whether in the States or overseas, you will probably find that you can do a far better job than any laundry when it comes to treating stains on your table linens. Therefore, it's usually a good idea to apply a commercial stain remover spray or stick to any tough spots and lipstick stains as soon as possible, and then follow up by washing those items before you send everything to the laundry. To combat the expense of sending your table linens to a commercial laundry, you might consider sending only your tablecloths, while you do the much-easier-to-iron napkins.

TREATING SPILLS

> **BLUF**
>
> Don't cry over spilled milk, but know how to get red wine out of your carpet or linens. Keep club soda handy!

Having company often means spilled drinks. Dealing with these spills quickly, effectively, and unobtrusively can be a challenge for even the most-experienced hostess. When a spill occurs, the best advice is to take care of it quickly, quietly, and then act as though it hadn't happened.

Drinks spilled on rugs and carpeting can be removed with hardly a trace if treated promptly with large amounts of soda water. Therefore, the smart host always keeps club soda on hand when entertaining, even if they don't expect anyone to drink it. Smart hostesses also never serve red wine at buffets, since red wine stains so much more than white. Another way to help prevent spilled drinks is to ensure that each guest has access to a table or flat, solid surface for drink glasses.

At the dinner table, spilled red wine can be more discreetly attended to by immediately blotting up any excess, then sprinkling enough salt on the stain to cover it. Later, any remaining spot should rinse out under cold running water.

SILVER-POLISHING GLOVES

Polishing silver pieces that are left out all the time is much easier if you routinely "dust" them with silver-polishing gloves, available at most jewelry stores. First, polish the silver thoroughly with a cream or liquid polish; then rub it well with silver-polishing gloves treated with a tarnish preventative. After that, "dust" (rub lightly) the silver once a week while wearing the gloves. You will find, much to your delight, that this keeps your silver looking great all the time, and eliminates that major chore when it comes time to entertain. (If possible, purchase a shaker jar of silver-polish powder sometimes available with these gloves and intended to replenish their tarnish-preventative properties; this will greatly extend the life of the gloves.)

CANDLE WICKS

Candles always come with long wicks because they are held by their wicks while being dipped. However, leaving these long wicks on the candles makes them unsightly and difficult to light. Whether candles are to be burned or are only put out for display, their wicks should be clipped (to about ¼-½ inch) and charred. Charring the clipped wick is done by lighting the wick momentarily and then blowing it out. This blackens the wick tip and makes the candle look more natural. It also makes the candle easier to light.

COOKING WITH WINES

When a recipe calls for red or white wine, you may substitute red Vermouth (called dark or sweet) or white Vermouth (light or dry). It's much cheaper than opening a bottle of wine for just half a cup or so, since Vermouth is reasonably inexpensive and, once opened, lasts a long time. Any subtle taste change in the recipe will not be noticeable.

SERVING WINES

White and red wines are treated differently before serving them to your guests. White wines are served chilled, allowing at least several hours in the refrigerator, longer if you have a number of bottles. Red wine is usually best served at room temperature, and it's generally believed that uncorking red wine at least one hour before serving improves its flavor.

BAMBOO LAP TRAYS

These are a boon to the Army host who entertains often with buffet dinners. However, they need to be protected from staining with a coat or two of varnish on the tops and bottoms. A coat of varnish is necessary to prevent spills from penetrating the bamboo, even if you always use cloth tray liners. Also, since spouses often lend trays to one another for large parties, write your initials on the bottom of your trays with indelible ink or a waterproof magic marker to ensure that they are returned to you.

BREAD-BASKET LINERS

Keeping bread or rolls warm by putting them in a napkin-lined basket is attractive, but hard on the napkin that absorbs oil from the rolls. Napkin cleanup is made easier and the bread kept warmer by lining the bread basket in the following manner; (1st) place an opened, cloth napkin in the basket; (2nd) place a sheet of aluminum foil, slightly smaller than a paper napkin, on top of the napkin; and (3rd) place a paper napkin, preferably the same color as the cloth napkin, on top of the aluminum foil. Once the bread is put into a bread basket lined in this manner, the paper napkin can't be seen. Then, as the bread is removed, the paper napkin hides the aluminum foil, which serves the dual purpose of reflecting both heat and grease.

GIFTS OF FLOWERS

Cut flowers are a special hostess gift, but they often require immediate attention. Although some flowers arrive pre-arranged in a vase or other container, some are presented as bouquets. The prepared host/hostess will have several sizes of vases available, and flower clippers (conveniently located in the kitchen) in case the stems need shortening. The vases need not be expensive; a leftover wine carafe makes an attractive substitute.

TUB FOR FLATWARE/SILVERWARE

When a host, who has no household help, takes the plates and flatware to the kitchen, they should put the flatware immediately to soak—to make washing easier and, in the case of silver, minimize tarnishing. The best way to accomplish this is by placing all of the flatware, handles up, into a plastic bowl or tub that has been partially filled with hot, soapy water. The bowl should be large enough to hold all of the eating and serving utensils, but small enough so that each piece remains in a relatively upright position. If using silverware, be sure to fill the bowl with only enough soapy water to submerge the knife blades, not the handles, since silver knife handles are hollow and should not be allowed to soak.

BARTENDER

When the party is large and you decide to hire a bartender, these hints will help.

- At the time the arrangements to tend bar are made with someone, confirm the duties, salary, arrival time, and appropriate dress. (Black slacks or skirt, with a white shirt or blouse and black bow tie, are standard—also a jacket, if the event is more formal.)

- The bartender should arrive early enough to help set up the bar area and attend to such duties as opening the wine and slicing lemons. Usually, ½-1 hour before guests are expected to arrive is sufficient. The host or hostess should show him/her where the bar supplies are kept, and where and how the bar is to be set up.

- Your soldier should remind the bartender before the guests arrive that he/she is not to drink any alcohol during the party. Suggest other beverages and, during the evening, offer something for the bartender to snack on. You might even offer something more substantial to eat before the guests arrive, or after they leave.

- Be sure to have plenty of ice, soft drinks, and juice on hand, as well as the alcoholic beverages.

- Place an empty cardboard box nearby in an inconspicuous spot for the empty bottles that will accumulate.

BACK-TO-BACK ENTERTAINING

If you find yourself faced with the situation in which you need to entertain a large number of people, more than can be invited at one time, consider back-to-back entertaining. If you are considering two dinner parties, why not have one on Friday night and the other on Saturday night? You only have to clean the house once, with a quick vacuuming in between parties. The same menu can be served both nights, with double quantities prepared at one time. The centerpiece and, hopefully, the tablecloth can be used for both events. The china, glasses, silverware, and serving pieces have only to be washed and replaced, not stored and later reassembled. Entertaining two nights in a row has one possible disadvantage—you may still be tired from the efforts of the first party by the time the second party rolls around. However, good prior planning and a little help in the cleanup department can overcome that problem.

NEW IDEAS

> **BLUF**
>
> "Don't reinvent the wheel! Utilize ideas from other sources. Add personal touches to make the event your own."
> *Angel Mangum*

As you become comfortable with tested menus and recipes, don't get into a rut and forget about trying new ideas. Styles of entertaining, food fashions, and party themes change with the times. A new item or two on your menu, or a change in setting, may be all that's needed to keep you current and prevent monotony. Don't hesitate to be innovative. A real boon, in that regard, for the Army spouse is the opportunity to live in many different parts of America and overseas, and sample the traditional recipes of those areas. Foreign specialties have always been a favorite, and American regional specialties have truly "arrived" in the last decade and promise to continue in popularity. Be sure to take the opportunity to become familiar with the cuisine of your area and find recipes that can be adapted to the way you like to entertain. These can add great variety and interest to your entertaining.

Watch the popular magazines for interesting recipes and new ideas for informal table settings, napkin folds and placement, centerpieces, and room decorations. That is where you will first notice new trends in entertaining. Clip out the recipes and pictures of table settings you especially like, and try them at the next opportunity. Your post library is also a good place to start looking for new ideas. It has subscriptions to numerous magazines, as well as the newest books on food and entertaining.

ENTERTAINING OUTSIDE THE HOME

Entertaining outside your home, at a restaurant or an Army club, is a wonderful way to treat your guest(s) to an enjoyable meal and a few hours of relaxation, without having to do the work yourself. It will probably cost you more than when you entertain in your home, but for the busy spouses of today, it is often a welcome alternative. More and more Americans use this form of entertaining, as do many other nationalities, especially Asians. Where you take your guests will depend on the type of meal you want, how much you are planning to spend and, perhaps, the number of guests.

You can make entertaining outside your home a special occasion by taking time to plan it in advance. If you will be entertaining in a restaurant, be sure to call for reservations and to make any other needed arrangements. If you know the layout of the restaurant, you can request a specific table or location in the room. When you have only a few guests who aren't arriving together, arrange to get there early and wait for them at your table. Decide before they arrive where you will seat everyone, being careful to give your guests the best seats. If you are entertaining in an official capacity, use the same seating protocol as you would in your home. (See Chapter 18, "Seating Arrangements.")

Creative personal touches will enhance any event hosted outside the home. Place cards are appropriate for larger groups. You might also add small favors to the table setting. If you plan a set menu, consider ordering a special dessert or wine you know your guests will enjoy. You might also bring some small items from home to augment the appearance of the table—a footed compote dish filled with fruit or candy, or perhaps a floral centerpiece or bouquet that might then be presented to the guest of honor at the end of the meal. Whether in or outside the home, it is the hostess/host who sets the tone of the event.

****

Hostess with the mostest, doesn't mean you have to spend the mostest!

SECTION SIX: SOCIAL FUNCTIONS

Chapter 20

Coffees

"As spouses, if we do not know one another, how will we be able to support one another, and stand together in times of need?"
Alma Powell, spouse of General Powell

The tradition of military spouses getting together for coffee dates back to the establishment of the first military posts. Wives of the frontier Army had to endure many hardships, and an hour or two spent sharing a cup of coffee with a friend must have been a welcome respite. Like so many of the social customs of the nineteenth-century military, this social tradition survived and evolved. Coffees became not just social get-togethers for all spouses, they offered an opportunity to develop friendships within the unit and a time to share information about unit and community activities.

This broader purpose resulted in an informal support network that, in turn, has served as the role model for today's command-sponsored Family Readiness Groups. These groups are not coffee groups, but offer spouse get-togethers and a support network to the unit families. It is important to note that, although some general information may be disseminated at a coffee, the primary function is a social one. The official information source for families in a unit is the Family Readiness Group (FRG). Because of these similarities to coffee groups, Family Readiness Groups will be discussed later in this chapter.

THE UNIT COFFEE

BLUF

Coffee, Tea, or FRG?

The evolution of coffees has seen a change not only in their purpose and function, but also in participants. After World War II, these get-togethers were primarily held exclusively by and for officers' wives. During that time, very few NCOs and enlisted men were married. Consequently, there were simply not enough NCO and enlisted spouses available to establish the tradition of coffees.

Today's Army is different. The all-volunteer Army has resulted in an enlisted force in which more than half the members are married. The increased number of working spouses in American society has resulted in officers' spouses no longer accepting the old-fashioned tradition of not working outside the home. The Army now has many more female members; some are single, some are married to civilians, and some are married to other military members. The American trends of later marriages, same-gender marriages, higher divorce rate, and more second marriages are being reflected in Army society. The result of all of these changes on coffees has been that coffees are no longer the private enclave of non-working officers' spouses.

Everyone who takes the opportunity to attend these spouse get-togethers on a regular basis soon learns the benefits of belonging to such a group. All unit coffees today serve the same useful dual purpose, which are:

1. They provide the spouses of the unit an opportunity to become acquainted in a comfortable, relaxed atmosphere. The friendships that result form the basis for an informal support system.
2. Coffees held on a regular basis give the coffee group an opportunity to exchange information about upcoming events in the unit and the community.

Coffees were once held only in the mornings, but today's lifestyles have changed that. Evening get-togethers are more convenient for the majority since many spouses work, and childcare is easier to find at that time of day. The term "coffee" doesn't even necessarily mean that coffee will be served since these get-togethers can take many forms. Thus, the evolution of coffees continues.

HOSTS AND HOSTESSES

Every member of a unit coffee group generally takes a turn serving as host or hostess. A sign-up sheet is periodically passed around at the coffees so that everyone will have an opportunity to volunteer. It's not a lot of trouble or expense to host a coffee, and everyone in the group takes a turn sharing in this responsibility. The group size will, of course, determine how frequently everyone needs to take a turn. It helps to plan ahead and decide which month or meeting would best fit your schedule so that when the list is passed around, you're ready to sign up.

If your coffee group is small enough to meet in the home, don't hesitate to volunteer to host the group because of the size of your home or apartment. It doesn't matter if it's small and some have to sit on the floor; everyone has lived in small places before. Nor is it important to have china and a coffee service; those aren't essential either. Getting together is what's important. Just clean the living areas, prepare your favorite snacks, and use whatever you have—without apologizing. Even paper plates and cups are fine to use; they may be a bit expensive, but cleanup is a snap.

If your coffee group is large and gets together at a unit or community facility, more than one hostess will probably be needed. Talk to a few of your friends about working with you to host one of these get-togethers. You can each be responsible for a different aspect of the event (invitations, program, refreshments, clean-up), or you can work together on everything. Either way, you'll find that co-hosting with your friends can be fun and an opportunity to strengthen your friendship.

Two Hosts/Hostesses, Rather Than One

It's *always* a good idea to have at least two people host each coffee, no matter the size. They share the work, the fun, and the cost of preparing for the get-together, as well as the responsibility. It's true that co-hosting does mean everyone's turn comes around twice as often. However, it's well worth it to the host or hostess because the work and expense are cut in half; it's worth it to the group because they'll never be left without a get-together if one host or hostess gets sick at the last minute.

GUIDELINES:
FOR HOSTS AND HOSTESSES
- *Day/Date and Time* - Coordinate the day/date and time with the coffee-group leader *before* sending out the invitations, unless the coffee group has a regular meeting date and time. Include the day of the week along with the date of the event; it prevents any miscommunication.
- *Special Guests* - Discuss with the coffee-group leader whether there are any special guests who should be invited. The leader may occasionally want to invite senior spouses from outside the unit, particularly when special programs are planned. The group might also consider inviting any fiancées in the unit who are living nearby or visiting; making them feel welcome is a wonderful introduction to Army life.
- *Invitations* - Send the invitations out approximately two weeks in advance, using R.s.v.p. with your phone number or email address. Call or email those who haven't responded a few days before the coffee. It is helpful to have a telephone number on the invitation as well as the GPS address, just in case invitees have difficulty finding the venue or have an unexpected delay or cancellation. (Alternatives to R.s.v.p. for very large groups are, "Please email if you can come," or "Email reply by ... if you plan to attend." However, with these requests you will only hear from those who are coming; you won't know who didn't receive an invitation.) Online invitations, such as Evites, are a great way of reaching everyone at little or no cost and allow the host or hostess to see who has viewed the invitation. They keep track of R.s.v.p.s, send reminders and updates, and possibly maps and directions for the event.
- *Door/Opportunity Prizes* - If you are responsible for providing door/opportunity prizes, remember to plan ahead. They need not be expensive, and home-baked goodies are always welcome. Gift-wrap anything other than plants or food; decorate those with ribbon. These prize-giving events are not legal in some states, so the best advice we can give is: If you are thinking of having a raffle or opportunity drawing at your event, check local regulations and policies. If you are the coffee-group leader, be sure to always get a legal brief.
- *Refreshments* - Plan to serve something nutritious, in addition to the usual sweets. The attendees may not have eaten breakfast before coming to a morning coffee, or dinner before attending an evening get-together. They might not have had time, or perhaps they're watching their calories. Although the coffee hostess isn't expected to provide breakfast or dinner, savory small plates or appetizers are a welcome addition to an array of sweets.

- *Type of Event* - Remember that a coffee doesn't really have to be a coffee. It can be any type of get-together: potluck, salad luncheon, picnic, baby shower, bowling, movies, wine-tasting, craft idea, short trip/sightseeing, or lunch or dinner at a favorite restaurant. Just be sure to coordinate your ideas with the coffee-group leader first.
- *Program* - If you plan to have a program or guest speaker, first discuss your ideas with the coffee-group leader and whether the program should take place before or after the business meeting. When making the arrangements with a speaker, be specific about how long the program can last and when it should begin. Coordinate everything early, preferably 3-4 weeks in advance. Then, a few days before the coffee, check back with the speaker to remind him/her and reconfirm the date, time, and place.
- *Name tags* - Be sure to use name tags. They help members learn and remember names, and make newcomers feel more at ease. The write-your-own, stick-on type is the easiest.
- *Table Setup* - If you plan to serve the food buffet style on the dining table, remove the chairs from around the table to make access easier, before everyone arrives. The chairs will probably be needed for seating elsewhere.
- *Candles* - Light the candles, which may be used on the dining table for late-afternoon or evening coffees, before the food is served. Candles are not usually used in the morning since their function is to provide light, but scented candles and candles used for ambiance provide a welcoming atmosphere.
- *When to Serve Food* - Coordinate when to serve the food (before or after the business meeting and/or program) with the group leader and any guest speaker.
- *Who Goes First* - Invite the senior spouse present to serve him/herself first when the food is ready. They may decline and suggest that someone else have the honor of going first, but it is correct for the host or hostess to recognize the senior spouse in this manner. The other guests will follow without any encouragement. (If someone has brought an older guest, it is appropriate in this instance to invite this older guest to go through the line first.)

FOR ATTENDEES
- *Invitation Response* - Respond to your coffee invitation within 24-48 hours.
- *How to Dress* - Wear something comfortable unless the invitation indicates otherwise; coffees are normally casual get-togethers, regardless of the time of day or evening they're held.
- *Transportation* - Try to arrange a carpool or, if invited, accept an offer to go with others. Sharing rides is more economical and another opportunity to make or strengthen friendships. It's also helpful when parking is limited.
- *Babies and Children* - Don't take your baby or child with you unless the hostess offers. Some groups make special arrangements for newborns. Normally though, adult social functions are not intended for children. Everyone occasionally enjoys having some time with only adults. Learn what your group guidelines are and be sure to follow them.
- *Expense?* - You'll want to take a little money along. Most coffee groups have a small

treasury collected from either nominal dues or the sale of door-prize chances. The money is used for good causes—possibly welcome plants, flowers for sick members, farewell gifts. Participation in the door/opportunity prize drawing is fun, but voluntary.
- *Visit with Everyone* - While it's fun to see and talk with your close friends at these get-togethers, always take time to chat with the host or hostess(es) and the senior spouse, as well as any newcomers, visitors, and guests of honor.
- *Hostess Gifts and Thank-yous* - Senior spouses often take a "standard" hostess gift and write a thank-you note on behalf of the group; therefore, it is not necessary for individual attendees to do either. However, telling the hosts or hostess(es) how nice everything was, and how much you enjoyed the coffee, is always appreciated.

FOR COFFEE GROUP LEADER

Coffees are primarily social groups. The commander's spouse normally serves as the leader and facilitates the group meetings. If, for whatever reason, she/he is not available to assume that position, someone else will need to lead the group. In most such instances, the commander will ask a particular person to take over the responsibility. It may be a field grade spouse (spouse of the XO, S3, or spouse of the chaplain). It is important for the cohesion and continuity of the group that this position does not go unfilled for more than a few weeks.

General guidance

- *Coffee Roster* - A voluntary up-to-date unit or coffee roster may be important for both you and all the members of the coffee group. It could include names, addresses, phone numbers, emails, and perhaps children's names and anniversaries. (The Privacy Act reminds us that no one should have personal information given out without his or her approval.) This roster should be regularly updated and kept by the senior spouse or group leader. The unit executive officer or S-1 (for officers' families), first sergeant or personnel NCO (for NCO/EM families), can keep you informed about new arrivals and planned departures. These rosters are strictly social and do not, in any way, replace those rosters held by the FRG.
- *Welcoming Committee* - Welcoming newcomers warmly is an important responsibility for any coffee group. It demonstrates that this is a caring, cohesive unit. Since welcoming all newcomers is normally too large a task for just one person, most coffee groups rely on a welcoming committee (discussed at length later in this chapter). A new group leader needs to ask the members of the welcoming committee if they are willing to continue to serve, or if they would like new volunteers to take a turn. The group leader also needs to ensure that the committee promptly receives the names and addresses of all newcomers, and that the committee understands exactly what the group expects of it. All members of the coffee group should help to keep the committee informed and actively reach out to make newcomers feel welcome.
- *Visiting the Sick* - Reaching out to any member of the group who becomes ill is an important service the group leader can provide. Taking time to visit someone in the

hospital shows that you feel they are a valued part of the group. Ask the group members to notify you as soon as they learn about someone who is ill or hospitalized. Be cautious of HIPA rules. Most commanders assign a Care Team (casualty-response group) to visit, arrange meals, or buy flowers for hospitalized members of their unit. Coffee groups in coordination with their FRG may supplement what the Care Team is doing.

- *Informal Fund (Dues/Opportunity Prizes/Farewell Gifts)* - An informal coffee group may need approval from the post commander to have an informal coffee group fund. This fund is typically used to purchase unit pins, farewell gifts, and token welcome gifts. It is recommended to have those in inventory rather than to have money in a coffee group fund.

  Whether or not the coffee group charges its members dues should be a group decision. Some groups have found that annual dues discourage membership, and the small amount of money needed isn't worth the trouble of keeping records on dues collection. If the coffee hostess(es) provides several door prizes/opportunities and most who attend buy chances, the proceeds are usually sufficient for the limited expenses of a coffee group, mainly the unit pins. Door prizes should always be inexpensive. A good guide to promote is: *Make it, bake it, or grow it*. If the informal fund runs low because of an unexpected expense, you can always pass a basket around and ask for donations. An added benefit to collecting money through door prize opportunities is that anyone can choose not to buy a chance without being noticed, a blessing to those who feel they just can't afford to participate this month. Some member of the group, other than the leader, should serve as treasurer; the leader should ensure that the treasury records are kept accurately.

- *Farewell Gifts* - These are a traditional part of unit coffee groups. They serve as mementos of the good times and friendships formed with the unit and coffee group. Therefore, the best farewell gifts are lasting in nature, related to the unit; i.e., engraved unit insignia/crest on an item or some memento of the area, but of reasonably low cost. The two basic issues to be decided are: What will the gifts be, and how will they be paid for? Both issues should be decided by the group, but take care not to let these decisions become major topics of discussion and dissension. A new coffee-group leader will want to find out if the group is satisfied with the current farewell gift and payment procedure. If a change is desired, ask for a volunteer committee to study various options and present them to the group for a decision. Traditionally, spouses would pay for their own "farewell gift." The treasurer would annotate that a spouse is paid up, and that spouse's farewell gift would be purchased by the coffee group leader or volunteer and given when that spouse is PCSing or departing that coffee group. It is unfair for a few spouses to pay for farewell gifts for those who do not attend the coffee on a regular basis. However, sometimes these gifts are handmade, token farewell gifts such as a coffee-group cookbook; then it is nice to share this memento with everyone.

Specific guidance for planning coffee get-togethers

The unit coffee is a regularly scheduled time for the group to socialize and share news. The group leader's responsibility is to oversee the scheduling and planning of the meetings, and to serve as leader of the meetings. The following suggestions will help the coffee-group leader as he or she works with the coffee host/hostess(es) and prepares for the business meeting:

- Coordinate with the coffee host(s)/hostess(es) early enough to avoid a conflict on date and time.
- Respond to the invitation within 24-48 hours, both to be correct and to set a good example.
- Ensure that all newcomers come with someone to their first coffee—usually the group member who lives nearest or someone from the welcoming committee.
- Gather information to share at the coffee that is of *real* interest or value to the group. Your sources are the senior spouse at the next higher unit level, your partner NCO (or officer's) spouse, the garrison commander's spouse, any senior spouses' group to which you belong, and any information available to you about unit and community activities. In order to keep the business meeting as short as possible, limit the information you discuss to that which is truly important or significant. (For example, the menu for the next spouse club luncheon is not significant enough to take time discussing at a coffee.)
- Prepare a written or typed list of important upcoming dates and events, and make a copy for everyone in the group. This could also be emailed to the group or included in a unit newsletter that you might prepare or contribute to. A shared online calendar of events can also be created to which everyone in the group has access.
- Be organized. Have an agenda for each coffee, use it at the meeting, and keep it for future reference.
- A typical agenda:
  1. Thanks to coffee host(s)/hostess(es). Present the token thank-you hostess gift.
  2. Introduction and welcome of newcomers and guests (Newcomers might introduce themselves and give a little background about their family and where they came from; guests should be introduced by the members who brought them.)
  3. Farewell to departing members (usually including the presentation of a farewell gift)
  4. Family announcements (babies, promotions, illnesses, birthdays, honors)
  5. Reports from committees or group representatives, if applicable
  6. Calendar of events (Pass them out, but discuss only those items of special significance in order to highlight them; explain any topics that require more details than can be written down.) Highlight events on the online shared group calendar; answer any questions anyone may have.
  7. Ask for any corrections or additions to the calendar
  8. Old business
  9. Any new business not covered on calendar
  10. Announce date and program for next month's coffee, if known.

11. Pass and review the coffee group roster for any corrections or updates. Ask the spouses to initial their contact information for verification.
12. Door-prize opportunity drawing
13. Program, if any

- The meeting should stay on track, but not be too business-like. Strive to keep the meeting fun and light. Also, keep it as short as possible, with the bulk of the time spent on socializing and any program that's planned.
- Let the coffee group decide if it wants to have programs and, if so, how often. Discuss what type would be of interest. Work with the coffee host(s)/hostess(es) well in advance to help select a short program or guest speaker, like the garrison commander or a local historian.
- Group members will be as active as their time, talents, and interests allow. As the group decides on projects, don't try to do it all yourself; use the committee system. Not only will this be easier on you, but it will help members feel more a part of the group. Guard against being a micro-manager.
- A thank-you note to the host(s)/hostess(es) after each coffee, written by you on behalf of the group, is appropriate and always appreciated.

ATTENDANCE/INCLUSION

Each coffee group is different in composition and interaction. Many factors can influence a group's membership and participation: location, group demographics, average age of members, family responsibilities, availability of childcare, and perceptions about the unit coffee group. False perceptions can cause some servicemembers to be uncomfortable with the idea of their spouses attending such unit-oriented get-togethers. Yet, the unit coffee provides an opportunity for Army spouses to be a part of their extended family in the military community. In rural areas of the United States and at overseas locations, the unit coffee group may provide the only link to friendship and informal support that its members have. Therefore, it's important for the composition, purpose, activities, and benefits of the group to become widely known.

BLUF

Coffee groups were traditionally comprised of spouses of officers, warrant officers, and senior NCOs of the unit.

The question of exactly who is eligible to join a unit coffee group does not have a simple, clear-cut answer today. Army society now consists of family situations that were almost unheard of a decade or two ago, and not all the answers have been worked out as to who gets invited to a coffee and who does not. For example, does a coffee group invite the unit's female officers or female soldiers (single or married) to its get-togethers? Does a coffee group invite a live-in fiancée or "significant other"? These questions are all being discussed and handled in a variety of ways throughout the Army; no single answer has evolved. It can truly be dependent on the unit and the commander's and spouse's preference, as they are most likely leading the group in one direction or another.

A fair way to start the coffee group is for the unit to host a Hail and Farewell. The Hail and Farewell group is usually created by the commander, command sergeant major, and the adjutant. From that list, the names of the spouses may be drawn to start the coffee group. It is beneficial to have the Hail and Farewell first. However, if you are not able to have a Hail and Farewell as soon as you arrive, the recommendation would be to start the coffee group small, then include more if desired. It is not wise to invite and then uninvite spouses to groups.

It is a good time to have the roster available at the Hail and Farewell to confirm email addresses and inclusion in the coffee group.

It seems most reasonable that a group's membership policy should be decided on by the members of the coffee group and their leader, in concert with the commander. The commander needs to endorse any decision of the group that involves military members in the unit. This membership-eligibility policy for the coffee group needs to be applied to all fairly, and carried out in such a way that it takes into consideration the feelings of those involved. For example, if the group decision is to offer membership to singles in the unit, the leader can explain the policy to them and ask if they are interested in receiving invitations to the coffees. If the decision is not to invite a girlfriend until she is officially engaged and the wedding date is set, then the leader might want to explain the group policy to any young man who has a "significant other," so they understand why she isn't being invited to the coffees when someone else's fiancée is. What works well in one unit might not work well in another. The bottom line is that, when these situations arise, they should be addressed, not ignored.

WELCOMING COMMITTEE

> **BLUF**
>
> "Be kind to one another."
> *Ellen DeGeneres*

- The welcoming committee of a coffee group is essential. Members of this group contact and visit the spouses who are newly arrived and eligible to join the coffee group. Letting others know that their arrival is noticed and that their presence in the group is welcomed goes a long way toward making them feel happy and proud to be a part of this unit. The size of the welcoming committee depends on the size of the group; some get along fine with only one person serving in this capacity, while others have frequent newcomers and need several on the committee. Whatever you do for one, you must do for all. If spouses are arriving from a distance (overseas), sometimes a welcome basket with nonperishable items or a "pounding basket" is greatly appreciated or perhaps a combination of both.

Here is a list of some of the helpful things that a welcoming committee can do for a newcomer:

- Take a small welcome gift, usually a small home-grown house plant or homemade goodie.
  - Pounding Basket: Traditionally, the pounding basket would consist of a pound of butter, a pound of flour and a pound of sugar for the new spouse's pantry. These days, a homemade item such as bread, cookies, brownies along with the recipe continues the tradition with an updated twist.
- Ask if any assistance is needed (directions, transportation, names of babysitters); arrange for what's needed.
- Check to see that the family has gotten the ACS and unit welcome packets; provide them, if necessary. Some units have welcome booklets or guides – "Out and About."
- Share dates of upcoming unit and post events.
- Invite new spouses to the next coffee group and/or spouses' club function; offer to pick them up and take them so they don't have to go to the first function alone.
- Leave a group roster. Point out the POC's name, number, and email in case they need to ask any questions.

The welcoming committee may be the first contact a newcomer has with the unit coffee group. This sincere welcome creates that important first impression of a friendly group, eager to extend the hand of friendship.

NEWSLETTERS

Newsletters and/or unit community calendars are great communication tools. They can be a vital link among the coffee group or Family Readiness Group (FRG) members. Email is the best way to ensure everyone gets a copy quickly and at no cost to anyone. Newsletters are usually compiled by the unit with input from several sources, including the commander and command sergeant major, and distributed via email to the entire FRG. The newsletters do not need a legal review; however, it would be wise to have JAG review at least the first one to ensure the new command team is on the right course.

Who types the newsletter and prepares it for distribution will depend on the size and composition of the group. For small groups, the group leader may serve as author, typist, and editor. Alternatively, there might be a budding journalist in the group who would enjoy taking on these tasks, with the group leader providing the information. For large groups, especially at higher levels, the leader probably will not have time to do any more than supply the key bits of information, and the bulk of the work will be shared by a committee. Either way, the newsletter offers a great opportunity to share information with the group members.

WELCOME COFFEE

The typical way for a group to welcome the spouse of a new commander or command sergeant major is with a welcome coffee. Depending on the size of the group, this can be a simple coffee in someone's home, or a larger affair held at the club or community center. It can be held in place of the regular monthly coffee-group get-together, or it can be a separate event. Either way, this special event serves to warmly welcome the new senior spouse to the group, gives the contemporaries and other appropriate senior spouses in the community a chance to join in the welcome, and provides the guest of honor with an early opportunity to become acquainted with the members of his or her new group.

The person who takes the lead in organizing such a welcome coffee varies with the group. Typically, to welcome a new commander's spouse, the spouse of the executive officer of the unit takes charge on behalf of all the officers' spouses in the unit. If he or she is not available to do this, the spouse of the S-3 is usually next in line. To welcome a new command sergeant major's spouse, the senior NCO's spouse of the unit and the commanding officer's spouse usually work together to take the lead on behalf of all the unit's NCO spouses. The expenses for a welcome coffee are usually shared by all of those in the hosting group.

The degree of planning needed for a welcome coffee will, of course, depend on the size of the event. For large welcome coffees, the planning guidance offered in Chapter 23, "Teas," would apply. For small welcome coffees, the planning would follow much the same course as for a regular monthly coffee. Additionally, the following suggestions will be helpful to the person taking charge of planning such an event:

- Tell the new senior spouse that the group would like to have a welcome coffee in their honor. He or she will probably be grateful, but you must obtain their permission before proceeding. If by some slim chance he or she declines, you must honor his or her wishes.

- Discuss with the honoree the date and time you are considering for the event, before you prepare any invitations. Also, discuss your proposed guest list, and ask if there are others whom the honoree would like to have invited.
- Those invited usually include: all the members of the group hosting the event, the guest of honor's contemporaries, and the next-senior leading spouse. For example, the guest list for a welcome coffee for a battalion commander's spouse might include all the battalion officers' spouses, other battalion commanders' spouses in the brigade, and the brigade commander's spouse. A welcome coffee for a battalion command sergeant major's spouse might include all the battalion NCO spouses, other battalion command sergeants major spouses in the brigade, the brigade command sergeant major's spouse, and the battalion commander's spouse.
- Send the guest of honor a written invitation, just as you do for the other guests, but omit or mark through the R.s.v.p. and write "To Remind."
- Arrange for a senior spouse in the hosting group to bring the guest of honor and take him or her home afterward or, if you will be free, plan to do this yourself. If he or she lives within walking distance of the welcome coffee, make plans to walk with the honored guest. Let him or her know the transportation or walking arrangements, and coordinate a pickup time.
- A small corsage, wrist corsage, or boutonniere for the guest of honor is appropriate, and can be presented to the guest of honor upon their arrival at the coffee. (See "Pinning on a Corsage" in Chapter 23.) There is no Army-wide traditional color of flowers associated with a welcome or a farewell corsage or boutonniere; however, you may wish to ask their favorite color, favorite flower or even use the unit colors.
- The size of the event determines whether or not you have a receiving line. For any but the smallest gathering, you will want a receiving line to facilitate introducing all of the guests to the guest of honor. The welcome-coffee planner serves as the host/hostess (even if the event is held in someone else's home) and stands first, followed by the guest of honor. The host/hostess greets the guests as they enter the door and shakes their hand, then introduces them to the guest of honor. For small gatherings, where a receiving line would seem too formal, these introductions are accomplished by the host/hostess as the guests arrive or by escorting the guest of honor around the room.
- Plan for everyone to use name tags and, if possible, have them available for the guests to put on before they go through any receiving line that's planned. This is especially helpful for the guest of honor as he or she tries to learn everyone's name, and for the host/hostess who must make all of the introductions.
- The food served at a welcome coffee is the same as that served at any coffee, except that the selection may be expanded and the table set with an especially pretty centerpiece. If the food is not made available until later in the coffee, as with a typical coffee, the guest of honor should be invited to start the line.
- The size of the event determines whether or not you need a pouring list. For large welcome coffees, pourers would be helpful; follow the guidelines given in Chapter 23, "Teas." For small affairs, everyone can serve themselves, although the host/hostess usually offers to get the guest of honor a cup of coffee, while the honoree remains to talk with the other guests.

- A few words of official welcome to the guest of honor, on behalf of the hosting group, are expressed by the person who planned the welcome, usually about three-quarters of the way through the coffee. This is normally after everyone has arrived, met and talked with the guest of honor, had something to eat and drink, and mingled with the other guests. You will have to position yourself in the room so that everyone can see you and get everyone's attention first. This is also an appropriate time to present the guest of honor with a small gift of welcome from the group; it might be a miniature unit crest pin, a new guest book that all of the guests signed as they arrived, or the centerpiece of fresh flowers arranged in a pretty basket that he or she is to take home when the coffee is over. However, a gift is not necessary.

The guest of honor may or may not care to say anything more than thank you after this official welcome. Some people feel very comfortable in such situations and speak extemporaneously, others will have thought about it in advance and have something prepared to say to the group, and a few prefer not to have to speak in front of a crowd. Let the guest of honor take the lead after you have made your welcoming remarks and presented any welcoming gift.

If this welcome coffee is being held in place of a regular monthly unit coffee, dispense with the business meeting, if at all possible. Email everyone to pass along the month's information. If something of an urgent nature must be shared with the group, keep this talk to a minimum. If the guest of honor will become the coffee group leader for this group, he or she may or may not want to conduct a short business meeting; that's their prerogative. However, the primary purpose for this event should be to welcome the new senior spouse.

FAREWELL COFFEE

A coffee can also be used to farewell a commander's spouse or command sergeant major's spouse, although at senior command levels a farewell tea is frequently planned instead. The person in charge of planning the farewell may wish to consult with the guest of honor regarding whether he or she prefers a tea or coffee. Certainly, he or she should be consulted about the scope of the guest list because the size of the guest list may influence where the farewell is held. Occasionally in small units, the guest of honor will prefer to limit their farewell to only those in the unit. They may feel that this is a very special time for saying goodbye to the friends they've made in the unit and that to invite outsiders would detract from that. However, usually for all but the smallest farewells, the guest list includes the guest of honor's contemporaries, senior spouses, and any other members of the community with whom they have worked closely. Of course, the guest of honor should always be asked for a list of any special guests they would like invited.

FAMILY READINESS GROUPS (FRG)

Separation is never easy for either the active duty member or the family. Field exercises, extended TDYs, unit deployments, and times of crisis require special efforts by everyone. Fortunately, the Army recognizes this and continues to try to reduce the level of stress felt by Army families, and to help them become more self-sufficient and self-reliant. In 1987, the Department of the Army took a giant step forward when it issued a regulation implementing command-sponsored Family Support Groups, now called Family Readiness Groups.

Unit FRG

Family Readiness Groups are sponsored by every unit. The Family Readiness Group (FRG) is an officially command-sponsored organization of family members, volunteers, and soldiers belonging to a unit. Together they provide an avenue of mutual support and assistance, and a network of communications among the family members, the chain of command, and community resources. FRGs help create a climate of mutual support within the unit and community. Basic FRG goals include supporting the military mission through provision of support, outreach, and information to family members. FRGs play an integral part in the unit, family, and soldier readiness.

They are a part of the commander's program, and every soldier (all ranks) in the unit and his/her family members are automatically a part of the group. Each unit commander is responsible for developing how the unit's Family Readiness Group will function, so each FRG may be organized and operated a little differently. Nevertheless, the principal purpose of each remains the same: to establish a system for the unit's family members to receive information, assistance with solving problems, and support during times of stress.

This purpose is accomplished in a variety of ways; some are group oriented and some individually oriented. Group-oriented events might include: unit briefings to which family members are invited, Family Day at the unit, and spouse get-togethers. Individually oriented efforts might include: unit welcome packets, monthly newsletters, information on what unit and post support is available and how to obtain it, and a list of what the family members should do to prepare for those times when they will be separated. All efforts are aimed toward ensuring that military members in the unit and their family members are informed about the unit, prepared for separations, and, most importantly, included in the "chain of concern."

"Chain of Concern"

This most important feature of Family Readiness Groups is a network for communication and support among the family members. In order for this network to be established, the unit needs to learn: who and where the family members are, their special needs, skills, and concerns. This information is strictly confidential and protected by the Privacy Act, but essential for contacting and assisting family members should the need arise. Each unit is divided into groups, for easier information dissemination. Each group is headed up by a contact person. A small pyramid, or "chain of concern," comprised of these groups is then established.

FRG Leader

At the top of each unit "chain of concern" is the FRG leader (often the senior commander's spouse, but not necessarily). Sometimes spouses of the commander and command sergeant major are now FRG Advisors. The commander will ask and appoint an FRG leader. ACS hosts a myriad of FRG classes to ensure FRG leaders and team members are well versed in family readiness and legally approved. When the FRG leader receives information from the unit to pass down the chain, he or she calls those on the next level of the pyramid, who call the contact people. Each of them calls everyone in their group, and thus, the information gets passed quickly to all the unit's family members. No one is left to wonder what's happening to the unit, believe the rumors they might hear, or feel left out.

Contact Person

Each group's contact person, in addition to calling everyone when information needs to be passed down the chain, also stands ready to offer support to their group members. They are trained by the unit in how to serve in this capacity. The first time a contact person telephones his or her group members, he or she invites them to call if they have any questions or concerns, especially when their sponsors are away. The contact person may not personally be able to solve all the problems that are brought to their attention, but they can be an important source of information about the support services available on post.

Family Readiness Groups can be more than just conduits for information and questions. Most of these groups also offer social get-togethers and group projects similar to coffee groups. In fact, some coffee groups have decided to merge their monthly get-togethers with those of the unit Family Readiness Group. The social activities these groups offer are especially important when the unit is away; they can fill empty hours with a little entertainment and fun, create a sense of belonging, and, most importantly, provide the opportunity for friendships to form. Most spouses realize that it is these friendships that will be their real support in a time of crisis.

* * * * * * * * * * * * *

You can find your "family away from home"
in your unit coffee group or Family Readiness Group.

Chapter 21

Showers

> They showered her with gifts and gave her a fine dowry.
> *Dutch Custom*

The tradition of the bridal shower is said to have begun many years ago in the Netherlands. A young Dutch girl fell in love with the village miller who was so generous to the poor and needy that he had few worldly goods. Her father told her she must marry a wealthy man of his own choosing or she would have no dowry. The village people who had eaten because of the miller's generosity felt sorry for the lovers. They didn't have much money, but each thought of a gift to contribute. They came to the girl in a long procession, each with a gift, and "showered" her with a dowry finer than her father's. Even the father came to wish them happiness, and the custom became a bridal tradition.

Bridal showers and baby showers are always enjoyable events. However, they take on special significance when some lucky young person is about to become an Army spouse, or an Army Family is planning on welcoming a new family member far from their family's home. If you've been invited to join the party and to help shower them with gifts as they begin this new and exciting chapter of their lives, you'll share more than fun—you will share in the camaraderie of Army spouses.

The shower may be for someone you know well, or it may be for a stranger. That occasionally happens in the Army; you get invited to a baby shower for a recent arrival whom you haven't yet had a chance to meet. That's okay! You and the others in your neighborhood or coffee group are "family" to one another, as you substitute for family and friends "back home." So, don't hesitate accepting an invitation to a shower for someone you don't know. A gift is expected, but it need not be large or expensive to be appreciated. You'll have a good time knowing that you are a part of the "Army Family" that takes care of one another.

BRIDAL SHOWERS

Helpful Tips for the Bridal Shower Hostess and Guests

- Friends of the bride or family friends usually give the bridal shower. Relatives of the bride or groom traditionally do not host a shower, though they might be invited as guests. However, today's more mobile lifestyle sometimes necessitates greater flexibility in this tradition.
- One shower is generally considered sufficient; surely, no more than two.
- Normally, the shower is scheduled with the bride, in advance of the wedding. However, occasionally, the shower is planned as a surprise—though great care must be taken to ensure that the bride attends. Under very special circumstances, a shower

might be held after the wedding.
- Cake is traditionally served at showers, but you may serve whatever else is appropriate to the time of day selected.
- "Co-ed" showers for the bride and groom are becoming very popular. They are usually held in the evening or on a weekend afternoon, and couples and single friends of both the bride and groom are invited. The theme of these co-ed showers should be appropriate to both of the guests of honor—perhaps equipping the household that today's busy couple will manage together, or furthering some mutual interest they might share, such as biking or hiking.
- Your bridal-shower gift can serve as a wedding gift, especially if you don't know the person well. Giving separate shower and wedding gifts is your choice. Alternatively, several guests may decide to pool their resources and buy one large shower gift.
- If you are invited to more than one shower for the same person, it is understandable that your gift at each shower may be small. Normally, only the bride's attendants are invited to more than one shower.
- If the hostess has not indicated on the invitation the type of shower (e.g., kitchen, linen) or the bride's color preference, guests may contact the hostess for suggestions or ask the website or location of a possible bridal registry.
- If this happens to be a shower given by Army spouses for a new Army spouse, you might consider sharing Army spouse customs and traditions; i.e., Newlywed Basket with etiquette and protocol guides, notecards, unit wine glasses, local-area coffee-table book. Make the basket fun and inviting! Being an Army spouse is special, but it can be intimidating, so help the new spouse feel empowered and excited about joining this new "tribe!"

Another idea is a newlywed "pounding basket" consisting of:
    bread (so the couple will never know hunger)
    bottle of wine (may they always have something to celebrate)
    box of salt (if they ever need to add spice to their life)

Traditionally, the "pounding basket" was a pound of butter, pound of flour, and pound of sugar so new spouses could stock their kitchen.

Tips for the Bride
- The bride should thank everyone personally at the party, and send each a short thank-you note as well. Many civilian etiquette books say that, when a gift is opened in front of the gift giver, a thank-you note is not necessary. This may be true for children's birthday parties, but it isn't true for showers.
- A special advantage of a co-ed shower to the bride is that she will have someone to help write the notes, and everyone enjoys receiving a thank-you note from the groom!
- The person or group who hosts a shower for you deserves a special and prompt note of thanks.
- All friends invited to your shower(s) should also receive wedding invitations.

BABY SHOWERS

Baby showers are usually given by friends or relatives of the expectant mother or father, or friends or relatives of expectant parents of those being blessed with an adoption. However, because of the Army lifestyle, a mother-to-be can move toward the end of her pregnancy and find herself in a new home, a different part of the country, or even overseas, without the support of her close friends and family. This is a time when a caring "Army Family" can make a difference. Neighbors may host a shower for the expectant mother, but more frequently this is a pleasure assumed by the coffee group.

Whether or not a coffee group has baby showers for all of its members who are expecting babies should be a group decision. However, every mother-to-be or new parent deserves to be recognized at this special time in her or his life. The groups that decide to have baby showers as a part of the unit coffee get-togethers ensure that no one is overlooked or excluded.

Decision to Have Showers

The leader of a coffee group that decides to have baby showers needs to be alert to who in the group is expecting and when—in order to coordinate with the appropriate month's coffee host or hostess(es). The shower can be a surprise, if you like, because the mother-to-be won't know exactly which month she will be honored. (Just be sure she is sent a plain invitation, while the others are sent invitations telling of the surprise shower.) To ensure the honoree's attendance, arrange for someone to bring her or coordinate with her spouse. Occasionally, combined showers might be necessary, when several of the group members are due at approximately the same time. That can be great fun, with each spouse sitting by their pile of gifts and all opening their gifts together.

For these shower-coffee combinations, the group leader needs to shorten the business meeting as much as possible. Eliminate or postpone anything that isn't absolutely necessary. Put as much of the information as possible to be disseminated on the "calendar of events" or newsletter.

Some coffee groups decide to have baby showers only for those members who are having their first child. For subsequent births or children, they prepare a "money tree." They feel that when you have one child, you already have many of the necessities, but there are always new things that are needed. With the gift of money, you can select what you truly need. Also, see "baby baskets" below.

Decision Not to Have Showers

The leader of a coffee group that decides not to have baby showers will want to keep track of which pregnant members have and have not been honored with a shower. If someone has been overlooked or is so new that no one knows her, the group leader can tactfully arrange for a shower. The point is that, whichever way showers are handled, no one should ever be forgotten, regardless of how new or shy she might be.

Difficult Situations

A difficult situation arises if there is an expectant mother who is eligible to be a part of the coffee group, but seldom or never participates. Should the group have a shower for her, should the group-leader help arrange one, or do you just "forget" this one, since she "doesn't seem interested" in the coffee group? A "baby basket" might be the answer—each spouse in the group can give a small gift (usually under a specified cost), wrapped and unsigned, to be put in a pretty basket or baby bathtub for the new baby. If the expectant mother doesn't come to the coffees, the "baby basket" can be taken to her, as a gift from the group. It should be given with genuine affection—to an Army wife who is away from home and family at a very special time in her life—along with a renewed invitation for her to attend the unit's coffee get-togethers.

Some coffee groups opt to have a "baby basket" as their shower-coffee tradition due to the dynamics of their coffee group. The expectant parent takes the baby basket home to open the thoughtful gifts with their spouse.

Advice for Invitees

A word about gifts—if you are invited to a shower but unable to attend, you need not send a gift. However, if you are a close friend of the guest of honor, you will probably want to do so. Another idea is for several friends to pool their resources and buy some especially needed baby equipment.

"Co-ed" Showers

As with wedding showers, co-ed baby showers for expectant parents are taking place today. This is an innovative and fun way to celebrate an expected birth that involves everyone. Such events are usually held in the evening when everyone can attend, and a light buffet supper, wine and cheese, or simply cake and coffee are served.

Advice for the Expectant/New Mother or Parents

Baby gifts should be acknowledged with a thank-you note. Even if you receive the gifts at a shower and thank the gift givers personally, a brief note is appreciated. This advice may differ from that found in civilian etiquette books; however, military manners call for the use of many more thank-you notes than civilians typically write, and this is one of those times. Of course, if you give birth shortly after a shower, no one expects an immediately written thanks.

WEDDING, BIRTH, AND ADOPTION ANNOUNCEMENTS

Announcements (wedding, birth, adoption, and graduation) are normally sent only to friends and relatives. However, in the military, announcements are often sent to the military member's commanding officer or supervisor as a courtesy. This is not a request for a gift, nor will the recipient regard it as such; you are simply sharing your good news. The announcement might be mentioned in the unit newsletter, or sometimes the unit presents baby certificates or baby cups through the Cup and Flower Fund. This fund is typically set up and maintained by the adjutant to cover hail and farewell flowers, soldiers' farewell gifts, and, if all are in agreement, baby cups for new babies and flowers for a death in the family.

For those receiving announcements of glad tidings, good manners do not require you to send a gift. Do so only if you know the individuals well and want to send a gift. However, it is always appropriate to send a note or card of congratulations.

* * * * * * * * * * * * *

Army weddings and births are truly significant events,
worthy of celebration.

Chapter 22

Spouse Clubs

"The cultivation of friendships is perhaps
the most worth-while thing you can do."
Eleanor Roosevelt

Most clubs that used to be called wives' clubs are now referred to as spouse clubs and their membership includes all spouses—female, male, same-gender spouses, and retiree spouses. Membership may also be open to active duty military and DA civilians. Spouse clubs have traditionally been an important part of the social life of Army spouses. They have provided an opportunity for spouses to come together to share friendship, mutual interests, and promote the good of the community. These goals haven't changed over the years, yet the means of achieving them have. Today, perhaps more than in the past, spouse clubs are being redefined by society's changes. Yet, as with all evolution, tomorrow's spouse clubs will be shaped in part by the past. Let's metaphorically open "grandmother's trunk" to see how it all began.

SPOUSE CLUB HISTORY AND TRADITIONS

Military spouse clubs have a long and proud history dating all the way back to the Revolutionary War. Esther Reed, along with thirty-nine other wives, formed the first military wives' club in May of 1780. Their purpose for doing so was to raise money among the citizens for the soldiers. Esther had been born in England, but married a young American lawyer who had gone abroad to study. Soon they moved to America, and five short years later her new country was at war with her native country. By that time, Esther's husband was a lieutenant colonel serving as George Washington's aide, and Esther had become a passionate patriot. From her vantage point, it was clear to her that there was a critical shortage of supplies for the Army. Soldiers were forced to take what was needed from the local citizenry, and the enthusiasm that sparked the Revolution was beginning to wane. Esther and a group of her equally dedicated friends formed what they called "The Association," and resolved to start a fund-raising effort throughout the country. Their plan was for the money to be given directly to the soldiers as a gesture of support. They were remarkably successful, raising $300,000 in Philadelphia alone. (When Congress devalued the Continental currency, this equaled $7,500—still not a paltry sum.) Their system of distributing fliers resulted in ladies throughout the country not only knocking on doors to solicit funds for the soldiers, but also making soap, knitting socks, and making shoes for them. Within just two months, the ladies had completed their fundraising, and Esther corresponded with General Washington. The general convinced Esther that the soldiers needed shirts more than money, and the shirts were made, many of them in Esther's home. This was the noble beginning of Army wives' clubs, now known as Army spouse clubs.

As our nation grew and expanded, soldiers and their wives went west. The frontier lacked many of the niceties of life they had grown accustomed to, so the wives did what they could to maintain as much normalcy and civility as possible. For example, they met socially to organize and perform their own amateur theatricals, read books, sew, and plan the social life of their small communities. Meanwhile, in the east, military wives formed societies, such as the West Point Ladies Literary Society. They met regularly, kept minutes, and developed organizational skills that were later transferred into trying to improve their communities. They became involved in some of the important social issues of their time. For example, we know that Mrs. William T. Sherman signed the 1892 Anti-Suffrage Petition. Many Army wives worked to obtain more adequate pensions for Army widows, while Navy wives helped to found Navy Relief.

In the years between the World Wars, more formally organized clubs developed. The Association of Army/Navy Wives was created during that time. When the Air Service, the forerunner of today's Air Force, was established in 1923, those young wives asked the chief of staff's wife for help in forming the Air Service Ladies' Club. World War II brought a change in focus of these clubs; social activities were abandoned as the women geared up for war work. Yet, the skills women had acquired in their socially focused activities were not wasted; rather, they were expanded for war work in support of national defense.

The model for the modern spouse club developed during the first decade after World War II. Club officers were elected from the general membership, and wives of senior officers took honorary and advisory roles. This became a period of institution building for wives' clubs. They organized and staffed the first post nurseries so that volunteers would have a place to leave their children; opened thrift shops, flower shops, and gift shops where these were needed; and published club magazines and even protocol guides. When noncommissioned officers' wives clubs and enlisted men's wives clubs began to form, they too focused much of their attention on supporting welfare and community activities. The 1970s was a period of restructuring for wives' clubs in response to the changes brought about by the all-volunteer Army. The early part of that decade was a high point for wives' club membership—more officers and enlisted men were married, greater numbers of families lived on post as more quarters were built, and wives were more willing to participate in club work as the strict formality of earlier years was relaxed. Consequently, the average age of board members began to decrease. These young professionals brought their education and enthusiasm with them, and soon many wives' clubs became "big business," with significant financial assets and responsibilities. Concurrently, clubs began to re-examine their structures—did they want to be strictly social clubs, or should their focus include both social and welfare interests? The decision was almost unanimously for the latter, and the club constitutions and structures of the 1950s and 1960s, no longer adequate, were rewritten. Wives' clubs were reorganized into two parts. The charitable side made incredible financial contributions to the community—from funding the construction of youth centers to donating scholarships, supporting youth activities, and other diverse welfare projects. Simultaneously, the social focus continued as the bonds of friendship among the club members were strengthened through their shared social experiences.

By the 1980s, wives' club boards were all composed of bright, well-educated, young women, many who brought previous professional experience to their clubs' management and activities. By this decade, many of the earlier, traditional welfare programs and volunteer activities of wives' clubs had been taken over by the Army: child care centers, Army Community Service, youth centers. Wives' clubs turned to other worthy, but usually less visible, welfare needs in their communities. With the increased presence and participation of "Army husbands," some officers' wives clubs changed their names to "officers' and civilian spouse clubs." In an attempt to keep the interest of their members, wives' clubs started offering more diverse activities: e.g., legislative awareness groups, investment groups, drug awareness classes, and craft fairs and bazaars where creative members could sell their wares. Today, clubs are known as spouse clubs and nearly all of these clubs are all-inclusive, meaning the members are spouses of both officer and enlisted personnel. Yet even with these changes, clubs in some areas declined.

It's difficult to predict what other changes will take place in spouse clubs as they continue to adapt to society's changes. Yet, as we look into "grandmother's trunk" and examine the "garments" placed there over the years since wives' clubs first began, it is a good bet that the "ball gown" of social interaction and the "apron" of social service will not be discarded, but carefully folded up and returned to the trunk for future generations of Army spouses to reflect upon and appreciate.

ARMY SPOUSE CLUBS

Today on most posts, you will find combined clubs for officers' and NCO/EM spouses. These clubs have addressed the problem of declining membership by combining to form spouse clubs for all ranks. Regardless of membership, the purposes of these clubs usually encompass such worthwhile goals as to organize and sponsor educational, charitable, cultural, and social activities; to provide information of interest to the members; and to foster ideals of charity and fellowship in keeping with those of the U.S. armed forces. Because spouse clubs operate on post, they must conform to the Army regulation that governs private organizations; they must also comply with the Internal Revenue Code.

Spouse clubs are led by a board of governors that includes honorary, elected, and appointed positions. The honorary positions are offered as a courtesy to the senior spouses; those who serve in those positions advise the elected officers who, in turn, manage the club's affairs. The appointed positions are held by club members who have been asked by the club president to be responsible for specific activities or tasks.

Prior to arriving at a new post, you may receive material on the spouse clubs in an ACS packet sent to you by your sponsor. Most spouse clubs have a bulletin board located somewhere in the club where you can learn about their upcoming events. You can also look online at the club's website for general information regarding the club in your area. Most clubs hold monthly functions throughout the school year, September through June. Be sure to find out when the next event is scheduled and plan to attend.

MEMBERSHIP

Membership in spouse clubs is open to various categories: active, associate, and honorary. Active membership is offered to spouses of active duty military. Associate membership is usually offered to spouses of retired military and reservists, spouses of foreign liaison soldiers assigned to that post in accordance with the club's constitution and bylaws. Honorary membership is a special category that may be extended to particularly distinguished men and women in the civilian community with an association or interest in the local military community. Active and associate members pay nominal annual dues to help defray costs.

Your degree of participation will probably vary, depending upon your "age and stage." Many find that they move in and out of active club involvement, based upon their current circumstances. It's interesting to note that many spouses of retirees continue to enjoy club work, and one of the wonderful features of spouse clubs is this capacity for building bridges between generations of military spouses.

ACTIVITIES

Social Activities - Spouse clubs offer more social activities than just monthly luncheons. Typically, most offer such diverse activities as seasonal formals, theme parties that raise funds for welfare, and the ever-popular "crystal bingo." Luncheon programs are carefully planned so as to provide great variety; they may be cultural, educational, humorous, or fashion oriented.

Recreation and Classes - Spouse clubs sometimes sponsor lessons and interest groups for a variety of sports—e.g., golf, tennis, aerobics, racquetball, riding, yoga. Other special activity groups might include bunko, bowling, book club, wine night, walking groups, gourmet cooking, and financial investments. Short classes are frequently offered, such as cake decorating, protocol and etiquette, or seasonal crafts.

Travel - Many spouse clubs, both overseas and in the U.S., have active travel programs that offer tours for the club members and their families. Some are purely for sightseeing; others are for shopping or taking classes, such as regional cooking classes offered at a hotel or resort.

Fund Raising - Spouse clubs raise money for welfare through such diverse projects as bazaars, art auctions, raffles, and fund-raising theme parties. Many clubs also periodically publish and sell cookbooks of the members' favorite recipes. Another important money raiser for some clubs is the ongoing, year-round project of managing and operating post thrift shops. Whether managed by the spouse clubs or independently as fundraising ventures, these second-hand stores are a vital community asset; they provide many welfare dollars (usually put into scholarships), as well as a place for ID-card holders to sell unneeded personal property and buy good used items at greatly reduced prices. Thrift shops realize a profit by taking a small percentage of the selling price of consigned items. They also sell for profit all serviceable, donated items and those that are left to become thrift shop property.

Welfare and Scholarships - Spouse clubs donate a great deal of money every year to their communities. As a result, everyone on post benefits from these efforts, either directly or indirectly. Requests for welfare grants submitted to spouse clubs' welfare committees are received from such diverse groups as Boy Scouts and Girl Scouts, post schools, hospitals and health clinics, and military units. In addition to responding to such requests, welfare committees also seek out worthwhile projects in the community that need additional equipment or funding. However, the one welfare project common to all spouse clubs is educational scholarships. The number and size of the scholarships, of course, vary with the size of the clubs. Some even offer adult and vocational scholarships, as well as the traditional college scholarships awarded to high school seniors and young college students.

Newsletters - Each spouse club usually publishes a monthly online newsletter during the club year. This can be found on the club's website and is an excellent source of information for members, especially those whose schedules don't permit regular attendance at the monthly luncheons. These publications provide a calendar of club and community activities, articles of interest to the general membership, and a place for members to advertise their crafts and businesses. The club newsletters also offer members who aspire to become writers and artists an opportunity for their work to appear in print—whether it's poems, prose, sharing customs and traditions, or pen and ink drawings.

SPOUSE CLUB COURTESIES

Just as with every other aspect of Army life, there are points of traditional protocol and etiquette associated with spouse clubs.

- Newcomers should never have to attend their first luncheon alone. If you know someone who is new in your neighborhood or unit and eligible to join your spouse club, invite them to go with you to the next spouse club luncheon.

- If you want to save a seat for yourself after arriving at the luncheon, put some personal item at that place or place the napkin over/on the chair. Don't tip the chair and lean it against the table. That looks unsightly, and people can trip over the extended chair legs.

- Talk with everyone near you at the table, not just your friends. If there are some you don't know, introduce yourself and take this opportunity to get to know them.

- When there is a head table, any guest of honor or guest speaker takes precedence over the senior spouse and is seated to the right of the club president.

SPOUSE CLUB BOARD

Spouse club boards serve for one year, and are usually elected in the late spring to take office before the end of the current club year in June. This gives them the summer months to organize, plan the next club year, and prepare a budget for presentation to the club members in the fall. It also allows the new club president time to select members for the various appointed positions on the board.

If you are asked to run for office or serve as a committee chair or committee member, consider this an honor. The nominating committee or president who asks you to volunteer obviously believes that you are the one best qualified for the job. You may not have previous experience in the position, but they believe that you have the right skills and personality for the job. There are definite advantages to accepting. Every spouse club board or committee position gives you valuable personal experience, as well as volunteer experience that can be included on your résumé. It fosters camaraderie and friendship, a greater sense of self-worth, and the knowledge that you are contributing to your community. Nevertheless, club work is time-consuming. Discuss the offer with your spouse and family. The one question you need to answer is, "Do I have the time?"

It's important for a club to transition smoothly from one board to the next. In order for this to happen, each outgoing board member should ensure that their files are complete and up-to-date, with the final after-action report included, as soon as possible after their term ends. Many club presidents plan a workshop and luncheon for the old board and new board a few weeks after the last club luncheon, in order to facilitate this transition. Such a meeting ensures that old board members take the time to share what they've learned with the new board members, and this also serves as a deadline for the transfer of files.

ELECTIONS

The election of spouse club officers is serious business. The elected officers accept significant responsibilities, including the management of large budgets—sometimes in the hundreds of thousands of dollars. For these reasons, the manner in which elections are conducted is particularly important.

Most spouse clubs used to offer a double slate for election of officers. However, with more spouses working and fewer available for the myriad of volunteer jobs in each community, a single slate for elections makes more sense than a double slate.

A single slate doesn't mean that the nominating committee decides who will be the club officers for the next year. Nominations from the floor are still accepted. What it does mean is that the valuable pool of capable members is not drained by having to provide two good candidates for each elected position. It also means that the spouse club doesn't divide into rival camps. Most importantly, it means that all of the willing-to-volunteer, best-qualified members are placed in jobs for which they're best suited, without having half of them discarded by the election process.

In order for a club to use a single slate, the club's by-laws have to allow for it. If they don't, they need to be changed before this nomination system is implemented. The next step is the selection of a strong nominating committee. It should be chaired by a senior spouse who is well-versed in club work and a member of the spouse club board. The honorary club president should serve as advisor to this committee. These two spouses, working together, select the other nominating committee members, seeking those who are: good judges of character, acquainted with all or a large segment of the club members, and probably not available for nomination because of their spouse level of assignment or anticipated reassignment. Nominating committee members need not be current board members. Among the committee members will probably be all of the leading spouses-of the major units on post, any available past presidents of the club, and select senior spouses or board members. (The current president should not serve on this committee.) It is advisable to have enough committee members to be representative of the club membership, but not so many that they impede brief discussion and prompt committee action. The committee must agree to keep their discussions absolutely confidential. A good nominating committee is the key to the success of a single-slate nominating system.

ADVISORS AND HONORARY PRESIDENT

Advisors and honorary president are courtesy positions on a spouse club board offered to the senior spouses and spouse of the senior commander, out of respect for their positions in the community but maybe more importantly, because of the experience and guidance they can offer to the club. These are not purely figurehead positions; those who are asked and have the time to accept these honorary positions are taking on special responsibilities. They are being asked to oversee the running of the club and give advice to the board members, as it is needed. Some combined spouse clubs have an honorary vice president. This position may be offered to the spouse of the command sergeant major.

To prepare for the position, advisors need to familiarize themselves with the military regulations that pertain to private organizations. They need to read the club's constitution and by-laws, and keep them available for ready reference. Part of their responsibility is to ensure that the club abides by all of these guidelines. Advisors and the president should have frequent, open, and honest communication. Advisors should attend all board meetings, and keep the honorary president and vice president in the loop, as honoraries may or may not attend board meetings on a regular basis.

As the new board begins to work together, the advisors need to make themselves available and observe. For the most part, they are seen and not heard, until they are asked for advice. The elected board should make its own decisions and steer the club as it sees fit, without undue control by the advisors or honorary president. There are occasions when advice needs to be given, without the advisor waiting to be asked, but those are rare. One significant point that must be made: Once the advisors and/or honorary president have voiced their opinions or given their advice, they must allow the voting board members to make the final decision—without getting angry or offended, if their advice isn't followed.

There are other significant ways in which advisors and the honorary president can be of help to the board and the club:

- Get to know the board members, especially the president, and let them know that you are available to help at any time during the month, not just at the board meetings.

- Look for opportunities to publicly praise individual efforts, committee work, and club activities that are especially noteworthy. Recognition of effort is the primary reward for volunteer work. Encourage and promote attitudes and activities that are fun! Board work is serious business, but it doesn't have to be all work and no play.

One of the best ways for a new board and its advisors to get to know one another well is to take a short trip together or spend the evening or afternoon together using a team-building event. There's much work ahead for a new board as it prepares for the coming year's activities. However, this work can be accomplished easier and more enjoyably when everyone on the board knows one another and has had time to become friends. It doesn't have to be a far-away, fancy, or expensive trip to create this unity of spirit—just time spent away together, team building, perhaps with a little shopping or sightseeing, away from the distractions of family and community demands. It takes a little planning and coordinating, but it can pay big dividends. Besides, volunteering to serve on a spouse club board should have some benefits!

* * * * * * * * * * * * *

Membership in a spouse club offers friendship, as well as an opportunity to contribute to one's community.

Chapter 23

Teas

"A tea, no matter how formal it pretends to be, is friendly and inviting."
Emily Post

One of those lovely, vintage "garments" in "grandmother's trunk" passed down to us by earlier generations is the formal tea. While many of our civilian counterparts have discarded this "frilly dress" as too time-consuming for their busy lives, Army spouses continue to use the formal tea for very special occasions. In fact, it is considered the dressiest daytime event and is traditionally reserved for welcoming and farewelling commanders' spouses, very senior spouses, and command-team spouses. Thus, to have a tea held on your behalf is a very great honor. Today's formal teas, patterned after the elegant tea parties of old, preserve a very stylized format designed to lend great dignity to these afternoon affairs.

If a commander and the command sergeant major arrive at the same time, it has become a tradition, if both spouses agree, to have a joint welcome. The spouses may or may not have a joint farewell, as spouses sometimes differ in how they wish to be farewelled—some prefer a small, intimate farewell (luncheon, dinner, coffee) and others a more traditional farewell tea. These welcome and farewell teas are usually hosted by the spouses of the unit involved. Many spouse clubs also have "teas" as their kick-off event or as one of their lunchtime events in order to pass down this enjoyable tradition to the newer spouses. What is important is keeping this wonderful tradition alive so that spouses can continue to experience it and share it at least once during an Army career!

Attending a tea can be a delightful experience. The tea table and room are beautifully decorated with flowers and candles. The table is laden with delicate bite-size morsels, both sweet and savory, with variety enough to tempt any palate. Often, soft music plays in the background to complete the relaxing atmosphere. After greeting the host/hostess and the guest of honor, the guests can spend an hour or so chatting with friends while enjoying the refreshments. Hopefully, they will also have an opportunity to spend more time talking with the host/hostess and the guest of honor. To complete the afternoon, the host or hostess or master/mistress of ceremony may say a brief welcome (farewell) and possibly present some appropriate gift to the guest of honor.

As relaxing and enjoyable as a tea is, many young spouses are apprehensive about attending their first tea because they don't quite know what to expect. Senior spouses often feel a little uneasy, too, the first time they are asked to pour. And the one person responsible for planning and organizing the tea has many things to consider, a daunting task if she or he has no experience or guidance. Hopefully, this chapter will help everyone enjoy the formal tea, and want to keep it in "grandmother's trunk" for future generations.

GUIDELINES FOR GUESTS

> **BLUF**
>
> "Military teas are not like a scene from Alice in Wonderland. They are steeped in tradition and protocol. It's best to know before you attend so you are not acting like the Mad Hatter!"
> *MAJ Cassandra Perkins*

- **R.s.v.p.** - Respond as soon as you receive your invitation.

- **Dress Up** - Teas are the fanciest daytime function that spouses attend. Therefore, wear a dress or a suit and heels. (Add a hat if you like, especially if the invitation states that.) Coat and tie would be appropriate attire for male guests. Basically, informal attire is best.

- **As You Enter** - As you enter, there are several things you'll probably need to do before going to the tea table: find a place for your coat, get your name tag, possibly pay your share for the event, sign the guest book, and go through the receiving line.

- **Receiving Line** - Receiving lines at spouse functions differ from those at receptions and balls, in that the first person in the receiving line is the host or hostess (not the announcer). Introduce yourself and shake his/her hand. You will then be introduced to the next person, who is usually the guest of honor. Everyone should go through the receiving line at some time—even those involved with putting on the tea, the pourers, and those handling the guest book and name tags.

- **Refreshment** - Once through the receiving line, you are free to enjoy the refreshments and mingle with the other guests. The coffee and tea may be on the same tray or on separate trays at opposite ends of the tea table. Punch will probably be available as well in another location. It may or may not be alcoholic; you may ask the pourer.

- **Standing vs. Sitting** - Army spouses traditionally stand throughout a tea, because this enables them to walk around and talk with different groups of people. However, a few chairs or small seating arrangements are normally scattered around the room for those who need to sit down. As the afternoon goes on and your feet tire, don't hesitate to sit for a while.

- **Chat with Pourers** - If you notice that one of the pourers is alone, be friendly and chat with them for a while. As the tea progresses, fewer guests continue having their cups replenished, so the pourers often find themselves alone while everyone else is talking and having a good time. A smile and some friendly words, even from a stranger, are most welcome.

- **Soiled Cup** - After finishing with your cup and saucer or punch cup, put them someplace other than the tea table. This leaves the table beautiful from the beginning of the tea until the very end, without unsightly, soiled cups and used napkins spoiling the effect. Normally, the planners will have placed a few small tables for "empties" strategically but unobtrusively around the room.

- **Thanks and Farewell** - Whether held in someone's home or at the club, a tea requires considerable effort. Take a moment to thank the host/hostess before you leave. If a musician has provided background music, you might thank him or her as well. Then be sure to say goodbye to the guest of honor.

GUIDELINES FOR POURERS

Being asked to pour is traditionally considered an honor. Therefore, most spouses graciously accept when asked, unless they have a medical problem or aren't certain they can arrive on time. As a rule, spouses are asked in advance to pour. Once a pourer has accepted, they should be told what, when, and how long to serve. They may also be told where the position will be located in the room, and whom he or she is scheduled to follow. If you are asked to pour and are not provided with this information, ask.

Pouring at a tea is an honor because it provides an opportunity to see and chat with a number of the guests as you pour for them. The beverages are ranked (coffee, tea, punch, in that order) because at the time this American tradition was established, coffee was the most popular drink. More guests would approach the lady pouring coffee than those serving tea or punch. While you're pouring, enjoy this opportunity to visit briefly with the guests you are serving. Pouring for others also models the idea of servant leadership, thus illustrating that the more senior people in an organization are not "above" assisting other spouses whose servicemember spouses are junior to the person pouring. It presents a subtle, but clear vision of the role of our senior spouses.

There are firmly established guidelines for pourers. They aren't difficult to master and will seem quite natural after a little practice. However, the main consideration is to relax and enjoy this honor. The following guidelines are simple and will help you prepare for pouring:

1. **Know what, where, when, and how long you are to pour.**

Arrive at your pouring position on time, and don't ask to pour longer or shorter than your allotted time. A pouring schedule is timed very precisely, and anyone appearing late or staying longer spoils the carefully planned schedule and inconveniences the other pourers. It is absolutely essential that the volunteers scheduled to pour first are in their places on time, or even a few minutes early. Since the most senior spouses are always scheduled to pour first, they should go through the receiving line first (if there is one), and then go directly to their pouring positions.

2. **Remain at your position until you're relieved by the next lady or gentleman.**

3. **Specifics for coffee and tea pourers:**

- Sit at the table in front of the tea service. If a cloth napkin is provided, place it in your lap. If you have a purse, put it either in your lap, under your chair, or under the table (not on the back of the chair where it can get knocked off).

- Examine the coffee or teapot to see if the lid is securely fastened.

- When someone approaches you, if you are serving both coffee and tea, ask which they prefer.

- Pick up the cup and saucer before filling it. Exception: If you discover the lid doesn't securely fasten to the pot, set the cup and saucer on the tray if there's room, or next to the tray if there's not, before picking up the pot with both hands.

- Ask if the guest would like cream or sugar (or lemon, in the case of tea). If sugar is desired and you have sugar cubes, ask, "One lump or two?" Anything added is put in the cup after pouring the coffee or tea. Sugar is put in before cream or lemon. A lemon slice, if desired, goes into the cup, not on the saucer.

- As a rule, the pourer does not stir the coffee or tea. However, if only one or two spoons are provided, you will have to do so. In that case, replace the spoon on the serving tray.

- Hand the cup and saucer to the guest with the cup handle to the guest's right. Spoons and napkins are usually arranged so the guests can take these for themselves. However, if these are placed very close to the pourer's position, put a spoon on the saucer (unless the coffee or tea is black) and pick up a napkin to hand to the guest.

- Sometimes a waste bowl is provided with the coffee and tea service. It is an empty, small-to-medium size silver bowl used for emptying the cold remains from a cup before refilling it.

- Keep an eye on your supply of coffee, tea, milk, cups, etc. If no one else checks on these periodically, you may need to ask someone to see that they are replenished before they run out.

- Pouring does not relieve you from the responsibility of going through the receiving line. If you can't do that before it's time to be at your pouring position, pour first and go through the receiving line afterward.

4. **Specifics for serving punch:**

- Stand by the table next to the punch bowl. If you have a purse, put it on the floor under the table. If you don't know whether or not the punch is alcoholic, inquire before you begin so that you can tell any guests who ask.

- When someone approaches to be served, ladle the punch into a cup held over the punch bowl. Be careful not to fill the cup so full that it's difficult to handle.

- If the outside of the cup gets wet as you fill it, blot it with a napkin. (Ladles seldom pour without dripping.)

- As you hand the cup to the guest, turn it so that the handle is free and in a position for them to take the cup from you easily.

- If no one checks on the supply of punch and cups periodically, you may need to ask someone to see that they are replenished before they run out. It is best to designate the person or people who agree to do this task prior to the event so things remain calm and relaxed, including the pourer!

5. **Regardless of what happens, don't get nervous.**

Everyone makes mistakes, even old-timers, and no one thinks anything about it; so don't get upset. If you inadvertently put a lemon slice into a cup of tea that has already had milk added, laugh at your lapse of concentration and pour a fresh cup. If you spill the coffee, dab it up and forget it. If the punch ladle causes the punch to dribble down the side of the cup (and they all do), just do the best you can and wipe the cup dry before handing it to a guest. No one expects you to be perfect!

PLANNING A TEA
1. Who's in charge?

One person must take on the responsibility to oversee the entire effort of the tea from the beginning of the planning until the tea is over, clean-up is started, and thanks have been expressed. This ensures unity of effort and prevents something from being overlooked because of a misunderstanding about areas of responsibility.

That responsibility normally is determined by the purpose of the tea and the hosting unit. If the tea is to honor a battalion or brigade commander's spouse, the executive officer's spouse is usually in charge. For a division or corps commander's spouse, the chief of staff's spouse is usually responsible. For a division or corps command sergeant major's spouse, the spouse of the senior brigade command sergeant major or senior subordinate division command sergeant major spouse normally assumes responsibility. If a unit is planning a tea for the other post spouses as a learning experience for the younger spouses, the first lady or gentleman of the hosting unit usually supervises the event. Teas that are hosted by spouse clubs can be planned by either the president, the program chairman, or someone designated for that event.

2. Set the place, date, and time.

First, decide where the tea will be held and contact the facility to find out what acceptable dates are available. Next, ask the guest of honor to select his or her preference from those available dates. Then, schedule the chosen date with the facility. Do all of this early enough so that invitations can be prepared, addressed, emailed or mailed, and received by the guests at least two weeks in advance of the tea date. Invitations for very large teas can be emailed or mailed as much as a month before the event. (Exceptions to these timing guidelines must be made when the person to be honored receives "last-minute" orders.)

As for the time, teas were traditionally held in the late afternoon, beginning around three o'clock. If many of the guests have a long distance to travel, it is a courtesy to them to begin earlier, possibly around two or two-thirty; however, with working spouses and childcare constraints, you may have to be even more flexible and lenient with times.

3. Prepare the guest list.

Usually, the person in charge of the tea prepares a tentative guest list, with input from the unit protocol office, if appropriate. In compiling this list, include all of the members of the hosting group, the guest of honor's contemporaries, and the spouse of the unit's senior commander. Consider the various boards, committees, and clubs to which the guest of honor may belong. Don't forget the members of their coffee group and any local civilians who should be invited. If the guest of honor is an officer's spouse, consider the senior NCO spouses with whom he or she works; if he or she is a senior NCO's spouse, think about inviting the officers' spouses with whom he or she has worked or will work.

Discuss this tentative guest list with the guest of honor, for their approval and any additional names of personal friends or family members whom they would like invited.

Compile a final list of the guests' names, to include their spouses' first names (for addressing the invitation envelopes), and the name each guest goes by (for response list and name tags), plus their addresses. This is a must if other people will be helping to address the envelopes.

4. **Decide on funding.**

Will the spouses of the hosting unit pay for the tea expenses or will everyone attending pay a pro rata share? If the latter method is selected, decide how collection will be made, but don't ask the guest of honor to pay.

5. **Decide on the invitations.**

Formal invitations may be typed or handwritten. Today, invitations are widely extended using electronic invitation systems as well. There are several ways to prepare invitations. Most popular choices are:

- Typed (ink jet printer or laser printer)—can produce beautiful invitations in a variety of fonts, especially pretty when linen paper is used

- Mass reproduction of a calligraphy or handwritten invitation—using a copier

- Handwritten—equally elegant, but labor-intensive, and requires quality control (correct color and type of pen, excellent penmanship, match writing on the invitation and its envelope, proper method of address)

- Email invitations—"e-vites" are the most popular way of sending invitations today

- Combination—handwritten or elegantly printed for senior ladies and special guests, and mass reproductions or evites for everyone else

- Telephone calls—used for really short notice

The traditional tea invitations were on single (not fold-over) cards, approximately 5¼ by 4 ¼ inches. However, style and content are more important than the size of the invitation.

6. **Select invitation wording and R.s.v.p. phone number(s) or email address.**

A tea invitation uses the third-person format of formal invitations. There is some flexibility in how the time is expressed; for example, half hours may be expressed as half after two o'clock, half past two o'clock, or two-thirty o'clock. Check and double-check the day, the date, and the spelling of the guest of honor's name. Use no initials in the guest of honor's name.

Decide how many R.s.v.p. telephone numbers are needed; even for large teas, two are usually sufficient. Email response is the most efficient way to keep track of responses. (It is not appropriate to ask the protocol office or any other military office to accept responses for a tea since this is not an official military function.) If you are in a foreign country and foreign guests are invited, consider asking spouses who speak the local language to receive the calls or emails.

> *The Wiesbaden Community Spouses' Club*
> *requests the pleasure of your company*
> *at a Welcome Tea*
> *honoring*
> *Mrs. Christopher M. Commander*
> *{or Jennifer R. Commander}*
> *on Monday, the thirtieth of June*
> *at half past two o'clock in the afternoon*
> *The Wiesbaden Community Club*
> *(GPS Address)*
>
> *R.s.v.p. by 25 June* *Informal*
> *xxx@gmail.com* *$7.00* in Advance*
> *(049) xxx-05440 Mr. or Mrs. Hooah* *Payable to: WCSC*
>
> **($5 will cover the cost of refreshments and $2 will cover the cost of a group welcome gift.)*

> *The Ladies and Gentlemen {or Spouses}*
> *of the*
> *1ˢᵗ Engineer Battalion*
> *request the pleasure of your company*
> *at a*
> *Farewell Tea*
> *in honor of*
> *Mrs. George P. Doe*
> *{or Elizabeth A. Doe}*
> *on Friday, the sixth of June*
> *at half after two o'clock in the afternoon*
> *Community Club*
> *(GPS Address)*
>
> *R.s.v.p. by 2 June*   *Informal*
> *Hooah@bellsouth.com*   **$10.00 cash*
> *678-xxx-xxxx Ms. Hooah*
>
> **($5 will cover the cost of refreshments and $5 will cover the cost of a group farewell gift.)*

Do you want to have an R.s.v.p. deadline? This can be especially helpful for the person responsible for the pouring schedule; they can begin asking the senior guests to pour after the R.s.v.p. deadline, without fear of late acceptances spoiling the schedule.

7. **Set up a tea committee.**

The number of helpers needed will depend on the size of the tea. The areas of responsibility and what each entails are as follows:

- <u>Invitations</u> - Prepare, address, and mail or email the invitations. Prepare an alphabetized R.s.v.p. list for those taking responses and, possibly, for those making name tags. Send a copy of the invitation and the final guest list to the guest of honor; if R.s.v.p. is printed on the invitation, strike through it and write "To Remind."

- Responses - Accept the R.s.v.p. calls and emails. Be prepared to give the names of the guests who will be attending to the name tag committee (if there is one), and the person preparing the pouring list. The tea planner will need to know the total number of acceptances.

- Refreshments -

    a) For catered teas—plan the food, beverages, and room arrangement with the caterer. Be specific about the size of portions and food presentation. Plan for coffee, tea, punch, and water. If it is a hot day, expect more attendees to drink punch.

    b) For homemade refreshments—coordinate with those preparing the food. Arrange for a nice selection of sweets and savories (all in finger-food size portions), brought on a serving plate or a tray with the donor's name taped underneath, and delivered early enough to be arranged on the table before other guests arrive. (Reminder: Paper doilies should not be used directly under food.)

- Room Decorations - Decide on a color theme. Then arrange for flowers for the tea table, a corsage or boutonniere for the guest of honor, and any other desirable room decorations. Consider coordinating napkins, candles, and the table underskirt with the color theme. Before ordering a corsage or boutonniere, find out what color the guest of honor plans to wear. For ladies, sometimes flowers for the hair, a lei, or wrist corsage might be preferred to a pinned corsage (See "Pinning on a Corsage" at the end of this chapter.)

- Name tags -

    a) For name tags made in advance—obtain the guest list from the invitation or response chairman; buy or make name tags; use calligraphy or very neat printing; be prepared to make more name tags at the tea.

    b) For name tags not made in advance—buy name tags and have pens available (small felt-tips, not ballpoints). The adhesive name tags or clip-ons are easier than those needing straight pins; have a small basket handy for disposal of any paper backings.

- Pouring List -

    a) Traditionally, the pouring list is based upon each guest's spouse's rank, job assignment, and date of rank. Compile the pouring list by considering either all the guests who will attend, or only those from the hosting group. The pouring positions begin with coffee in position #1, tea in position #2, and punch in position #3. When there will be two silver services, both with coffee and tea, then the one nearer the door is considered position #1.

    b) For a tea that includes significant numbers of both senior officers' spouses and senior noncommissioned officers' spouses, two pouring lists may be used in order to incorporate both groups. This is best accomplished by assigning all the senior officers' spouses to pour coffee and tea, and assigning all the senior noncommissioned officers' spouses to pour punch. Prepare each list using the order of precedence described in (a) above.

c) Ask the pourers in advance. It is more polite to telephone or email those whom you are asking to pour than to include a written request in with their invitations. Besides adding a more personal touch, the telephone call or email allows you to tell the particulars of their assignment to those who accept—especially important if they need to arrive early.

d) Provide each pourer with this information: his or her position, time to start, length of time to pour (5 minutes for senior spouses, 10-15 minutes for all others), and whom he or she will follow. Those in the receiving line should not be asked to pour. However, if there is no receiving line, the guest of honor should be given the opportunity to pour first, if he or she would care to.

e) Put a small card inconspicuously by each pouring position with the schedule for that position. Be sure a chair and napkin are provided for the coffee and tea servers, and a napkin for the punch server.

f) Keep a master copy of the pouring schedule handy, and watch to ensure that each pourer is relieved on time.

g) Be flexible. If a spouse scheduled to pour doesn't arrive, ask someone to pour a little longer or make an appropriate substitute.

h) It's not necessary to have pourers available until the very end of the tea. As guests begin to depart and the crowd thins, it is fine to stop the pouring (even if others are on the schedule). Leave the coffee, tea, and punch on the tables for the remaining guests to help themselves.

- Music - It's nice if you can arrange to have live music (usually piano only). Check in advance to see if the piano needs tuning. Supervise the positioning of the piano in the room. Consider a flower arrangement or candles for an upright piano that coordinates with the color theme.

- Guest Book - Purchase a guest book or borrow the departing guest of honor's book, if he or she received one on arrival. Prepare the title page and heading of the tea page, preferably in calligraphy. Have a table near the entrance, usually positioned before the receiving line, for the guest book. Provide a nice pen for guests to use in signing the guest book. Decide if you want signatures on both or only one side of a page. Someone should be asked to stand or sit by the guest book as guests arrive to ensure that everyone signs, and that they sign where you want them to sign. Another idea for a guest book is a scrapbook. A page could be decorated for the tea where the guests would sign. The farewell scrapbook would capture their servicemember's time in command along with all the programs of changes of command/responsibility, balls, formals, menus, and FRG events they will have attended.

- Pictures - Take pictures of the entire event, but especially of the guest of honor, the tea table with the tea committee, guests greeting those in the receiving line, and any presentations. After the tea, ensure that selected pictures are provided to the guest of honor, the members and head of the tea committee, and the hosting group (if a scrapbook is kept of such events) on a CD or thumb drive.

- Gift - The tea planner usually decides who might want to contribute and how money can be easily and inconspicuously collected (if that's necessary). Contributions toward group gifts are limited to an unsolicited amount per giver. With tea planner's assistance, decide on an appropriate gift; allow sufficient time for any engraving needed. Be certain that the value of the gift is not so great that it will be an embarrassment to the guest of honor. The tea planner decides who will be the master/mistress of ceremony and present the gift. If a microphone will be needed, arrange for and test the speaker system. Remember, the gift should not exceed the legal limits.

  o A welcome gift need not be expensive. It is a token of friendship; a welcoming embrace to their new unit. Some examples of welcome gifts are a coffee table book of the local area, bling unit pins, unit tie tacks, stoneware crocks with personalized bases/posts. A tradition for a brigade gift is a bracelet made with all the battalion unit crests. These unit crest pins are sometimes presented during a pinning ceremony; i.e., crests pinned to a ribbon dangling from the corsage.

  o A farewell gift is a meaningful reminder of friendship and sincere appreciation. Guidelines within the sections of the Joint Ethics Regulation (DOD 5500.7-R) establish a limit on the total value of gifts to a departing superior. As spouses, we, too, should comply with the guidelines. Some examples of farewell gifts are homemade unit quilts or wall hangings/afghans with unit crests, unit stained glass panes, a watercolor of the honoree's house, unit engraved trays, vases or pitchers.

- Hospitality -

  a) Arrange for the guest of honor's transportation, or for someone to walk with him or her to the tea. (Never let a guest of honor come by themselves; even if he or she is only coming from the house next door.)

  b) Decide if you want to have individual escorts for each senior spouse, or simply have hosts or hostesses stationed at strategic points to direct guests and/or take their coats. (It's always thoughtful to assign two hosts or hostesses to the same location, or near one another so that during a lull they do not have to stand there alone. Also, if one is called away for a few minutes, the position is not left vacant.) Escorts are discussed later in this chapter.

- Security and Parking - Make any arrangements necessary for the safety and convenience of your guest of honor and other guests.
- Clean-up - This may not be necessary in a club, but in a home it should not be overlooked. In a home setting, spouses should be asked well in advance to assist with this task. Those who've agreed should be scheduled to remove the soiled cups, saucers, and napkins periodically. A clean-up committee made up of volunteers should remain after the other guests have departed.
- Publicity - If publicity is desired, before and/or after the event, prepare and submit an article to the local or post newspaper. This responsibility might be combined, or at least coordinated, with the person taking pictures.

8. Decide on receiving line.

Plan this in conjunction with the guest of honor. Remember that the shorter the line, the better—preferably only the host or hostess, followed by the guest of honor. In this type of receiving line, there is no announcer; therefore, the guests shake hands with everyone standing in the receiving line. If this is a spouse club tea being hosted at the senior spouse's home, the receiving line would start with the spouse club president and then the senior spouse (who usually is the honorary spouse club president).

9. Arrange the tea table.

- *Tablecloth* - A tea table always has a tablecloth and, sometimes, a skirt. An undercloth of a contrasting color can be used to add interest to a lace or cut-work tablecloth. The tablecloth may be additionally decorated with the use of swags or corner arrangements of flowers and/or bows.
- *Centerpiece* - A full floral centerpiece is standard. Try to provide one that is tall and large enough so that the height and size of the candles, silver service, and punch bowl don't overshadow it.
- *Candles* - Candles may be used if the day isn't too sunny or the curtains are closed. Since candles are meant to add light, they are not used unless extra light is needed. However, if candles are on the table, they must be lit. Clip the wicks to about ¼" beforehand, and have matches handy.
- *Silver Service* - If only one silver service is used, place it on the end of the table nearer the door, with the punch bowl on the opposite end. If two silver services are used, one goes on either end of the tea table, with the punch bowl on another table.
- *Cups and Saucers* - Do not stack cups and saucers, or punch cups. The only exception would be for extremely large teas when they simply cannot be replenished fast enough for the crowd. In such a case, small stacks of cups and small stacks of saucers may be placed around and behind the silver service. Always line up the cup handles so that they appear orderly.

- *Pourer's Chair* - Place a chair in front of each silver service. Provide a napkin for each pourer, including the punch server.

- *Milk, Sugar, and Lemon* - Use milk rather than cream, because the tannic acid in tea may cause the cream to curdle. Use lump sugar rather than granulated, if possible, because it's easier for the pourer to judge the amount of sugar being added. Don't forget lemon slices for the tea. At a small, private tea, you may want to offer artificial-sweetener tablets, but this is too involved for the pourer at a large tea.

- *Food Trays* - Arrange trays of food on the table so that the appearance is balanced, and the savories are interspersed among the sweets.

- *Replenishment* - Trays should be replenished out of the guests' view. As the tea draws to a close, trays that are beginning to look sparse may be combined (out of sight of the guests) in order for the table to continue looking reasonably attractive.

9. Write an after-action report.

Anyone having the responsibility of putting on another tea after you will greatly appreciate any information provided concerning your planning, cost, numbers, and suggestions for improvement. Write your report as soon after the event as possible, while the details are fresh in your mind.

Coffee and tea on one end, punch on the other

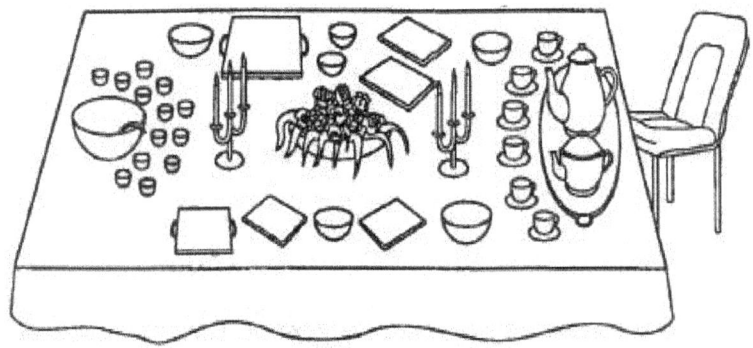

Coffee and tea on both ends (or one on each), punch on a separate table

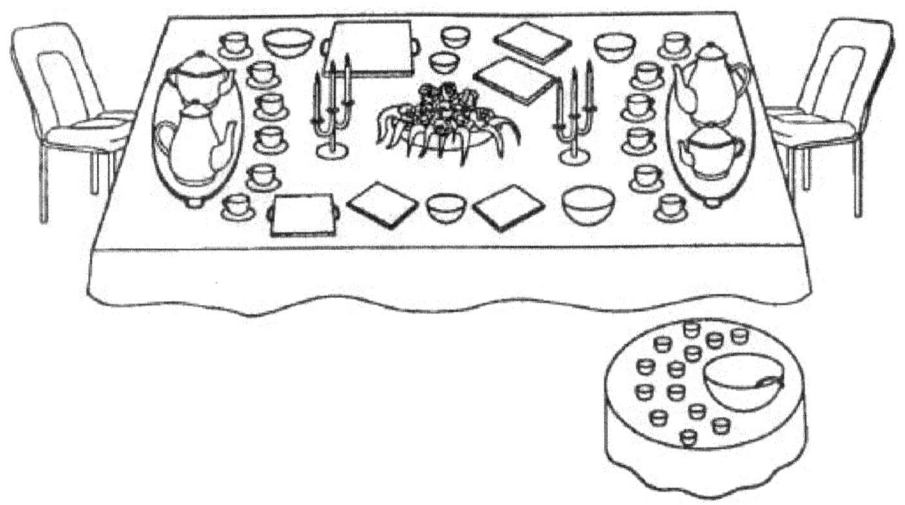

T-shaped table to accommodate two coffee services and one punch bowl

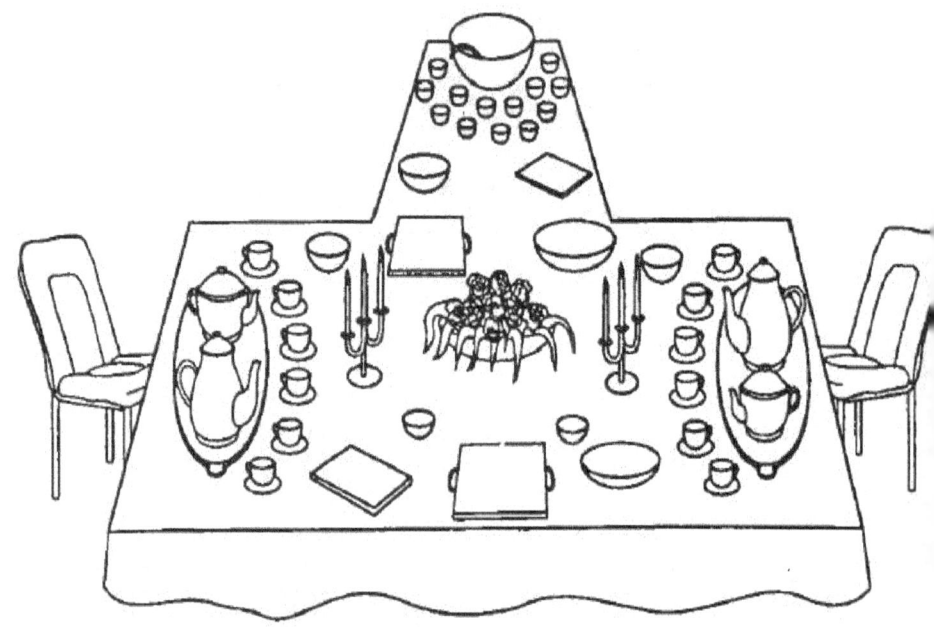

ESCORTS

If the decision is made to have escorts for senior guests and you are asked to serve in this capacity, the following guidelines will be helpful. To begin, it greatly helps to be able to recognize the senior spouse you have been assigned. If you don't know them by sight, try to find someone who does and get a description, because your first responsibility is to meet the spouse at the door and welcome him or her to the tea. This, by the way, requires you to be there early.

Alternatively, you may be asked to drive the senior spouse from their home to the tea. Call early to coordinate this with them, confirm where they live, and set the time when you will come by for them. Allow sufficient time to get them to the function a few minutes early because, as a senior spouse, they have probably been asked to pour during one of the early time slots. Before you go, take time to clean out the clutter from your car; it doesn't have to be spotless, but a mud-spattered car with gum wrappers and discarded sweaters on the front seat isn't very inviting. If, for convenience, you are also driving several other spouses, be sure to stop for the senior spouse last and take them home first. The right front seat should be left vacant for them. When you arrive at the function, if you wish to let the senior spouse out at the front door and then park the car, be certain that there are "greeters" standing by at the front door to welcome and talk with him or her until you can join them.

When the senior spouse arrives at the tea, hang up their coat and give them a name tag. (If you are there early, have their name tag in hand; if you arrive together, get your name tags at the same time.) Confirm that they know when and where they are scheduled to pour. (Check with the pouring chairperson in advance for this information.) Escort them to the guest book and the front of the reception line. (If you arrive together, inquire about the location of these beforehand.) Any spouse senior enough to have been assigned an escort is senior enough not to wait in the reception line. Excuse yourselves to those waiting near the front of the line for asking to go ahead of them. Follow the senior spouse through the receiving line so that you can stay with them. Some senior spouses will prefer not to "cut the line" and wait with the rest of the spouses. It is their prerogative, so no need to insist they move ahead!

During the tea, your responsibility is to see that the senior spouse has a nice time. When they are scheduled to pour, escort them to the correct location and check that everything is in order for them. During lulls when no one asks them to pour or is chatting with them, take the opportunity to get to know them better. Don't let them sit or stand there alone! When the senior spouse isn't at their pouring assignment, offer to get them something to drink, and later, ask if you can replenish their cup. There will be many who will want the opportunity to meet and chat with the senior spouse, so the courtesy of your providing their beverage allows them more time to talk with other guests.

When introducing others to the senior spouse, always name the more senior spouse first. (Review Chapter 2, "Introductions.") Try to add some informative details in each introduction. "Mr./Mrs. Senior, this is Mike/Mary Junior. His/her spouse is our new battalion S-3, and they just arrived here last month from an exciting tour in Alaska." Such an informative lead-in provides the basis for conversation and helps everyone feel comfortable.

Some escorts have the mistaken impression that they need to stay with their senior spouses throughout the entire tea. This is not necessary or desirable. You do want the senior spouse to meet many of the other guests and have time to chat with them, but this can be accomplished without your being there every minute. It does require that you keep alert to the senior spouse's situation, but don't "hover." If the senior spouse is engaged in conversation, you can excuse yourself and chat with your friends. If he or she has been talking with the same group for some time, you might offer to introduce them to another group. Alternatively, you might encourage others to walk up and introduce themselves. Let the senior spouse set the pace. They may already know many of the other guests and enjoy chatting with them on their own. They may even suggest that you don't need to watch out for them, that they are fine on their own. However, watching out for them is exactly what you should do, but do so unobtrusively.

At the close of the tea, the senior spouse will want to find the host or hostess and any guest of honor to say their goodbyes. Then you escort them out, retrieve their coat, and say farewell (unless you are taking them home). Although escorts are not used as often as they once were and many senior spouses would prefer to manage on their own, if you are assigned as an escort for a senior spouse, it's important that you do it well. Your position is that of a substitute host/hostess, because the host/hostess can't see to all of the needs of their very special guests, especially if there is a guest of honor. You can represent the host or hostess and your unit well by taking your assignment seriously.

PINNING ON A CORSAGE

A corsage or boutonniere is often given to a guest of honor at a tea. For a lady, unless she has chosen to wear it in her hair, or on her purse or wrist, someone will be asked to pin the corsage on for her. Knowing how a corsage or boutonniere should be worn and how to pin it on quickly and securely are easy skills to learn.

A corsage or boutonniere is worn "as the flowers grow"—in other words, stems down. Usually, it is placed on the wearer's upper left side, though the style of the dress or personal preference of the wearer may cause that to change.

To pin a corsage or boutonniere on well requires a steady hand and a little patience, but a special technique in the use of the corsage pin will help. That technique is this: While holding the corsage or boutonniere in place, stick the pin in and out of the garment on one side of the corsage or boutonniere, run it over the flower stems (not through), and then in and out of the garment on the other side. The pressure of the pin against the flower stems is what holds the corsage in place, not actually pinning the corsage to the garment. (Your fingers slipped underneath the garment where the pin will enter and exit can help prevent sticking anything besides the garment.) For large, heavy corsages, two pins may be needed. Boutonnieres are usually worn by inserting the flower stem through the left-lapel buttonhole. They may or may not require a pin to hole the flower in place. If the buttonhole has not been sufficiently opened to allow the flower stem to pass through, then the pin will definitely be needed.

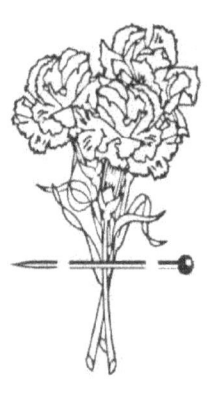

* * * * * * * * * * * * * *

A tea is intended to be an elegant, enjoyable affair for everyone—guests, pourers, and planner. Relax, mingle, and have fun!

Chapter 24

Receptions and Receiving Lines

"Receptions of various kinds are a most popular form of official entertainment."
Mary Jane McCaffree and Pauline Innis, Protocol

Receptions are lovely, stand-up affairs held in honor of a special occasion or an honored guest. When held alone and not in conjunction with another event, a reception is considered to be more formal than a tea or a cocktail party. Accordingly, it is an adult gathering, not intended for children. Receptions usually begin with guests going through a receiving line to be greeted by their host and hostess and to meet any guests of honor. This is followed by an opportunity to mingle with the other guests and enjoy the beautiful array of food and wide variety of drinks. Receptions seldom last more than a couple of hours, so that guests depart while they're still having fun.

When you receive an invitation to a reception, respond promptly, as with any other type of invitation. Many guests are usually invited to a reception and one might think that an exact number doesn't matter; however, those responsible for the arrangements need to know exactly how many to expect. Nearly all receptions are catered, and the planners have a deadline by which they must tell the caterer the number of guests who will attend.

ATTIRE AND ARRIVAL

Receptions may be either informal or formal affairs—the later the reception begins, the dressier it gets. The style of attire should be stated on the lower, right corner of the invitation and, for a military event, this will usually be a particular uniform or choice of uniforms. The uniform is determined by the time of the reception.

Spouses are expected to take their cue from the formality of the uniform stated on the invitation. For example, an invitation to a reception that states "Duty Uniform" implies: for a female spouse, a simple dress or suit; for a male spouse, a sports jacket or suit. "Army Service Uniform with four-in-hand tie" is a bit more formal and calls for a dressier dress or suit for the female spouse, and a suit for the male spouse. "Army Blue Mess" equates to Black Tie and "Army Blue Evening Mess" equates to White Tie, so they are the most formal options.

For a female spouse, that means a really fancy, street-length dress or suit if the reception begins before five o'clock, and usually a tea-length or long dress if the reception begins at or after five o'clock. The generally accepted rule on skirt length is "no long, dressy skirts before five o'clock." The later the event the longer the dress. For a male spouse, those formal uniforms mean a dark suit, or possibly a tuxedo for an evening reception.

Since a reception usually begins with a receiving line, it's important to arrive punctually. If, by chance, you are unavoidably detained and miss the line, seek out the host, hostess, and any guest of honor *as soon as you arrive* "to pay your respects." In other words, say hello or introduce yourself, make your apologies for being late, and chat for a minute or so. It is your responsibility to do this immediately upon your arrival—even before you get a drink, have anything to eat, or mingle with the other guests.

THE RECEPTION LINE

As soon as the guests arrive and their coats have been hung up, they join the *reception line* (those waiting to greet the host, and any other special guests), to go through the *receiving line* (those waiting to greet the guests). Women keep their purses with them as they go through the line. For any wearing long or short evening gloves, today's etiquette allows for them to keep them on as they go through the line, regardless of what the women in the receiving line are wearing. Occasionally, for a very large brigade or division reception or ball, guests may be directed to go through the receiving line at a certain time prescribed for their individual units. Though this seems to be an unnecessary extension of militarism, there are some practical reasons for this. It helps with timing so people aren't waiting in long lines; it helps those in the receiving line, who can't possibly know all of the personnel in their sub-units or other organizations, mentally connect those guests to a unit; and it also makes the line a bit more manageable. At such a function, when guests do not immediately line up to go through the receiving line, they are free to have a cocktail before their appointed time. However, once in line, it is not appropriate to drink or smoke. (Note that on rare occasions when the line is exceptionally long, waiters may serve drinks to those in the reception line, and a table is placed near the receiving line for their empty glasses.)

As couples join the reception line, they line up in pairs, with the military sponsor standing on the side nearer the receiving line. This puts the sponsor, typically a male, in the position of standing closer to the aide or adjutant (the announcer) who is standing at the beginning of the receiving line to take your names. The sponsor's guest, typically a female, then steps ahead of the sponsor as they proceed through the line. This follows the Army's traditional courtesy of putting "ladies first." However, today the composition of the military has changed sufficiently that this protocol is evolving. The Army still puts forth the "ladies first" policy, but some locations simply follow the procedure of having spouses/guests go first, regardless of gender. In couples with both in the military, the one who is the member of the hosting unit usually stands nearer the receiving line (announcer). And some locations are no longer concerned with this detail and have guests go through the line in whatever order they happen to join the line. Because of this evolution, it would be helpful to determine the preference of your host and what works best at your location.

A few more tips: (1) If, as you line up, you cannot see the receiving line for a guide as to where to stand, simply follow the lead of those ahead of you. Then, as soon as the receiving line comes into view, you can change your positions if necessary. (2) While waiting in line, if a very senior officer/NCO comes up and stands in line behind you, it is proper to invite that officer/NCO and his/her spouse or guest to precede you. In this way, they get passed along to the front of the line—a simple courtesy.

There is no ironclad rule for the formation and location of receiving lines; often times you will have to make a case-by-case judgment depending on circumstances. Typically, the receiving line is formed first by an adjutant, followed by the commander, the commander's spouse, the command sergeant major, and the command sergeant major's spouse. It is acceptable for a female to stand at the end of the line.

It's important for guests to understand that the first person they come to as they approach the receiving line is not actually a part of the receiving line. He or she is the "announcer." Usually the aide or adjutant is given this task. They are there to hear the guests' names and introduce them to the first person in the receiving line. Do not shake the hand of the "announcer"! In international settings, you may see two announcers: one who speaks English, and one who also speaks the host nation language.

As each couple steps up to the announcer, the gentleman (or military sponsor) gives their spouse's (date's) name and then theirs, "Mrs. Smith, Lieutenant Smith." It is appropriate to give only the title and last name. Even if the servicemember is a good friend of the announcer or even if they are wearing a name tag, they still identify their spouse (date) and themselves by name. The announcer can have a momentary memory failure, so they shouldn't be put in the difficult position of having to remember names or read name tags. After the soldier has given the names to the announcer and the person ahead has moved a step away, if this is an Army reception, the spouse or date steps ahead of the soldier and goes into a single, rather than double, line.

Not all receiving lines follow the "ladies first" rule, as the Army traditionally has done. When invited to a reception or ball hosted by another military service or to a diplomatic reception, it's appropriate to follow the protocol of the hosting group. Foreign countries have differing protocol regarding receiving lines; check with the local protocol office or embassy for guidance.

Traditional Receiving-Line Protocol

Army and civilian receiving lines: Females precede their accompanying gentleman guests.

Air Force, Navy, and Marine Corps receiving lines: Sponsors precede their spouses or dates, or the servicemember to whom the invitation was addressed precedes his/her spouse or date. However, some hosts of these services prefer the "ladies-first rule." If the couples ahead of you are following that courtesy, you might feel more comfortable doing the same.

White House and diplomatic receiving lines: Traditionally, invited guests precede their dates. For today's evolving protocol, see "An invitation to the White House" in Chapter 9.

GOING THROUGH THE RECEIVING LINE

As a guest, you may not know the people standing in the receiving line waiting to greet you. Therefore, it is helpful to understand a bit about how a receiving line is set up. The first person (not counting the announcer) in the receiving line is the most important person there. He or she is not necessarily the highest ranking in attendance, but for this event he or she is the most important. This usually is the host, and that person stands in this position of honor. However, if the guest of honor is very senior or very important, he or she may be placed first. In most Army receiving lines, spouses usually stand beside their military sponsor; however, you may occasionally see the male guest of honor standing beside the host or hostess, followed by the two spouses. If there are more than two individuals or more than two couples standing in the receiving line, they are usually positioned in descending order (e.g., brigade CO, battalion CO, brigade CSM, battalion CSM).

As the announcer gives your name to the first person in the receiving line, you simply step in front of them, turn to face them, shake their hand, and exchange a brief greeting: "It's nice to meet you," "Good evening," or a similarly short greeting. Remember that all of those people waiting in line behind you hope you will hurry; save any extended conversations for later in the evening. Give a light, but pleasantly firm, handshake and look at the person's eyes.

Continue down the receiving line, greeting each person in turn. As you proceed, you should continue to be introduced to the next person in line by the one whom you have just met. If this doesn't happen or your name gets mispronounced, introduce yourself to each person or correct the mispronunciation as you shake hands. Don't hesitate to do this or feel badly if it's necessary. Those in the receiving line will appreciate your help. They try hard to hear every name correctly and pass it on; but, with so many names and the noise of the crowd, that's not always easy and mistakes can occur. Once through the receiving line, it's time to join the party.

Who goes through the receiving line? Every guest, except those who are actually in the receiving line and the aide or adjutant who serves as the announcer should go through. Even those guests who are helping with other details of the event, such as taking coats or giving out name tags, should find time to go through the line. If your soldier is not available to go through the receiving line with you, don't hesitate to go through alone. However, in that case, you will give the announcer your name, and it is helpful to tell the first person in the receiving line (not the announcer) who your soldier is and what he or she does.

Guests should try to find an opportunity to chat, at least briefly, with the host and hostess, and any guests of honor, after the receiving line ends. Other than that, their only other responsibilities for the remainder of the reception are to enjoy the food and drink, mingle, and chat with the other guests.

SITTING

Although receptions are intended to be stand-up affairs, there are always a few chairs available for those who need them. Certainly, pregnant women should not feel that they need to stand the entire time if they are uncomfortable. The object of standing is to mingle, but it should not be done to the point of discomfort. Receptions are meant to be enjoyed!

SETTING UP A RECEIVING LINE

Location - Ordinarily, a receiving line is formed to the left of where the guests will enter the room. If this is not convenient, it may be formed to the right. The receiving line is usually situated near the door or entrance, but may be placed in another location in order to balance the room's appearance or to permit the flow of the reception line. Consider: the desired "traffic pattern," space for those waiting to go through the line, and easy movement from the end of the line to the refreshment area.

Flags - Flags are placed directly behind the receiving line. The American flag is always to the "right of the flag line," meaning "marching right." Therefore, as you face the line of flags, the American flag will always be on the extreme left side. Next come any appropriate foreign national flags (placed in English alphabetical order) or state flags; organizational flags (in order of echelon); and, finally, general officer flags (in order of rank). Usually only flags of general officers in a participatory role are displayed. Flags are not usually displayed for those merely in attendance. At very large gatherings, only the star flag of the most senior general present is used. When two or more services are present, organizational and star flags for each service are used.

Red Carpet - The purpose of a red carpet is: (1) to honor those in the receiving line, and (2) to provide a softer surface for them to stand on. Therefore, place a short, red carpet immediately in front of the flags so that it will be used by those standing in the receiving line. (This red carpet is not for those going through the receiving line!) If you want to honor the other guests as well, or protect them from slipping on a bare floor, use another long runner where it's needed, perhaps at the entrance to the club or ballroom. Note that any receiving line may be honored with a red carpet, regardless of the rank or gender of the participants.

Forming the Receiving Line - The host (servicemember) and spouse of host are responsible for deciding how the receiving line will be formed, often with input from the adjutant or protocol office. The guiding principles should always be: (1) keep the line as short as possible, and (2) position the most important person for that particular event immediately after the announcer. Typical receiving line arrangements used by the Army are as follows:

Most common arrangement:
    (announcer) host/ spouse of host/ guest of honor/ spouse of guest of honor
Alternate arrangements:
    (announcer) host/ guest of honor/ spouse of guest of honor/ spouse of host
    (announcer) host/ guest of honor/ spouse of host/ spouse of guest of honor
Arrangement for a very special, high-ranking guest of honor:
    (announcer) guest of honor/ spouse of guest of honor/ host/ spouse of host

Occasionally, another gentleman is asked to bring up the rear of the receiving line so that a lady does not have to be last. However, this is not required and should be done only if the gentleman is an appropriate guest to be honored—for example, the unit's command sergeant major. To choose someone such as the youngest lieutenant in the unit for this honor is not in keeping with the purpose of the receiving line. An alternative to adding a gentleman to the end of the receiving line is to place a small table with flowers there as a graceful transition.

Courtesies - A small table placed near the receiving line is very handy for the women's purses, a water pitcher and glasses, hand sanitizer, and any gifts that might be received. (Gifts are sometimes presented by foreign guests.) If the receiving line is expected to last a long time, it's also nice to place a few chairs nearby for use during short breaks.

RECEIVING-LINE GUIDANCE

BLUF

Don't act like a politician working a rope line—shaking hands with one person while smiling and talking to the next person down the line.

- The host decides whom they want in the receiving line with them and asks them in advance. Preferably none of the other guests in the receiving line outrank the guest of honor.

- The host and spouse of host arrive early.

- The host and spouse of host or their representative greet the guests of honor at the front door when they first arrive, take their coats, and escort them to where the receiving line will be formed.

- Guests of honor arrive at least ten minutes early in order to be in place at the start time.

- Guests of honor look to the host and his or her spouse for guidance concerning where to stand in the receiving line.

- Remove any rings from the right hand before standing in a receiving line. If you forget, the first few handshakes will remind you.

- Once the line begins, listen carefully to each person's name. Say a very brief greeting, and then introduce the guest to the next person in line. Use the simplified method of introduction, saying each person's name only once: "Mr./Mrs. Jones, this is Colonel Brown," or "Colonel Brown, Sergeant Smith." A servicemember standing before his or her spouse in a receiving line should simply say, "Sergeant Major White, I'd like you to meet my husband/wife." If you aren't given the guest's name or don't hear it clearly, ask the guest for his or her name.

- Avoid giving the impression of "pulling the guests down the line" with your handshake.
- Remember to smile.

RECEPTION/RECEIVING LINE DIAGRAM
The reception line and receiving line typically look like the following diagram when the line is set up to the left of the entrance. The line is a mirror image of this diagram when it is formed to the right of the entrance; however, the flag line remains the same as shown.

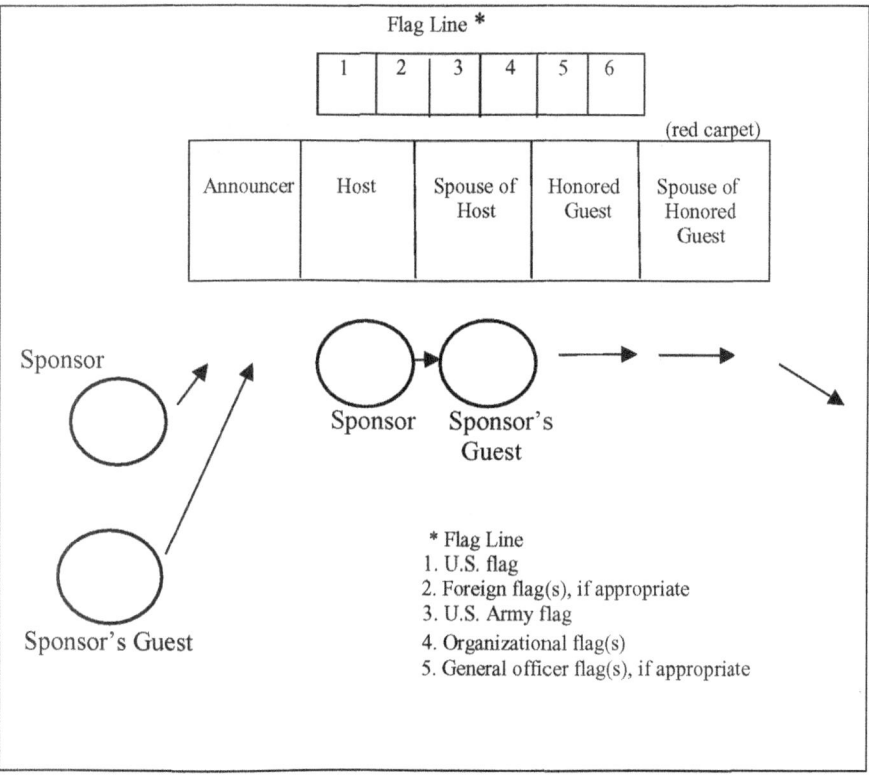

NEW YEAR'S RECEPTION

The commander's New Year's reception has special protocol associated with it. It is considered a "command performance" event, meaning all invited military members are expected to attend, unless they are ill or out of town. The military member should attend even if his or her spouse cannot; on the other hand, a spouse should regret if the military sponsor cannot attend. Hostess gifts are not appropriate. For a reception held in the commander's home, guests may leave their calling cards (see Chapter 5 for guidelines); however, calling cards are never left at a reception held in the club. Accepting an invitation to a commander's New Year's reception does not result in any social indebtedness. Usually the invitation will give both a beginning and ending time. Be sure to depart on time, as there may be another group invited to arrive a few minutes later.

* * * * * * * * * * * *

All Army spouses deserve the "red-carpet treatment" but, at a reception, only the people in the receiving line stand on it.

Chapter 25

Formal Functions

"A ball is the only social function in the United States
to which such qualifying words as splendor and magnificence
can with proper modesty be applied."
Emily Post, 1927

Some of the most elegant Army traditions that set us apart from civilian society are those associated with formal military affairs. Many of our civilian counterparts rarely have the opportunity to attend a formal function, yet most Army spouses are invited to at least one formal ball or dining-out each year. And who doesn't enjoy dressing up for a fancy party? Of course, knowing what to expect at these events makes them even more enjoyable.

FORMAL BALLS
Dress - Formal balls are usually held to celebrate special military occasions or holiday seasons. By definition, the proper dress for these events is always formal. For officers, that means either Army Blue Mess uniform or Army Service Uniform with bow tie. (Army Blue Mess or Army Service Uniform may be worn year-round, while Army White Mess may be worn in the summer, or year-round in tropical climates.) For NCOs and enlisted, formal also means Army Blue Mess or Army Service Uniform with white shirt and bow tie. Usually a military invitation will state the desired uniform; however, if it simply says "Formal" or "Black Tie," military members are expected to wear the appropriate uniform. The Army expects name tags to be worn on all uniforms except the Army Mess uniform.

Male civilians should wear appropriate black-tie attire. Women's fashions, however, are continuously changing, and formal dress styles are no exception; of course, long formal gowns are always in style for formal balls. In addition, short formals, tea-length gowns, and evening pants with fancy tops are currently in vogue. One custom that seems to have faded from fashion is that of a gentleman giving his escort a corsage, though if you receive one, wear it with pride. So respond to your invitation early, decide what you will wear, and get out your dancing shoes—you're going to a ball!

Arrival - The evening of the ball, be sure to arrive on time. There probably will be a receiving line and you shouldn't miss it. If the reception line is already forming, join in. Otherwise, you might have a drink and mingle with the other guests. Don't be tempted to look for a place to sit down; this is a time to socialize. At some time before dinner is announced, you or your spouse should locate the seating chart (usually placed in a prominent spot or by the door to the ballroom) and find out where you will be sitting. If the seating has been left open, places may be reserved by leaving some personal item at your selected place, but not by tipping the chair. Dinner may be announced by a bugler sounding Mess Call, or simply by the doors to the ballroom being opened.

Receiving Line – A receiving line is an efficient and gracious way to allow the honored guest(s) to meet all guests. Oftentimes, the receiving line forms immediately prior to entering the ballroom. There will be a table located near the start of the line where drinks can be placed. Ensure your purse, gloves and anything else is tucked away so that your hands are free to shake. The woman precedes the man EXCEPT at White House or Air Force Functions. The first person you encounter in the receiving line is the aide or adjutant. The servicemember gives names to the aide and introduces their spouse or date. Do not shake hands with the aide. To keep the line moving, make sure you speak only briefly and shake the hand of those in line. For a more detailed explanation of receiving lines, see Chapter 24.

Seating - Once you find your seat for dinner, don't automatically sit down. Check to see what the spouses at the head table are doing, or listen for instructions. If those at the head table are already sitting down, you should follow their lead. Alternatively, you may be requested to remain standing or you may be asked to take your seats and await the entrance of those who will sit at the head table.

There usually is a printed program at your place that will indicate the order of events. First listed is normally the color guard, which will march in and post the American flag and the organizational colors. During this ceremony, military members stand at attention and civilians stand quietly with their hands at their sides (not on the back of their chairs). Keep your eyes on the American flag as long as you conveniently can, turning your head or body slightly if necessary. Don't pirouette or turn in a complete circle to "follow the flag." It's all right for the flag to pass behind you; simply follow it with your eyes once it moves into view. If a band is present, the "National Anthem" may be played after the colors are posted. As a civilian, it is customary to put your right hand over your heart at this time. Next, the chaplain may offer a short invocation, and then toasts are made to a variety of groups. (A discussion of the etiquette of toasting appears later in this chapter.)

The Fallen Comrade Table (also known as Missing Man/Soldier Table) - This is a beautiful and solemn observance in honor of our Fallen Comrades and POW/MIA. It is a moving tribute to our servicemembers who have given the ultimate sacrifice in honor of our country. Variations may occur, but here is one adaptation: A small round table is located in a place of honor; it symbolizes they are there in spirit. These items are symbolic: The small round table represents everlasting concern for our soldiers and their families with a complete setting for one with the wine glass inverted, which symbolizes their inability to share this evening's toast. The tablecloth is white, symbolizing the purity of their motives when answering the call to duty, with a single red rose and yellow ribbon to remind us of the life of each of the fallen and their loved ones; a slice of lemon to remind us of the bitter fate of those who died; and a small amount of salt to symbolize the tears shed by families and friends who lost their loved ones. A small bible, which represents the strength gained through faith that sustained our soldiers lost in battle while fighting for our country, founded as one nation under God, is placed next to the rose. The chair remains empty as a reminder of the loss of a hero. The attendees raise their glasses but do not drink as they observe a moment of silence in honor of their fallen comrades.

After the toasts, everyone sits down and dinner is served. Later, there will possibly be a speech or entertainment, followed by the retiring of the colors, and, finally, dancing!

General Guidance
- Talk to the people sitting around you at the table, especially the ones on either side of you. It may happen that you don't know them but, as a courtesy, one should chat with his or her dinner partners. Couples are often not seated together. Look at this as a good opportunity to make new friends. Traditionally, one was expected to talk with each dinner partner in turn, as courses were changed. This stilted procedure faded into history long ago, but its mere existence indicates how important proper dinner-partner courtesy is considered.
- If there is a standing ovation given to the guest speaker and you feel the speech warrants such an honor, feel free to join in. (A reminder for the speaker's spouse: Applaud and stand with the audience; not to do so implies that you are accepting some of the tribute for yourself.)
- When everything but the dancing is over, you may leave the table briefly to mingle with the other guests. As a courtesy, use this time to go up to the head table and speak to the host, hostess, and guests of honor.
- Wear comfortable shoes so that you won't be tempted to take them off as the dancing warms up. Taking your shoes off is not considered polite or proper. Ladies, you can bring ballet shoes or fancy slippers to change into.
- If the room gets extremely warm, hope that the senior servicemember present will remove their coat (and tie); if they don't, it's not appropriate for the other servicemembers to do so.
- If possible, try to stay until the senior person leaves. Hopefully, they will recognize that others are waiting for them to depart, and won't stay until everyone "turns into a pumpkin." If you absolutely must leave before they do, it is proper for you and your spouse to say goodbye to the ranking person and make your apologies for leaving early.
- The last thing to do before leaving is tell the host and hostess (or the senior person of the hosting unit) what a nice party it was and how much you enjoyed it. Consider how much effort went into the preparation; they deserve your thanks.

FORMAL FAREWELLS

Formal balls are frequently held to farewell senior commanders. They follow the procedure just described, with the addition of entertainment (frequently quite humorous) that is centered around the departing commander, and provided by members of the hosting group. The hosting group of a formal farewell is normally the departing commander's own unit. However, occasionally a senior commander prefers to ensure that all of their senior subordinate commanders receive comparable farewells by hosting them, with the assistance of his staff and the departing commander's staff.

The person responsible for making up the guest list is determined by which group hosts the farewell. If the senior commander hosts the event, then that person decides on the guest list. If the departing commander's unit hosts the farewell, then the departing commander decides on the guest list. This includes deciding whether or not to invite their contemporary commanders, any senior commanders, and their spouses. They make this decision based upon their personal preference, local tradition, and any guidance that might have been given by their senior commander. Probably the only ones who are never included on the guest list of a formal farewell are the incoming commander and spouse. This is time for those who have known the outgoing commander during their command to celebrate their accomplishments, and not a time to welcome a newcomer.

Seating for a formal farewell, when the departing commander's contemporaries and their spouses are included on the guest list, offers an opportunity to deviate slightly from the guidance given in Chapter 18 under "Head-Table Seating." Although the contemporary commanders may warrant head-table seating due to their rank, they may instead be scattered throughout the audience and seated in places of honor, with the various subordinate unit commanders and their spouses. For example at an aviation brigade farewell, other brigade commanders of the division may be seated at the senior end of a table across from the aviation battalion commander whose unit supports that brigade. This is usually preferable to seating all of the contemporary commanders together at one table and separating them from the hosting unit's subordinate commanders.

No one needs to be concerned that the presence of any senior guests might overshadow the person being farewelled. On the contrary, their presence should be considered an honor to the one who is departing. The guest of honor is "center stage" throughout the evening. He or she and their spouse will be first in the receiving line, they will be seated at the center of the head table, the spouse may be given a small gift, the entertainment will be centered around the guest of honor, words of praise will be spoken about them, and the departing commander will receive one or more farewell gifts within statutory guidelines. Check with your local Staff Judge Advocate for legal limits and Joint Ethics Regulation. In other words, the entire evening revolves around the departing commander, and no one can overshadow that!

The formal farewell of a commander is perhaps the most bittersweet event of their command. The hard work is over. Many lasting friends and wonderful memories have been made. This is the last time before they turn over command that they will be able to enjoy the company of all of the members of their unit.

THE DINING-OUT

Two similar terms you will hear in the Army are *dining-in* and *dining-out*. A *dining-in* is a traditional formal dinner for military members only. A *dining-out* is a relatively new version of this tradition that includes spouses and guests. The purpose of the dining-out is to bring members of the unit and their guests together in an atmosphere of camaraderie, good fellowship, fun, and social rapport.

History - The tradition of *dining-in* probably dates back to the Vikings, who held formal ceremonies to celebrate their victories and feats of heroism. It is believed that this custom later spread to England—first to the monasteries, then the early universities—and, with the appearance of the officers' mess, the British military also adopted this custom. Later, World War II brought American and British military closely together and exposed our officers to this ancient tradition of dining-in. Quickly seeing its benefits for the units, we adopted it as our own, and today both officers and noncommissioned officers in the American Army hold dining-ins. These formal dinners are regarded as excellent opportunities to build *esprit de corps* and educate young unit members in the ways of formal social traditions. Dining-ins also provide the unit with special occasions to recognize unit achievements and honor distinguished visitors. Occasionally, a unit will decide to invite spouses and dates to its formal dinner; this is known as a dining-out, meaning people outside of the unit are invited.

Dress and Arrival - Dining-ins and dining-outs are always formal events. The desired uniform will be stated on the invitation, usually Army Blue Mess or Army Service Uniform for officers, noncommissioned officers, and enlisted personnel. Male civilians should wear appropriate black-tie attire. Female civilians wear evening clothes—long gowns, short or tea-length formals, formal evening pants.

Be sure to arrive on time; if there is a guest of honor and/or special speaker, there may be a receiving line. There is always time for cocktails and socializing prior to the dinner hour. A seating chart is usually located in the reception area, and place cards will be on the tables. When it's time to move into the dining room, any unfinished drinks should be left in the reception area. There probably will be a printed program at everyone's place that details the schedule for the evening, outlines the mess "rules," and provides a biography of the guest of honor. Prepare yourself for an evening of ceremony and tradition filled with good fun and camaraderie.

Ceremonies - Dining-ins and dining-outs are extremely ceremonial in nature, and these ceremonies differ slightly with our various military services. Following are the Army traditions and etiquette important for you to know.

- *Principal Officials of the Dining-out*
  - President: Usually the commanding officer of the unit is the central figure of the dining-out. He/she introduces the speaker and honored guests, proposes the first toast, and presides over the ceremony.
  - Mister Vice/Madam Vice: The person responsible for the evening is a member of the unit, usually selected for possessing wit and the ability to speak. Mister/Madam Vice sits at the end of a table or alone at a small table on the opposite side of the room from the President of the Mess. Members and guests must be prepared to follow instructions given by either the President or Mister/Madam Vice.
- *Toasting* - The custom of toasting at formal events is universal. At the opening of the dining-in or dining-out and during the evening, a number of toasts will be made. (See the following section on "Toasting.") The appropriate responses may be printed in your program.
- *The Punch Bowl or Grog Ceremony* - A Master of the Punch or Grog may be appointed when this ritual is a unit tradition. The punch is prepared ceremonially before attendees. The Master selects one assistant for each recipe ingredient. As each assistant adds an ingredient to the bowl, he or she makes suitable remarks. The ingredients are usually selected and reflect the unit's long history. Once the punch is prepared, the President is offered a taste to see if it is ready for consumption.

- *Smoking Lamp* - A symbolic ceremony at a dining-in or dining-out occurs when, at the close of dinner, the President of the Mess asks Mister/Madam Vice to light the smoking light. After this has been accomplished, the President formally announces, "The smoking lamp is lighted." With the current trend of smoke-free environments, many halls and community clubs are non-smoking facilities. There may be an area for cigar smoking outside. The lighting of the Smoking Lamp signifies the end of the official portion of the Dining-In/Dining-Out.

Traditions - It is the traditions associated with dining-ins and dining-outs that make them so much fun. However, these traditions are carried out in utmost seriousness.
- Throughout the dinner, various members of the unit may request permission from Mister/Madam Vice to address the Mess and their guests. Often this is to report some humorous "infraction of the rules" by other members of the unit, for which a small fine may be levied. The money collected is used for some worthy purpose, such as a charity. It's all done in the spirit of fun and the fines are nominal, so don't worry if your spouse gets fined. (But they should remember to wear their dog tags that evening, because those who don't are sure to get caught!)
- Try to avoid going to the restroom during dinner. It is not appropriate for you to leave the dining room, since the members of the unit are not allowed to leave without Mister/Madam Vice's permission. Your departure would be "noticed" and, in fun, someone might call attention to your "infraction of the rules." Note the time dinner is to begin and make any necessary stops before then. Also, watch the head table. If the spouses sitting there leave the table, then you can do so without worry of being chided.

A large part of enjoying dining-outs is waiting to hear the next infraction of the rules that's called to everyone's attention, and hoping that it isn't something you or your spouse has done, or failed to do. Remember, though, this is all done in a sense of camaraderie and fun.

TOASTING

Toasting is the ancient tradition of drinking together in honor of someone or some group in order to show respect or appreciation. It is believed that this custom came into wide acceptance after the effects of poisons were discovered. When two persons, who might be antagonists, drank from the same source at the same instant and suffered no ill effects, a degree of mutual trust or rapport was established. Today, toasting is a gesture to honor the person or group being recognized. It is not necessary to drain the glass, or even to sip the wine or water; a mere touch of the glass to the lips satisfies the ceremony. Toasting is almost always a part of the ceremonial beginning of Army formal dinners and balls, and frequently a part of small dinner parties.

Toasting Etiquette

There are two important rules to toasting etiquette:

1. *Do as the toastmaster does.* The toastmaster is the person offering the toast. If he or she stands to propose a toast, then everyone joining in should stand as well. If he or she

remains seated, often the case in home settings, then everyone remains seated. However, one point needs clarifying. While traditionally, only men stood during toasts, this is no longer true. Unless the toast is to the ladies or spouses as a group or to you personally, you should stand with the servicemembers to drink a toast, if the toastmaster is standing. The one time everyone should stand is for a toast to the President of the United States.

2. *Never drink a toast to yourself.* Since toasting is done to show respect or appreciation, it is inappropriate to drink when a toast has been proposed for you personally, or to all of the ladies, spouses, or guests as a group. Therefore, even when the toastmaster is standing, if the toast is to you, you should be seated. That is why you will often hear the toastmaster request, "Gentlemen, please seat the ladies," or "Soldiers, please seat your guests," a signal that a toast to the ladies, spouses, or guests is about to be offered. If a toast to you is proposed without the request to seat you first, then sit down quickly (and quietly). That says, it's important to listen very carefully when a toast is offered. It also says that the toastmaster should state at the beginning the name of the person or group to whom the toast is being proposed. To begin simply by saying, "I would like to propose a toast," leaves everyone wondering what to do.

Traditional Toasts and Responses

Tradition also prescribes the manner in which we respond to a toast. When the toastmaster proposes a toast, he/she will name the individual or group being recognized. The others present, excluding the individual or group being toasted, repeat the name in unison before sipping from their glasses. Sometimes, if the title of the individual or name of the group is very long, it may be shortened. For example, an appropriate response to a toast proposed "To the President of the United States" would be "The President." Other services may have slightly different traditions. For example, the last toast given at an Air Force formal function is "To our honored guests," to which the response is "Hear, hear." The printed program for the evening's events may list the toasts that will be offered as well as the responses expected of the guests. There are variations to the MIA and Fallen Comrades toast, but it is customary to observe a moment of silence in their honor and this should be stated in the program. Most also follow the tradition of raising their glasses, but not sipping from them, since the Fallen Heroes can no longer join in the toast.

The toasts usually proposed at a formal Army dinner or ball proceed in the following manner: (In international settings or when foreign officers or dignitaries are present, appropriate toasts to their heads of state should be included.)

        To the President of the United States
            (or To the Commander in Chief)
        To the United States Army
        To the (unit)
        To All Our Missing and Fallen Comrades Who are
            Unable to Join Us Tonight (moment of silence)
        To our Guests (or To the Spouses)

At official military functions, toasts are generally offered after the colors are posted and before the guests are seated. This is most helpful in a crowded room where the closeness of the tables makes it difficult for guests to move their chairs to rise in response to a toast. However, if people are already seated at an official function, the toastmaster usually provides instructions, such as "Will everyone please rise, and join me in a toast to...." If in doubt, take your cue from the spouses seated at the head table.

To acknowledge a toast that has been made to you personally, remain seated and don't drink the toast, then respond immediately by saying a few words and offering a return toast to the person who has just toasted you.

When a toast is offered to an individual, the spouse joins everyone in drinking the tribute. To refrain from doing so looks as though he or she is accepting some of the group's respect or appreciation that is being offered to another. The same is true when the group is applauding or giving a standing ovation. Unfortunately, some spouses refrain from toasting, applauding, or standing for their spouses from a misplaced sense of modesty.

When is it appropriate to offer toasts during dinner? That can vary. Toasts at American formal military dinners and balls always occur after the colors are posted and before dinner. At small informal dinner parties, toasts may be offered any time after the wine is poured. The only time it is inappropriate to offer a toast is after the liqueurs have been served, because toasts are never drunk with liqueurs. In international settings or with foreign guests, be sensitive to their customs that may be different from our own. For example, the British Army toasts after dinner. In Germany, a toast is almost always made after everyone's glass has been filled for the first time, but before anyone has taken a sip. Even at a very large gathering, such as a German beer fest, no one in a group will begin drinking until everyone in the party has raised their glasses together, saying "Prosit" or the universal "Cheers."

The final question for some is, "What if I don't drink alcoholic beverages and a toast is offered?" The answer is simple. The spirit of the toast is served equally well by lifting a glass to your lips without drinking, or by drinking whatever beverage you have. Don't hesitate to join in the toasting, even if you prefer not to drink any alcohol.

* * * * * * * * * * * *

A toast to every Army spouse—May you always have
a loving family and dear friends, now and forever, until life ends.

SECTION SEVEN: MILITARY FUNCTIONS

Chapter 26

Parades, Changes of Command, Changes of Responsibility, and Retirements

"... every Citizen who enjoys the protection of a free Government,
owes ... his personal service to the defense of it . . . ."
George Washington

To both new and experienced Army spouses, a military parade epitomizes the very essence of the Army. The soldiers standing at attention looking their very best, the flags and guidons flapping smartly in the breeze, the band playing John Philip Sousa music—whose heart doesn't swell with pride and patriotism at such a sight?

Just as the movements of the troops on a parade field follow a prescribed tradition, so too do those of the audience. Therefore, attending a parade becomes much more than just watching the pageantry when you understand what is happening on the field and what is expected of you as a spectator.

A parade, as an all-inclusive term used by Army spouses, can take several different forms. A "review" consists of troops of an entire major unit or division parading together for "review" by their commander or a more senior officer. A change of command parade honors an outgoing commander as he or she turns over command of the unit to the incoming commander. A change of responsibility ceremony honors the senior noncommissioned officer that is leaving a leadership position as a new senior noncommissioned officer takes their place. A "retreat" ceremony is held in the late afternoon and features the lowering of the American flag. Whichever form an Army parade takes, it will always be exciting. Never miss an opportunity to go to a parade!

ON THE FIELD

The entire orchestration of troop movements during an American military parade dates back to 1778. General George Washington saw the need in his troops for organization, control, discipline, and teamwork; he asked a Prussian officer, Baron Fredrich von Steuben, for assistance. Von Steuben went to Valley Forge, saw the disarray of the troops, and quickly set to work. He developed drill movements, taught them to a model company of 120 men, then convinced the troop leaders to incorporate the drills into their daily training. From this, our parades of today have evolved. Following is a description of the sequence of events that occur at a typical parade.

The parade begins with the "formation of troops." The units on parade are either already standing in their places on the field, or will march on at this time. At the appointed starting time, the adjutant announces, "Sound attention," and the band plays a four-note chord. The units are brought to attention by their commanders, as the adjutant marches to the center of the field. The commander of troops and his staff also march to the center of the field, and the troops "present arms" in a salute to their commander. Soldiers return to "order arms" (position of attention with a weapon) and the staff marches to the other side of the commander of troops, who turns and faces the reviewing stand. The "formation of troops" is now complete.

The reviewing party is announced and steps forward to the reviewing stand. The troops on the field "present arms" (salute). If there is a general officer in the reviewing party, the band commences playing "Ruffles and Flourishes" and "The General's March." Simultaneously, a cannon salute is fired, if guns are present. Following the last cannon shot or note of music, the troops are brought to a position of "parade rest."

The "inspection of troops" is accomplished by the reviewing party moving to the center of the field where it is met by the commander of troops. This group proceeds to inspect the troops by passing along the front line of the troops and returning behind them. The reviewing party then returns to the reviewing stand. (At a large review, this inspection may be accomplished by vehicle or on horseback.)

Next, the unit commanders, the units' colors, and the national colors are brought forward in preparation for the playing of the "National Anthem." If this parade is taking place in a foreign country or any foreign units are present on the field, their national anthems are played first, followed by the U.S. national anthem.

If this parade is being held in honor of a change of command, presentation of honors, or retirement, that special event takes place at this time. The principal participant(s) move to the center of the field directly in front of the reviewing stand. The officiating officer stands in front of the participant(s), with his back to the audience, to conduct the ceremony. Following the passing of colors from the outgoing commander to the incoming commander (signifying the actual change of command), or the presentation of honors to the honoree(s), the principal participants return to the reviewing stand to make public remarks.

Remarks by one or more of the reviewing party are addressed to guests and the troops on the field. Following that, the commanders and colors return to their original positions. The band and troops begin to "pass in review," that is—each unit marches past the reviewing stand, salutes the reviewing party, and marches off the field. The announcer may name each unit, its commander, and command sergeant major or first sergeant as they pass. The audience often applauds to commend the soldiers for their part in the ceremony. After the last troop unit has passed, the band plays "The Army Song." The commander of troops announces to the reviewing officer, "Sir, this concludes the ceremony." The parade is now over.

The playing of "The Army Song" to close the parade is a fairly recent addition to the ceremony. In the early 1960s, the Army adopted the Field Artillery song, "The Caissons Go Rolling Along," as its own, after changing the title and words. (Artillerymen love the fact that everyone stands up for "their song.")

A typical parade program for a review:

> Sequence of Events
>
> Formation of Troops
> Arrival of Official Party
> Honors*
> Invocation*
> Inspection
> National Anthem*
> Remarks
> Pass in Review
> The Army Song*
> Conclusion
>
> * Please stand

On the program for a change of command/change of responsibility, presentation of honors, or retirement parade, the appropriate entry would appear between the National Anthem and Remarks. Note that not all programs will contain each component listed in the above Sequence of Events example. It will depend on the type of parade. Additionally, for changes of command/responsibility and retirement parades, a one-page biography is often included with the program of the new and old commanders/senior noncommissioned officers, or for the retiree(s).

On occasion, there will be minor deviations from the typical parade just described. For example, after the formation of troops, the band may be directed to "Sound off." At this command, the band marches the length of the parade ground and returns, playing music all the while. The use of "Sound off" began during our Civil War when the band played popular music to entertain the troops.

Company, Battery, or Troop-level Change of Command/Change of Responsibility

The change of command/change of responsibility for a company, battery, or troop-level unit differs significantly from the typical parade just described. Because this event is held for a smaller-size unit, there will be fewer personnel on the field, and probably no band or general officer present. Nevertheless, the importance of such a change of command/responsibility should not be overlooked. At this level, the commander/first sergeant has a greater impact on the soldiers in his/her unit. Therefore, these smaller changes of command/responsibility are more important to the Army and the nation than are the larger, higher-level changes of command/responsibility, even though they are not accompanied by such pomp and ceremony.

For the company change of command, the "formation of the troops" is accomplished by the unit's first sergeant. He takes the report from each platoon sergeant that "All present or accounted for," makes any necessary announcements to the assembled troops, and turns the unit over to the outgoing commander. Unlike higher-level changes of command in which the outgoing commander is a part of the reviewing party and first comes on the field to "inspect the troops," the outgoing company (battery/troop) commander takes his position on the field in front of his unit and guidon bearer, facing the audience and his senior commander.

At the appropriate time in the ceremony, the senior commander who will conduct the change of command and the incoming company (battery/troop) commander come onto the field to a position directly in front of the unit. The outgoing commander moves to one side and slightly in front of his senior commander, and the incoming commander does the same on the other side. The unit's first sergeant steps into position in front of the senior commander. This completes a diamond-shaped formation, with the first sergeant, outgoing commander, senior commander, and incoming commander each standing at one point of the diamond, facing toward the center. The first sergeant takes the unit guidon from the guidon bearer, passes it to the outgoing commander, and he relinquishes it to the senior commander, who in turn presents it to the incoming commander. Thus, the change of command is accomplished. The new commander, after passing the guidon back to the first sergeant (who returns it to the guidon bearer), takes his new place in front of his troops facing the audience, and the outgoing and senior commanders leave the field.

Remarks may be made by all the principal participants at this time. Because of the small size of the unit, the outgoing commander has usually already taken the opportunity to congratulate the soldiers in his/her unit on their performance and restricts this farewell to a few brief words of praise and thanks. Unlike the large change of commands, the senior commander normally says more, using this time to thank the unit for its past efforts, describe the qualifications of the incoming commander, and request continued cooperation and support from the unit for the new commander. The incoming commander may make brief remarks to the unit, although most will wait for a more relaxed atmosphere to do so and simply end the ceremony by turning the unit over to the first sergeant.

The company change of responsibility is very similar to the company change of command. The primary difference during the ceremony is that the commander will pass the guidon from the outgoing first sergeant to the incoming first sergeant symbolizing the transfer of the responsibility of the keeper of the colors and the senior noncommissioned officer's responsibility for the unit.

> **BLUF**
>
> Outgoing on the right and Incoming on the left.

SEATING

Seating at a typical parade falls into two categories: designated and undesignated. Those guests who receive and accept invitations to attend a parade will have designated seats reserved for them; all others are welcome to attend, and may sit in open seating to the side of the designated seats.

The protocol office or unit adjutant is responsible for organizing the seating, and does so according to the unofficial order of precedence (as shown in Chapter 18, "Seating Arrangements"). The commanding officer should review the seating plan because his/her more mature view, especially with regard to the importance of civilian guests, may differ from that of a young protocol officer or adjutant. In overseas areas, it is always wise to seek the advice of the protocol office on seating foreign military and civilian guests.

The seating for military guests is normally based on job assignment, rank, and date of rank. Spouses are seated beside their sponsors. If the sponsors cannot attend or are on the parade field, the spouses are seated where their sponsors would have been seated. (If the spouse is also in the military but invited as a spouse, she/he should be accorded the seat of higher precedence—based on either the spouse's rank or the sponsor's rank.) Invited civilians and any special guests are interspersed throughout, in positions appropriate to their importance for the event. All guests who are invited as members of a special group are generally seated together with the other members of their group.

It must be pointed out that, while there are many key military personnel who serve in important positions, both officer and NCO, not everyone can be accorded prime seating at a parade. Designated seating is best allocated according to the order of precedence, which everyone understands. The only exceptions typically made are for changes of command/responsibility, when key people are accorded a higher position in the order of precedence because of their relevance to the event. Many units seat their command sergeants major and their spouses next to their unit commanders and spouses at parades to show unity with their battle buddy/command team. The visual impact of soldiers seeing their command team seated together is very powerful. The commanding officer will need to use creativity in order to ensure that all guests are treated with respect to their rank and position. Ultimately, the decision of who sits where rests with the commander.

Typical Parade Seating - Designated seating is usually divided into sections, right and left of the reviewing stand. The aisle seat on the front row of the right side, as you face the field, is the seat of honor. The highest-ranking officer is normally seated there, with his or her spouse seated next. The second-ranking officer is normally seated on the left side, front row, aisle seat. Continuing down the list of guests by order of precedence, seats are assigned on alternating sides—first to the right, then to the left—starting in the center and moving outward. When each row is filled, begin again at the center.

Typical Parade Seating

The seating arrangement for a typical parade with two sections of designated seats looks like this:

```
                        Parade Field

                    ┌─────────────────┐
                    │ Reviewing Stand │
                    └─────────────────┘

    8    7    5    2         1    3    4    6
   16   14   12   10         9   11   13   15
   24   22   20   18        17   19   21   23
   32   30   28   26        25   27   29   31
```

1. Highest-ranking officer and spouse
2. Second-ranking officer and spouse
3. Spouse of CSM (Unit CSM will be on the field.)
4. Third-ranking officer and spouse
5. Fourth-ranking officer or Distinguished Visitor (DV)
6. Senior officer or DV and spouse
7. Senior officer or DV and spouse

8-32. Other invited guests by order of precedence, alternating sides

Note: There are usually other sections for family, friends, and other attendees.

Change-of-Command/Change-of-Responsibility Seating

The order of precedence for change-of-command/change-of-responsibility seating is different from that just described for the typical parade because of the importance accorded the spouses and families of the outgoing and incoming commanders and outgoing and incoming command sergeants major. Typically, the outgoing commander/command sergeant major's spouse and family sit beside the highest-ranking officer and spouse on the right of the aisle; the incoming commander/command sergeant major's spouse and family sit to the left of the aisle (as you're facing the field) on the left of the second-ranking officer and spouse. Also, the spouses of the senior commanders on the field and the commander of troops are accorded higher precedence, and are typically seated behind the family of the outgoing commander/command sergeant major. Other senior guests are seated beside and behind the incoming commander/command sergeant major's spouse and family, or they may be seated on the outgoing commander/command sergeant major's side based on the number of attending guests, configuration, and available seating. Often times, an organization or unit command sergeant major and spouse will be afforded special ceremonial protocol precedence and seated before the chief of staff or deputy commander or the row right behind their counterpart commander if the first row is taken up with the family members. You will note that for changes of command/responsibility, retirement ceremonies, and presentations of honors, not only the spouse(s), but also the children and other close family members of the principal(s) being honored are included in the order of precedence and accorded special reserved seating.

As you can see, it's difficult to depict seating for a change of command/change of responsibility in a simple progression, as was shown for the typical parade, because of the unknown number of family members to be accommodated. However, the order of precedence for those involved and their possible seating positions for a change of command in a two-section seating arrangement are shown in the next diagram.

Please remember that these are approximate positions that can vary significantly, not only with the wishes of the highest-ranking officer, but also with the wishes of the outgoing commander/command sergeant major, number of family members who need special seating, seniority and number of other guests, and available seating capacity and its configuration.

Typical Change-of-Command Seating

```
                        Parade Field

                     ┌─────────────────┐
                     │ Reviewing Stand │
                     └─────────────────┘

   6    4    4    2      1    3    3    5
   9    9    9    8      7    7    7    7
   9    9    9    9      9    9    9    9
  10   10   10   10     10   10   10   10
```

1. Highest-ranking officer & spouse
2. Second-rank officer & spouse
3. Outgoing commander's spouse & family
4. Incoming commander's spouse & family
5. Spouse of parading unit's command sergeant major
6. Third-ranking officer & spouse
7. Spouses of: senior commanders on the field & "commander of troops" (by order of date of rank)
8. Fourth-ranking officer & spouse
9-10. Other invited guests, by order of precedence

GENERAL GUIDANCE

Invitations - You don't need an invitation to attend a parade. Everyone is welcome. Those who receive invitations (and respond) will have designated seats reserved for them, and are usually invited to any reception that is held afterward. However, the parade itself is open to everyone.

Children - Children love a parade. If you will be sitting in the general seating (not designated), plan to take your children. Once children are old enough to sit still and remain quiet, they are welcome at parades. However, parades are dignified events, and the other spectators want to hear the announcements and speeches. If your baby cries or your child gets restless, courtesy demands that you get up and move far enough away that your child does not disturb others.

If you receive an invitation to a parade, the chances are your children are not included in the invitation unless the parade is being held in your spouse's honor. Should they wish to attend, make arrangements for them to sit with a friend in the open (undesignated) seating, since you will have a designated seat in the reserved section.

How to Dress - A parade is a very special military event, requiring considerable time and effort from the soldiers involved, for both the preparation and execution. It is expected that the members of the audience will honor the dignity of the event by dressing nicely. For gentlemen, a business suit or sport coat and tie is appropriate. For ladies, a dress or suit with heels is most appropriate. A casual skirt and blouse, pantsuit, or simple dress with flat shoes may be suitable depending on current fashion, local customs, and the location of your seating (designated or undesignated). Slacks and pants suits are not usually worn by those in the designated seating area, but others may wear them during extremely cold weather. (Suggestions for really cold weather: undershirt, knee-length "long johns," pocket warmer, and even a small blanket or lap robe.)

When hats are in fashion, they may certainly be worn to a parade, though a large hat that obscures the vision of those seated behind you would be inconsiderate. Whether or not the senior lady or female guest of honor is wearing a hat should have no bearing on what other ladies choose to wear.

Arrival - Arrive early enough to park your car and find your seat before the parade begins. Parades always begin on time, so be there a few minutes early. Once the parade begins, it is not polite to talk or smoke. Please silence all cell phones. Your attention should be on the activities on the field.

INVITED GUESTS

1. Respond to a parade invitation just as you would any social invitation. Since you are an invited guest, a seat will be reserved for you if you are able to attend. The protocol office or adjutant is waiting for the responses of the invited guests in order to plan and reserve the seating. Additionally, if the invitation includes a reception after the parade, those making plans need your response in order to plan for refreshments.
2. If, after having accepted an invitation to a parade, you find that you cannot attend, by all means, call and cancel. Reserved seating is always limited. Also, it's an embarrassment to the hosting unit to have empty reserved seats at a parade. Even as late as the morning of the parade, regrets can be accepted and adjustments made in the seating, though earlier notice would be preferable.
3. Find your seat as soon as you arrive. There is usually a seating chart near the entrance to the designated seating area. Additionally, ushers may be available with a list of invited guests and their designated-seat locations. It isn't necessary to sit down right away, but do find out where your seat is located so that you can move quickly to it before the parade begins.
4. It is nice to use the few minutes prior to the start of the parade to chat with the other invited guests. At a change of command/change of responsibility, also use this time to seek out the spouses and families of the outgoing and incoming commanders/command sergeants major, and briefly greet them. If the group of invited guests is very large, just speak with those seated near you, especially any civilian guests.
5. After a change of command/change of responsibility of major units, the Army tradition is for the outgoing commander/command sergeant major and family to depart the post, while the incoming commander/command sergeant major is welcomed at a reception

held in his/her honor. (Although some units do not follow this tradition, it was the standard.) Therefore, as soon as the change of command/change of responsibility is over, the departing commander/command sergeant major and his/her spouse and family usually stand in front of the reviewing stands, and the invited guests go by to say their final farewell.

6. Receptions after parades are usually only for the invited guests. If you are invited to a parade, read your invitation carefully to see if it includes a reception. These are adult gatherings and not intended for children.
7. At the reception following a change of command/change of responsibility, there will be a receiving line. If you arrive at the reception location before the formation of the receiving line, you will have to wait to enter. Once the line is formed and the most senior officer and his or her spouse have gone through to congratulate and welcome the new commander/command sergeant major, then everyone else follows. (See specific guidelines in Chapter 24, "Receptions and Receiving Lines.")

WHEN TO STAND

Knowing when to stand and when to sit at a parade is the true test of an experienced Army spouse! Of course, some units thoughtfully have the announcer ask the audience to stand and sit at the appropriate times. Others use the printed program to indicate by asterisk (*) those times when the audience is expected to stand. But you can't count on having this guidance. When no other help is provided, watch those who are seated on the front row of the designated seating; they are the most experienced and should know what to do. The best advice, however, is to learn for yourself what is expected. The times to stand are as follows, in order of their occurrence:

1. If a general officer is in the reviewing party, the band will play "Ruffles and Flourishes" followed by the "General's March" as soon as the reviewing party steps onto the reviewing platform. Simultaneously, a cannon salute may be fired.[1] (If cannons are positioned nearby, prepare yourself and your children for their firing at this time.) The audience stands at attention from the first note of "Ruffles and Flourishes" to the last cannon shot, and then is seated.
2. Stand at attention when our national anthem, other national anthems, and "To the Colors" are played. ("To the Colors" is normally played at retreat ceremonies and is accorded the same courtesy as our national anthem.) In addition to standing at attention when our national anthem is played out-of-doors, civilians and military in civilian clothing should put their right hand over their heart as a sign of allegiance. Legislation was passed in 2009, however, stating that veterans and military not in uniform, may render the military salute, if desired. Indoors, civilians

---

[1] A brigadier general gets an 11-gun salute, major general—13, lieutenant general—15, general—17, general of the Army—19, and the President of the United States—the ultimate 21-gun salute.

should place their right hand over their heart. Military in uniform without headdress simply stand at attention with hands at their sides.

Civilians do not put their hand over their heart for the national anthems of other nations, even though military members in uniform salute for all national anthems. A salute is rendered to show respect; a hand over the heart shows allegiance.

3. Some parade programs include an invocation, for which the audience is expected to stand. If an invocation follows the "National Anthem" (check the printed program), remain standing after the music ends.
4. As the flags pass in review, stand at attention when our American flag and other national flags are six paces before you, and remain standing until they are six paces past you. Thus, those in the audience will stand and sit at different times, producing a ripple effect. When indoors or outdoors, place your hand over your heart as the American flag passes.

At a parade in which troops from other nations participate, the foreign national flags may be carried in more than one location (e.g., both with the color guard and in front of the foreign troop units). In these cases, stand when each national flag passes, following the same rule of "six paces before until six paces after."
5. At the end of the parade, the band will always play the "Army Song," possibly preceded by a division or regimental song. The audience stands at attention for this music. Be prepared to sing along. Many times the words to the songs are included in the program. Following that, everyone remains standing until the commander of troops announces to the reviewing officer, "This concludes the ceremony."

CHANGES OF COMMAND/CHANGES OF RESPONSIBILITY

A change of command/change of responsibility is a very special occasion in the lives of the outgoing and incoming commanders and command sergeants major and their families. For the outgoing commander/command sergeant major and family, it is a bittersweet time—bitter because they have to leave; sweet because of the many experiences, friends, and memories the command has brought. The incoming commander/command sergeant major and family also have mixed emotions, probably best described as a mixture of pride and apprehension. The audience needs to appreciate the emotions of the occasion for both families, and help make it the memorable occasion it should be.

- Make the incoming commander/command sergeant major and his/her family feel welcome. This starts with their arrival on post, not with the departure of the outgoing commander/command sergeant major.
- Prior to the parade, there may be an award presentation to the outgoing spouse. The unit is responsible for ensuring all appropriate paperwork has been submitted in a timely manner so the award is ready at the time of the ceremony. This may be an end-of-tour award as well as a branch award.
- At the parade, both should be extended equal courtesies. If you bid goodbye to

the outgoing commander/command sergeant major and their family (and you should), then also take time to greet the incoming commander/command sergeant major and their family. If flowers are presented to the outgoing commander/command sergeant major's spouse, then the incoming commander/command sergeant major's spouse should also receive equally lovely flowers or perhaps a unit coin (if personally procured), unit crest hat or cigar for a male spouse. (No rule specifies which color or type of flowers each should receive, but tradition dictates red, fully bloomed roses for outgoing and yellow buds for incoming.) This is the traditional time for the outgoing spouse to present the incoming spouse with a unit pin/tie tack or lapel pin.

- If your soldier is the senior officer who passes the flag from the outgoing commander/command sergeant major to the incoming, urge him or her not to make a long speech. The parade is for the outgoing commander/command sergeant major; let them have center stage; let their speech be the highlight. As the senior commander, your soldier might have other opportunities to express appreciation and admiration for the outgoing commander/command sergeant major and spouse's accomplishments. (If all else fails, you might gently remind your soldier that, regardless of how great an orator he or she is, on this occasion, the audience didn't come to hear your soldier speak.)

- The outgoing commander's/command sergeant major's remarks can be long, but not too long; three minutes seem plenty to the soldiers on the field. In bad weather, the remarks should be even shorter. In a foreign country, very brief remarks in the host language or in the foreign languages of the troops on the field are appropriate. However, extensive remarks in a foreign language are not necessary, especially if the commander/command sergeant major has not mastered the proper pronunciation. Certainly, they should never translate their entire speech. (Expansive remarks about the loving and supportive spouse and family are best saved for their retirement parade.)

- The incoming commander's/command sergeant major's remarks should be the briefest of all. Opportunities for expressing their thoughts and philosophy of command will come later.

GUIDELINES FOR THE SPOUSE OF INCOMING AND OUTGOING COMMANDER, COMMAND SERGEANT MAJOR, AND SENIOR ENLISTED ADVISOR

- You and your soldier should prepare your personal guest list early—as soon as the date is established for the change of command/change of responsibility. Also, the incoming commander/command sergeant major should be asked to compile their portion of the guest list at that time.
- Ask to see the proposed seating chart early enough so that changes can still be made. Ensure that reserved seating for all family members of the incoming com-

mander/command sergeant major is provided beside the incoming commander/command sergeant major's spouse. See that the spouses of the senior troop commanders on the field, commander of troops, and unit's command sergeant major have special seats, with an unobstructed view for picture-taking.

- Discreetly inquire about flowers. If you are to receive them, make certain that the incoming-commander/command sergeant major's spouse does as well. Note: Since it is a courtesy to rise when someone is presenting something to you, be prepared to stand as the flowers are presented. Confer with the incoming spouse so they are prepared to stand as well. This also allows the audience to see you receive the flowers.

- You and your soldier set the tone of the welcome for the new commander/command sergeant major and his/her spouse. Remember how you felt when you first arrived, and treat them the way you were or would like to have been treated. Welcome them warmly and encourage others in the unit to do the same; sponsor them yourselves or ensure that they are provided with an energetic sponsor; include them in appropriate unit functions once they arrive; inquire about a welcome function to ensure it is being planned; ensure that all the arrangements for the reception after the change of command/change of responsibility are being made. If you set this example, the unit will follow, making the transition smooth and easy for everyone.

- The change of command/change of responsibility is an appropriate time for the outgoing commander/command sergeant major's spouse to present the incoming commander/command sergeant major's spouse with a unit pin.

- As soon as the change of command/change of responsibility parade is over and you hear "This concludes the ceremony," you and your children should immediately go stand beside your soldier in front of the reviewing stands. If you delay even one minute, the crowd will engulf your soldier and you separately, and the people in the audience will have to work their way to two separate locations to say their last farewells. It is very helpful if an escort has been assigned to move you quickly into place beside your spouse.

- On those occasions when the commander/command sergeant major's spouse cannot be present, an adult child or teenager may represent the spouse if the command approves.

- Although invited guests should not take their children to the reception unless they are invited, children of the incoming commander/command sergeant major are welcome and may stand in the receiving line.

- Incoming Spouses: Give yourself grace and time to adjust. Realize that the unit spouses are going through a transition that day also. If you arrive early, explore the environment, know your seating, and review the program/sequence of events for the ceremony. Be prepared to stand when receiving flowers or token of welcome. (Check with the outgoing spouse.) Leave immediately after the ceremony to get to your receiving line. Try to have someone at the reception assist with the set-up while you are at the ceremony. Enjoy the day and the momentous ceremony. Please refer to Chapters 30, 31, and 32 for additional information on advice for spouses of commanders, NCOs, and general officers.

RETIREMENTS

Retirement parades are especially moving ceremonies. They are a tribute to the military member's contributions and service to the Army and the United States of America. These retirement parades follow the same pattern as a typical parade described earlier in this chapter except that, after the playing of the "National Anthem," the retiree(s) moves to the center of the field for the retirement ceremony. If married, the spouse may join the retiree at this time or just prior to the presentation of awards and certificates.

The actual retirement at a retirement parade is comprised of the following parts: a brief review by a narrator of the highlights of the military career of the retiree(s), the presentation of any end-of-career award(s), the reading of the retirement order(s), and a few private words of appreciation from the senior officer conducting the retirement. If the retiree is married, the spouse is presented with a certificate of appreciation from the Secretary of the Army and may also receive an award for service provided to the nation and the Army. Following this, the retiree(s) returns to the reviewing stand, where the presiding officer makes brief remarks concerning the retiree's contributions and service and invites the retiree to make any remarks he/she would care to. (If more than one retiree is being honored, they may not be given this opportunity to make remarks.) During these final remarks, the retiree usually acknowledges those who have helped him or her along the way, with a special tribute to their spouse and family. It is always a moving and meaningful time for everyone. If the retiree is a general officer, following their remarks, their general officer flag may be furled and encased for the last time.

Retirement ceremonies, for both officers and noncommissioned officers, can also be held without the accompanying parade or review. These may take place in any appropriate location, outdoors or indoors. Regardless of where a retirement is held, any time a military member has served his/her nation long enough to be eligible for retirement, it is a very significant event.

Guidelines for Invited Guests - Respond promptly to your invitation, and plan to arrive a few minutes early so that you may be seated. Whether outdoors or indoors, there will be designated seating for family members and distinguished visitors, but other guests are free to sit in any undesignated seat. If this is a retirement parade, you will probably be given a printed program that lists the sequence of events and when to rise (similar to a typical parade). If it is an indoor retirement ceremony, the procedure may be very simple: Rise when the presiding officer enters and is announced. If the reading of the retirement orders is prefaced by the narrator saying, "Attention to orders," the military will stand, and, out of courtesy, civilians should as well. For all other portions of the ceremony, the audience remains seated.

Reception - If there is a reception following the retirement, be sure to take time during this event to congratulate the new retiree(s) and their spouse(s) and wish them well. When other family members are present, especially children or out-of-town relatives and guests, it's courteous to talk with them as well. When there is no reception, the retiree(s) and their spouse(s) usually form a receiving line in front of the ceremony location and those in the audience line up to walk by the retiree(s) and their spouse(s) to shake their hands and say a few words of congratulations. Retirement may be the end of a military career, but it is the beginning of a new chapter in life—it merits words of appreciation and encouragement.

* * * * * * * * * * * * *

Army tradition governs the movement of the troops on a parade field, as well as the audience's response—thus, everyone becomes a participant in the drama of this impressive military ceremony.

Chapter 27

Military Weddings

"How do I love thee? Let me count the ways."
Elizabeth Barrett Browning

A military wedding is a beautiful introduction for any bride to her new life as an Army wife. The scene includes the military members of the bridal party, handsomely dressed in their uniforms and providing a colorful contrast to the bride and her attendants. Memorable military traditions, such as the saber arch and caisson ride, may be added to accentuate the fact that the groom, and possibly the bride, is a member of the Army. What more could any bride ask for, in order to make her wedding the enchanting ceremony of which she has always dreamed?

Planning a military wedding isn't that different from planning any other wedding. In fact, because innumerable books have been written about civilian weddings, this chapter focuses primarily on the differences between military and civilian weddings, and special points of interest for guests who are invited to a military wedding. Anyone planning a military wedding should supplement the information provided here with that found in the books written on how to plan a civilian wedding.

The main difference between a military and civilian wedding is that, in a military wedding, the military members of the bridal party wear uniforms. In addition, one or more of the following military touches may be included:

- The wedding may be held in a chapel on an Army post.
- An arch of sabers may be formed for the couple after the wedding ceremony if the bride or groom is an officer.
- The bride or groom's unit may arrange for some special military transportation from the chapel to the reception, such as a horse-drawn caisson.
- The reception may be held at the community club on post.
- A military sword may be used by the couple to cut the first piece of wedding cake.

A military member who is married in a civilian church and has the wedding reception at a hotel or restaurant may still elect to have a wedding with military traditions. It is not the location of either the wedding or the reception that creates a military wedding; it is the military uniforms.

> **BLUF**
>
> This is how you start your Army Walk of Life!

UNIFORMS

The bride and groom determine the formality of their wedding by their attire. The other members of the wedding party dress accordingly. The wedding guests, who seldom know in advance how the bridal party will be dressed, tend to dress more formally later in the day.

The military wedding party's attire can be classified as informal or formal. Those terms imply the following:

Informal Wedding

- Bride wears short dress, or very simple long dress with short veil.
- Attendants and mothers of the couple wear short dresses.
- Military wear Army Service Uniform, with four-in-hand tie (long).

Formal Wedding

- Bride wears long dress with train and veil.
- Attendants and mothers of the couple wear long or tea-length dresses.
- Military wear Mess Uniform or Army Service Uniform, with bow tie.

If the bride is in the military, she may choose to wear civilian attire or her uniform. If she selects a uniform and the groom is also in the military, he wears the same type of uniform as the bride. Other military members of the wedding party dressed in uniform should wear the same type of uniform or, if in a different service, a uniform of comparable formality.

Female military members of the wedding party may wear uniforms or traditional civilian clothing, whichever the bride and groom prefer.

Ushers may be in different services and, thus, in different uniforms. Any of the bridal party not in the military should wear the appropriate civilian attire for the formality of the wedding. If there is to be an arch of sabers, the ushers may either wear the sabers while ushering or not, depending on the chaplain's preference—though most chaplains feel that arms should not be worn inside a chapel.

Military guests may choose to wear their uniforms (same type as wedding party) or civilian clothing. However, the bride and groom planning a military wedding frequently prefer for their military guests to wear their uniforms. If you are invited to a military wedding, you may want to ask their preference, as well as which uniform would be appropriate.

For the couple planning a formal wedding and wanting to let their invited guests know, they may want to add the words "Black Tie Optional" on the bottom of the invitation.

MILITARY CHAPEL

As with civilian churches, a military chapel must be reserved as soon as possible. (The West Point Chapel is sometimes reserved a year in advance.) The chaplain should also be consulted at that time. He may wish to arrange counseling in preparation for the marriage and discuss the many details to be arranged, just as with a civilian wedding. Couples preparing for a wedding in a military chapel should consider the following questions:

- Do they want their civilian clergyman to assist the chaplain? If so, he should also be consulted early.
- Is there a military organist assigned to the chapel, or is it necessary to arrange for a civilian one? The chaplain may be able to suggest someone. Do they want a soloist?
- Does the chapel furnish any decorations, flowers, or candles? Usually not. If not, are there vases or stands available for use? Which size (diameter) candles fit the candelabra?
- Do the bride and groom want to light a unity candle?
- When would the rehearsal be convenient for the chaplain, and when is the chapel available?
- If a saber arch is planned, does the chapel have sabers available or, if not, can they be borrowed from another chapel or the local ROTC unit? Does the chaplain prefer the arch be held indoors or outdoors?
- Which rooms are available for the bride and bridesmaids, and for the groom and best man before the wedding? Are they large enough for the bride to dress there, or should she do so before coming to the chapel?

There is no charge for the use of a military chapel or for the services of a military chaplain; however, a donation to the chapel fund is traditional. A check made out to the "Non-appropriated Chaplain's Fund," not to the chaplain, should be given to the chaplain or chaplain's assistant sometime prior to the wedding. No set amount is expected. As a guideline, civilian fees usually range from $100-150. If civilian clergy assists, they should receive a fee for their services, usually presented to them in a sealed envelope by the best man prior to the ceremony. The groom traditionally pays these expenses.

The organist, unless a servicemember assigned to the chapel, receives a fee for the ceremony, an additional sum for the rehearsal, and more if he or she rehearses with and accompanies a soloist. A soloist receives compensation as well. Neither the organist nor the soloist is really necessary for the rehearsal, but arrangements need to be made with them well in advance as to the music desired. These fees are paid prior to the wedding, traditionally by the bride's family.

A wedding coordinator is a reasonably new participant in the planning of a wedding. This is a person hired to assist the couple in planning the details of their wedding, such as: reserving the reception site, contracting for the flowers and wedding cake, and supervising all the last-minute details of the wedding. Many civilian couples, especially those who are planning very large weddings, hire a wedding coordinator to assist them with this myriad of tasks. While Army chapels do not keep a list of wedding coordinators to recommend, they certainly are prepared to work with anyone, whether a professional or simply a friend of the couple, who is working with the bride and groom as they plan for "their big day."

ARCH OF SABERS

An arch of sabers is not necessary for a military wedding, yet it does make a memorable ceremony all the more unforgettable. (Sabers are curved, one-edged swords.) If the bride and/or groom is a commissioned officer, and they contemplate this ceremony as a part of their wedding, the following guidelines may be helpful in their planning. (The duties of saber bearers and ushers are discussed later in this chapter.)

- On most Army posts, at least one chapel usually has sabers available for military wedding ceremonies. Another possible source is the local ROTC unit.

- Customarily, six to eight ushers in uniform create the arch. (It is not necessary to have an equal number of bridesmaids.) The ushers may be commissioned officers from different branches of service and, thus, in different uniforms. If that many ushers are not needed, other military guests may be asked in advance to assist the ushers in performing this service.

- The chaplain should be consulted on the location for the arch: inside the chapel, in the vestibule, or on the steps outside.

- If the chaplain prefers that sabers not be worn inside the chapel, they may be left in a side room until after the wedding ceremony.

- Only the bride and groom pass under the arch of sabers.

INVITATIONS

Traditional Wedding Invitations - Traditional wedding invitations are considered formal invitations; they are worded in the third person and each line is centered on the page. They may be engraved, thermographed, commercially printed, or handwritten in black ink on white or ivory stock. The traditional wording of the bride's parents inviting the guest to be present at the marriage of their daughter to her fiancée is considered standard. An invitation to a reception may be included in the wording of the wedding invitation when all guests are to be invited to both, or a separate reception card may be used. Many people now also include a response card to encourage guests to respond.

Following is an example of a basic formal wedding invitation and enclosures:

> *Mr. and Mrs. Theodore Martin Turner, junior*
>
> *request the honour of your presence*
>
> *at the marriage of their daughter*
>
> *Teresa Marie*
>
> *to*
>
> *Major Harold Hunter Christianson*
>
> *United States Army*
>
> *on Saturday, the eighth of June*
>
> *two thousand and nineteen*
>
> *at three o'clock*
>
> *Main Post Chapel*
>
> *100 Washington Road, Fort Carson, Colorado*

Reception
immediately following ceremony
The Community Club
150 Constitution Drive
Fort Carson

The favour of your reply is requested
on or before May 20, 2019
Name _____
Number of persons accepting ___
Number of persons regretting ___

Wording Guidelines

- When the bride and/or groom has the rank of captain or above, the rank appears before the full name, with the branch of service on the line beneath.
Captain John Joseph Jones
United States Army
- For lieutenants, their full name appears on a line by itself, with rank and branch of service on the line beneath.
John Joseph Jones
Lieutenant, United States Army
- Both first and second lieutenants of the Army are designated simply as Lieutenant, but those in the Air Force or Marines use their complete title.
- For noncommissioned officers and enlisted military members, rank is usually omitted. Their full name is written on one line, with the branch of service underneath.
John Joseph Jones
United States Army
- A father-of-the-bride or mother-of-the-bride who is a retired officer uses their military rank. They do not note on the invitation that they are retired, unless they are a widow/er and issuing the invitations in their name alone.
- If either the groom or father-of-the-bride issuing the invitations in his name only is retired military, that status is noted on the invitations in this manner:
Lieutenant Colonel John Joseph Jones
United States Army, Retired

Non-Traditional Wedding Invitations - Many couples choose to vary some of the traditional wedding customs, and invitations are no exception. Certainly, these non-traditional wedding invitations are just as appropriate for military weddings as for civilian weddings.

Non-traditional wedding invitations are less-formally worded, and often printed in colored ink on pastel paper with an embossed design. They may even reflect the colors chosen for the wedding. A lovely example of non-traditional wording is, "Please share with us our joy, and celebrate the marriage of our daughter...." Another recent innovation is for wedding invitations to list the names of both the bride's and groom's parents, appropriate if they are sharing the cost of the wedding. (This would have been regarded as the worst of manners only a few decades ago, but is now more accepted as wedding expenses escalate). The current trend toward later marriages is often reflected in invitations that are worded to indicate the bride and groom are sponsoring their own wedding. In other words, the bride and groom invite you to their wedding. On wedding invitations for same-gender couples, they will need to decide which name goes first, perhaps alphabetical.

Invitations to a small wedding may be handwritten by the bride and/or groom in the form of an informal note. An example of such a non-traditional wedding invitation follows. Whether traditional or non-traditional, the only important considerations when deciding on wedding invitations should be that they are in good taste and in a style that pleases the bride and groom.

Dear Jerry,

Sharon and I will be married at five o'clock on Saturday, the ninth of November, at Chapel One. Sharon's parents will host a buffet dinner at the Community Club following the ceremony.

We would love to have you there for both events. Please let us know if you will be able to come.

Sincerely,
John

Responses - Responding to a wedding invitation is just as important, if not more so, than responding to any other type of invitation. In fact, the only time a reply to a wedding invitation is unnecessary is if you are invited to the ceremony only, and not to the reception afterward. When the invitation is written in the formal, third-person format, the response is worded in the same manner. A reply to a non-traditional wedding invitation not worded in the third-person format may be written in the form of a short, personal note. Following are examples of a handwritten acceptance and regret to a formal wedding invitation: (Note that the acceptance mentions the time, and the regret does not.)

> Captain and Mrs. Ross William Robins
> accept with pleasure
> Colonel and Mrs. Crowell's
> kind invitation for
> Saturday, the seventh of June
> at four o'clock

> Captain and Mrs. Ross William Robins
> regret that they are unable to accept
> Colonel and Mrs. Crowell's
> kind invitation for
> Saturday, the seventh of June.

The practice of including response cards with wedding invitations to encourage and assist guests in their responsibility to respond has developed because so many people are negligent about responding to any type of invitation. While response cards are certainly not required, most people have grown to believe that they are an essential part of the wedding invitation—and they do make it more likely that guests will respond. If response cards are used, most wedding guidebooks recommend that their return envelopes be self-addressed and stamped.

With or without response cards, wedding and reception invitations should be responded to as promptly as possible. Knowing how many guests to plan for is extremely important for those planning the wedding reception, especially when dinner is being arranged.

ADDRESSING THE ENVELOPES

Wedding invitations are traditionally enclosed in two envelopes, with the inner envelope left unsealed. This custom probably dates back to the time when invitations were hand-delivered, as correspondence that is hand delivered is properly not sealed. Later in our history, with the advent of mail service, the unsealed envelope was placed in another for mailing. Today, we continue this practice because the inner envelope offers a place to write the names of all invited family members, escorts, and children. However, in the interest of conservation, it is perfectly acceptable to use only one envelope.

Names and addresses written on wedding invitation envelopes should be done in black ink. No abbreviations are used in the name, rank, or street address. The outer envelope is addressed using the guest's full name. (Omit the middle name if space is a problem.) The inner envelope, which remains unsealed, is addressed with the guest's rank or title and last name only; first and middle names are not used here. Examples are: outer envelope—Captain and Mrs. John Joseph Jones; inner envelope—Captain and Mrs. Jones. When the guest is retired military, this fact does not need to be noted. When the guest is a single friend who is being invited to bring a date, the inner envelope might say, "Miss Jones and Guest." Close relatives may have their more familiar names written on the inner envelope, such a "Grandmother" or "Aunt Margaret and Uncle Hugh." Anyone over eighteen should receive his or her own personal invitation. If you wish to include young children in the invitation, their names should be added on the inner envelope only, under their parents' names, as in:

Captain and Mrs. Jones
Stephanie and Steven

What goes into the inner envelope along with the wedding invitation varies, but typically the following might be included: a sheet of tissue, an invitation to the reception, a response card and return envelope, and perhaps a map (inserted in that order). The small sheet of tissue paper to cover the face of the invitation, once used to keep the engraving from being smudged, is no longer needed but still used by many. The inner envelope is inserted into the outer one so that the addressee's name faces the back flap of the outer envelope.

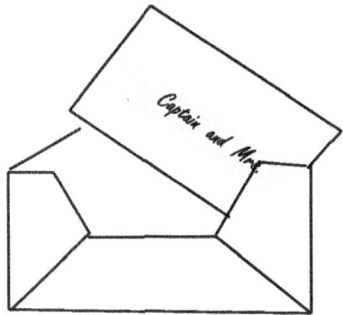

Special Invitations

- The bride's mother should send invitations to the groom's parents and all the bridal party, as keepsakes. No response card is included.
- It is also courteous to send an invitation to the chaplain or person performing the ceremony and his or her spouse.
- When a military member gets married, it is a courtesy to send an invitation to the commander and spouse. If the groom's (or bride's) parents cannot attend the wedding, the commander and spouse may be invited to sit in their place. Otherwise, the commander and any very senior officers may be escorted to seats beside or directly behind the family in the first few pews.

Announcements

Wedding announcements may be used in addition to wedding invitations. These announcements may be sent to friends and relatives who live too far away to attend the wedding (often the case with our military friends), and to local friends and relatives when the wedding and reception list is too small to accommodate everyone. Announcements carry no obligation for a gift; however, courtesy requires at least a warm note of congratulations. Usually, wedding announcements are mailed the day after the wedding, certainly never before. The announcement resembles the wedding invitation in every way except wording. Typically, the names of both the bride's and groom's parents are given, as shown in the following illustration, since this is not an invitation issued by those hosting the wedding.

> Mr. and Mrs. Harold Andrew Carter
> and
> Colonel and Mrs. Bryan Thomas Burns
> announce the marriage of
> Priscilla Anne Carter
> and
> Harold Andrew Carter, Jr.
> Lieutenant, United States Army
> Saturday, the fourteenth of December
> Two thousand and nineteen
> First Baptist Church
> Clarksville, Tennessee

GIFTS

Gifts are usually sent to the bride at her home before the wedding. However, they may be taken to the wedding reception where a table for them is usually provided. Gifts also may be sent after the wedding ceremony. There is no real time limit after which wedding gifts are inappropriate. However, if circumstances have prevented you from sending a gift for a good part of a year, you might consider making your gift an anniversary present.

CHILDREN

Children and babies should not be taken to a wedding or reception unless their names are included on the invitation. Some couples even state on their invitation "Adults only affair." Children who are invited and attend should be well behaved. Even at the reception, children should not be allowed to run about freely, nor should they be allowed to join in with the young single adults when they gather and attempt to catch the bride's garter or bouquet. Those brides and grooms who want to invite their friends to bring their babies and small children may hire someone to provide nursery service in a separate room at the church, and note this information on a card in the invitation envelope.

REHEARSAL AND DINNER

- The wedding rehearsal is usually held the evening before the wedding. All members of the bridal party and the bride's parents attend; the groom's parents may attend if they care to, but it isn't necessary.

- At the rehearsal, all of the movements of the procession are practiced, including walking down the aisle, which is done at a slow, natural pace—not a hesitation step.

- If the couple plans to light a unity candle, this procedure is rehearsed for the benefit of the bride's attendant who will help her with her dress as she moves to light the woman's candle.

- After the rehearsal, a dinner is traditionally hosted by the groom's parents. If they do not do so, the bride's family or a close friend may.

- All members of the bridal party, their spouses or fiancé(e)s, close family members and any other close friends the couple desires are invited to the rehearsal dinner. The chaplain and his wife may be included—definitely if the couple knows them well or they are good friends of the family.

WEDDING PHOTOGRAPHS/VIDEOS

Taking photographs has long been a cherished part of wedding activities. This practice is now frequently supplemented by videos. Whether pictures are taken by a professional photographer/filmmaker or an amateur-photographer friend, the bridal couple should review their personal preferences in advance with the photographer. There are no specific rules for how or when members of the wedding party and their families are photographed; the wishes of the couple and their families should serve as the photographer's guidelines. However, the chaplain should be consulted for any requests or suggestions. Certainly, everyone concerned will want the photographer and/or video operator to remain as unobtrusive as possible.

USHER AND SABER BEARER DUTIES

- An usher's main duty is to seat the wedding guests. He usually asks if the guest is a friend of the bride (to be seated on left facing the altar) or a friend of the groom (to be seated on right). He then offers his right arm to the lady and escorts her down an aisle. If a couple is together, the gentleman follows. If several ladies are together, the usher escorts the senior lady and the others follow. If one side of the chapel fills more rapidly than the other, the ushers may ask late-arriving guests to sit on the other side. Guests should never be seated during the solo.

- Since early arrivals should have the best seats, this can be accomplished by having the ushers escort the guests down the outside aisles (reserving the first pews for family and any very senior officers or commanders). Thus, the guests enter the pews from the outside aisle and proceed to as close to the center aisle as possible, giving them the best view. Seating from the outside aisles also precludes later guests having to squeeze past those who are already seated.

- The head usher escorts the bride's mother to the first pew on the left just before the wedding begins. She is the last person seated.

- At the conclusion of the ceremony, as the bride and groom (followed by the bridal party) leave the sanctuary, the last two ushers escort out the bride's mother (the bride's father follows) and, then the groom's mother (the groom's father follows). The chaplain will follow, and then the guests.

- The specifics of how, when, and where sword bearers form up depend on the plans arranged in advance between the couple and the chaplain. Wherever formed, they do so in pairs, drawing their sabers on command and holding them together over the aisle for the newlyweds to walk under. Traditionally, Army and Air Force sword bearers rotate their sabers so that the cutting edge is up, thus forming a more-true arch.

- If an arch of sabers is planned for the vestibule or outside, the ushers direct the guests where to stand. If the arch is to form on the outdoor steps, the bridal party usually stands on both sides of the door, with the guests along the steps and sidewalk. The best man should notify the bride and groom, who have gone to a side room, when the arch of sabers ceremony is ready. If sabers aren't available or can't be borrowed, the ushers may still form a double line for the couple to pass through.

- It is traditional, as the couple proceeds through the arch of sabers and approaches the last two saber bearers who lower their sabers in front of the couple, that they will be detained there until they kiss. The sabers are then raised and the saber bearer on the right, with the flat side of his saber, gives the bride a gentle "swat" on the rump and utters, "Welcome to the Army, Mrs. ___." This step is omitted if the bride is in the military.

- Cutting the cake: The senior saber bearer presents the groom with his saber to cut the first piece of the wedding cake. The groom then hands the saber to his bride and with his hand over hers, their first piece is cut.

GUESTS' PARTICIPATION

Guests should respect the religious practices of the congregation—standing, sitting, or kneeling with the audience. If you prefer not to kneel, at least bow your head while the others are kneeling. It is not necessary for a non-Catholic to make the sign of the cross at a Catholic ceremony; however, men of all faiths are expected to wear a yarmulke at most Jewish ceremonies. Skull caps are usually available in the vestibule for those who do not have their own.

There are two times when the guests at a wedding (other than Jewish) stand:
1. At the beginning of the ceremony, when the bride and her father (or her escort) start down the center aisle. Usually on the first notes of the "Wedding March,"

the bride's mother stands and turns to face in the direction of the bride, and everyone else does the same.
2. At the conclusion of the ceremony, when the couple kisses and turns toward the congregation.

RECEPTION

There are many types of wedding receptions. All serve the enjoyable purpose of allowing the guests to help the bride and groom celebrate their marriage. Food is always served, though it may vary from the simplest of finger foods served buffet-style to an elaborate sit-down dinner. Often music is provided, though it too may vary according to the formality of the reception.

Guests frequently wait outside the chapel to see the bride and groom depart, before going to where the reception will take place. However, if wedding photographs have not been taken before the ceremony, the bridal party may be delayed in the chapel. In that case, the wedding guests should proceed to the reception. It's nice, under these circumstances, if arrangements have been made with the club or caterer to open the bar and make available a portion of the food. A friend or family member should be asked to greet guests and invite them into the reception area. When the bridal party arrives, they can form the receiving line, if they choose to have one, possibly near the side of the room where the wedding cake is located, instead of near the door. The following points of etiquette concerning wedding receptions are important for everyone to know:

- A receiving line is not considered necessary at small receptions. Guests will have the opportunity to speak with the bridal couple and other members of the bridal party as they move around the reception area.

- If there is a receiving line, the bride and groom always stand together, with the bride on the groom's right. The parents of the couple, the maid of honor, and the bridesmaids may or may not be a part of the line, but never the best man (who stays available to assist the couple in any way) or the ushers (who should circulate and serve as unofficial hosts).

- Guests should congratulate the groom, but not the bride. The bride is wished happiness and complimented on her beauty. (To congratulate the bride implies congratulations on "catching" a husband.)

- The first toast is always offered by the best man to the bride and groom. Other appropriate toasts are to the bride and the bride's mother, offered by the groom. The bridesmaids may be toasted as a group. Those being toasted remain seated during the toast and do not participate, other than to smile and nod in acknowledgment of the honor.

- When it's time to cut the wedding cake, if the groom has a sword, he unsheathes it and hands it to the bride. She holds the sword, the groom places his hand over hers, and they cut the first piece together. If no sword is available, a silver cake knife may be used, possibly decorated at the handle with streamers of ribbons.

- Guests do not have to wait until the newlyweds depart before leaving the reception. However, they should wait until the wedding cake has been cut, and they have said goodbye to the bride's mother and father.

- The tradition of throwing rice on the departing couple has given way to throwing birdseed (out-of-doors, of course). It's best to check with the church, chapel, or venue on their allowed practices.

* * * * * * * * * * * *

A military wedding is a beautiful beginning for an Army spouse.

Chapter 28

Military Funerals

"Day is done, gone the sun;
From the lake, from the hill, from the sky.
All is well. Safely rest. God is nigh."
Taps "Day is Done"

The most solemn and moving military tradition one can observe is the military's laying to rest one of its own. Whether the deceased soldier was on active duty or retired, the military pays its last respects in gratitude for the service and sacrifice rendered to the nation.

For the spouse of the deceased, the military will provide a casualty assistance officer or NCO to assist with arranging the funeral and attending to the other important details that require immediate action. When the military member was on active duty at the time of death, the assistance officer or NCO will be automatically appointed by the unit. When a retiree dies, the Casualty Assistance Command at the nearest military facility must be notified for a military funeral to be arranged. If there is any question about whom to contact, call collect to the DA Casualty Operations Center for guidance.

Since the spouse of the deceased will be given a full explanation of the available options for a military funeral, this discussion is not intended to be complete. Rather, it is provided primarily for the friends of the family and first-time observers of a military funeral. It will help them understand the significance of the military traditions involved, and prepare them for the unexpectedly sharp report of the three-rifle volleys and the heart-rending playing of "Taps" that come at the end of the burial service. This chapter also includes a few points of protocol for the bereaved family and suggestions for those wanting to help.

BLUF

Honor and Remember Always

HONORS

Military funerals can be divided into three categories: full honors, simple honors, and simple burial. All military members are entitled to simple honors. These include a military chaplain to conduct the service (if requested by the next of kin), a U.S. flag to cover the casket, rifle volleys, and the playing of "Taps." Military members who served as sergeant major and all officers are authorized full honors.

Full honors include all of the simple honors just described, plus the use of a caisson (or similar vehicle), active pallbearers assigned to this duty by the military, an escort commander, a color guard, troops, and a band. The size of the troop detachment and band will vary with the rank. When the deceased was a colonel or general officer, a riderless horse with boots placed backward in the stirrups may be added to the procession. General officers may also have their personal star flag carried in the procession, and cannons may fire a farewell salute after the benediction.

Regardless of the honors authorized, the family always has the right to choose a less-involved service—even a simple burial service conducted at the graveside by a military chaplain, with none of the other honors. The non-availability of troops, band, caisson, or horse at the desired time may also cause the family to accept a military funeral with fewer honors than are authorized.

Irrespective of the honors rendered, military funerals can begin in different locations, and the details change slightly with each location. If there is to be a chapel or civilian church service, the mourners meet there. After the service, everyone walks or drives to the grave site. If no chapel service is planned, the funeral procession can form at the entrance to the cemetery, or simply assemble at the graveside.

CHAPEL SERVICE

The funeral escort for a full-honors funeral is formed in front of the chapel or civilian church prior to the service. This escort consists of troops, band (as available), colors, escort commander, chaplain, active pallbearers, and possibly honorary pallbearers. Just before the service is to begin, the hearse or caisson bearing the casket arrives at the chapel. As mourners, including the immediate family, arrive, they should walk around the troop formation, through the aisle of honorary pallbearers leading to the chapel, and proceed inside to be seated before the service begins. The immediate family sits in the first pews on the right (facing the altar). The two front pews on the left should be reserved for the honorary pallbearers (if any), who will follow the casket into the chapel. When there are no honorary pallbearers and the active pallbearers are friends of the family (rather than service-members assigned to this duty), they may occupy these front left pews. Otherwise, seats are reserved for the active pallbearers at the rear of the chapel. The band (which remains outside with the troops, colors, and escort commander) will stop playing when the casket enters the chapel.

After the service, the chaplain leaves the chapel first, followed by the honorary pallbearers, and the active pallbearers with the casket. They are followed by the immediate family members, who wait at the entrance of the chapel until the casket is secured on the caisson (or other vehicle). This may mean that the majority of the mourners have to remain inside the chapel during this time. Once the honorary pallbearers have broken ranks, the members of the immediate family, followed by everyone else, move to their places in the procession.

FUNERAL PROCESSION

A funeral procession accompanies the casket to the grave site. The front part of the procession consists of the escort commander, band, troops, colors, and chaplain (in that order), who march in front of the honorary pallbearers and hearse (or caisson). They are followed by the active pallbearers, the personal flag and the riderless horse (for authorized officers), and then the immediate family. Other mourners follow. The band plays solemn music, and the procession marches very slowly until it has left the chapel area. The band then stops playing and the marching speed increases.

As the procession nears the grave, the hearse (or caisson) halts and the military escorts and chaplain take their positions. The honorary pallbearers form an aisle toward the grave, just as they had formed an aisle into the chapel prior to and following the chapel service. The band plays a hymn, preceded by "honors" when the deceased was a general officer. The active pallbearers remove the casket from the hearse (or caisson) and follow the chaplain to the grave. As soon as the casket passes between them, the honorary pallbearers turn and follow. Having waited for this transfer of the casket to take place, the family and mourners now follow and are seated.

WITHOUT CHAPEL SERVICE

When there is to be a funeral procession but no chapel service, everyone meets near the entrance to the cemetery. The casket is brought to the cemetery by hearse. The hearse may join the procession, or the casket may be transferred to a caisson (or similar vehicle). If the casket is to be placed on a caisson, the family and mourners remain in their cars while the casket is being moved, and then they join the procession.

GRAVESIDE

If there is no funeral procession planned, the mourners go directly to the grave site. The military escort will probably be in place when the mourners arrive, and the casket will be in place at the graveside. If mourners do arrive while the casket is being moved, they should wait in their cars during that time. At the grave site, the chaplain may or may not remove his hat; if he does, all others in uniform (except ceremonial troops) should remove theirs as well.

Once the active pallbearers have placed the casket over the grave and the graveside service begins, they raise the American flag from the casket to a position waist high, taut, and level. The flag is held in this position until "Taps" is played at the end of the funeral.

After the chaplain conducts the graveside service, he renders the benediction. A cannon salute may be fired if the deceased was a general officer. The firing party then fires three volleys of blank cartridges over the grave. Immediately after the final volley, the bugler plays "Taps." Several versions of words have been written for this last bugle call of a soldier's day. Following are what are thought to be the original words, and a more modern version:

Fades the light,	Day is done
And afar	Gone the sun
Goeth day	From the earth
Cometh night;	From the sea
And a star	From the sky;
Leadeth all,	Rest in Peace,
Speedeth all	Soldier brave,
To their rest.	Rest in Peace.

At the conclusion of "Taps," the active pallbearers fold the flag in the prescribed triangular shape (representing a cocked hat of the American Revolution), and the chaplain or the escort commander presents it to the spouse or next of kin. These words, or something similar, are usually expressed by the one making the presentation:

"This flag is presented on behalf of a grateful nation as a token of our appreciation for the honorable and faithful service rendered by your loved one."

The funeral escort then forms up and marches off.

CREMATION

There is little difference between a military funeral in which the deceased's body has been placed in a casket and one in which the deceased's cremated remains have been placed in an urn. The primary difference is in the number of active pallbearers assigned and in their duties. When there is no casket to carry, fewer active pallbearers are required. One man is designated to carry the urn and four more will follow to carry and handle the flag. The funeral may be with full or simple honors, with or without a chapel service or funeral procession.

SYMBOLISM

- The American flag covers the casket to symbolize the service of the deceased in the Armed Forces of the United States. It is placed on the casket so that the field of blue covers the left shoulder of the deceased. This custom probably originated on the battlefield where caskets weren't available; a dead soldier might, instead, have been wrapped and buried in a flag which served as a makeshift pall. (A pall is the cloth covering for a casket and is usually black; in military funerals, the U.S. flag serves as the pall.)
- The use of a caisson in place of a hearse also dates back to early battlefields

when that was the most expedient means of moving a casket. Today, if a caisson is not available, some other modified military vehicle may be used to represent it.
- The three volleys of rifles originated in ancient symbolism to scare away evil spirits. Today they are thought of as the final tribute to the military member.
- "Taps," the last bugle call a soldier hears at night, concludes the military funeral and symbolizes the beginning of their last, long sleep. Important, but unexpressed, is confidence in the ultimate reveille to come.
- The boots placed backward in the stirrups of the riderless horse, often referred to as the "caparisoned" horse (meaning ornamentally covered), are a symbol of a fallen leader. Army officers who were colonels and above are authorized this honor.
- When the deceased was a military pilot, aviation participation might be provided as a part of the military funeral. In that case, a tactical formation, less one aircraft to symbolize the missing pilot, flies over while the casket is being moved to the graveside.
- At Arlington National Cemetery, every military funeral has an "Arlington Lady" present to represent the Chief of Staff of the Army. She is there to ensure that no soldier is buried at Arlington without at least one mourner to bid him or her farewell. As the representative of the Chief of Staff, she presents a note of condolence to the spouse or next of kin. Arlington Ladies are military wives who live in the Washington, D.C. area and volunteer their time to perform this very caring and meaningful service to our deceased military.
- It is important to note that, except for the few honors reserved for very senior soldiers, a military funeral for a private does not differ much from that of a general officer. That says a great deal about the respect our nation has for all of its soldiers, regardless of rank.

FUNERAL COURTESIES
- At a military funeral, all mourners in military uniform should face the casket and salute during the following times (listed in order of occurrence):
  - honors (played for a general officer)
  - when the casket is being moved
  - cannon salute (fired for a general officer)
  - three volleys
  - "Taps"
  - when the casket is lowered into the grave (if the mourners are still present)

During these times when the military in uniform are saluting, civilians and military members in civilian dress should stand at attention. Civilian men wearing hats should remove and hold them over their hearts.

- Prompt messages of condolence are very important to the family. A brief note, a telegram if you are far away, or a phone call if you feel close to the family is appropriate and greatly appreciated. In each case, brevity and simplicity (as well as speed) are called for—write from the heart, express your sympathy to the bereaved, make a kind comment about the deceased, and possibly offer to be of assistance. If you send a note, write it by hand, preferably on plain paper using dark ink. A commercial sympathy card does not have the same touch of personal warmth or sincerity as a handwritten note, but will suffice if a personal note is added.

- Send flowers for a funeral to the chapel, civilian church, or funeral home where the service will take place. For a graveside service, send the flowers to the funeral home. You may also send a small floral arrangement to the home of the bereaved. Enclose your personal or joint calling card, or sign a plain card provided by the florist. Sign your full name, unless you are extremely close to the bereaved and are certain that no mistake or doubt could arise from a shorter signature.

- If the family requests no flowers, you may send a donation to the charity they recommend. You may mention your donation in a note to the bereaved, without giving the amount, or you may prefer not to mention it at all. The charity should notify the family of your contribution if they receive sufficient information to do so.

- The family of the deceased should write brief notes acknowledging all messages of condolence, floral or charity contributions, gifts of food to the family, and any special assistance. A sentence or two is sufficient, written on plain white or dark-bordered paper. If the family chooses to use the pre-printed cards sometimes provided by the funeral home, they should add a brief, handwritten, personal note.

- Although the services of a military chaplain are free, the family should send him or her a brief note of thanks after the service. If a civilian clergyman is involved in the funeral, it is customary for the family to give this individual honorarium, either handed to them before the service or included in a letter of appreciation later. However, it may be a part of the bill from the funeral home.

MEMORIAL SERVICE

Often a memorial service for a deceased soldier will be held at the chapel on the post of his or her last assignment, or a memorial service may be held for a prominent senior military commander. Such a service, separate from the funeral and internment (which may be held in a distant city or national cemetery) is planned and conducted by the unit and provides colleagues and friends an opportunity to pay their last respects. Possibly the immediate family will not be present; however, this does not preclude the community from gathering to honor the departed servicemember. The memorial service is similar to a chapel funeral service, but without the presence of the casket, band, or troops. Occasionally, the combat boots and helmet of the deceased will be placed on a table in front of the altar to symbolize the soldier's presence. The "Battlefield Cross" is made up of the soldier's pair of boots, their rifle with dog tags hanging and their ballistic helmet placed atop the rifle. Generally, a few hymns are sung and a memorial eulogy presented. If the family, or a family representative, is present, you may take this opportunity to offer condolences.

When the deceased was a military pilot, a flyover may occur at the conclusion of the memorial service. In that case, after the family has been escorted outside, guests leave the chapel and remain outside for the flyover.

HOW CAN YOU HELP?

If you feel close to the family or your spouse worked with the deceased, you may want to help in a more personal way than just sending a note of condolence and, possibly, flowers. Everyone is apprehensive in such a situation because it's hard to know how to help without being a burden. Consider these suggestions:

- Help relieve the family of the burden of preparing meals. Take food that is ready to eat or that only requires warming. Try to use disposable containers, so that no one will have to keep track of which dishes belong to whom. Organize other friends and neighbors to do the same, supervising what's needed, so that complete meals are provided. Coordination can avoid overwhelming the family with food, plan for dietary considerations/needs, and take into account the tastes of any children in the home. If you are helping at the home of the bereaved, keep a record of who brings what and leave it on the kitchen counter for future updates, so that the family can later send their notes of thanks.

- Don't just offer your assistance; look for the areas where you can help, then step in and quietly do it. Friends can assist with the burdens of routine living, such as—keep the living area and kitchen picked up; shop for essentials like milk, bread, and toilet paper; wash and iron clothes; babysit or entertain the children; drive a carpool if older children need transportation to and from lessons or meetings.

- While at the home, answer the telephone and front door, and keep a list of calls and callers. If the bereaved wants to answer the phone or see visitors, that's fine, but that may not always be the case. In any event, be prepared to answer calls if the bereaved is resting, or during particularly busy times.

- Continue to support the family as the months pass. Sending flowers weeks after the funeral, with a simple "Thinking of you and your family" note, is one way to express your continuing concern. Keeping in touch, helping where needed, and staying as involved with the bereaved as you were before—these spell out your support in deeds, rather than just words. As the anniversary of a death approaches, remember that this date is a special time for concern and caring; a note, card, or possibly a visit at this time will be most appreciated.

WHAT TO SAY AND NOT SAY

What to say to the bereaved is often our biggest concern. No one wants to worsen the situation by saying the wrong thing, and yet even those who have experienced a close death in their own families don't necessarily know how to comfort others. At such times, it's best to turn to the experts.

These professionals describe the initial reactions to grief as numbness, shock, denial, and isolation. During this time, the greatest solace can come from a trusted friend, not a lot of friends—that comes later. The bereaved may appear disorganized and confused. Their emotions may become volatile as they progress through feelings of anger and guilt. They may do or say things totally out of character. On the downswing, they may express, or at least be experiencing, feelings of utter helplessness or fear that they are losing their mind. Depression is accompanied by feelings of tiredness and a general lack of interest, often accompanied by crying. At this stage, patience is needed, and the opportunity for the bereaved to talk, talk, talk. You don't really need to say anything.

Deaths involving active-duty military often occur suddenly, with no preceding illness. The survivors, having had no time to prepare, usually react in one of two ways—either stoic silence or hysteria. Those in the first category need to know that it's all right to express what they're feeling. Those in the second category need constant, calm companionship—and not be told to "Get hold of yourself!" or "Try to calm down." Give them time to let their feelings out.

Clichés, such as "I know how you feel" or "It's the will of God," are not helpful. If the bereaved expresses strong feelings—anger and resentment toward the deceased and others, even God—never respond with "You don't really mean that." Instead, bolster these mixed emotions by assuring them that, "It's natural to feel the way you do." What the bereaved need most is a trusted friend to talk to. If that someone is you, be there when you're needed. You don't have to do or say much of anything; just let them talk.

Later in the grieving process, during the depression phase, good friends need to stay close, offer specific help, and be persistent. Offers of general help—"Let me know if there's anything I can do"—are meaningless, and even specific offers may be repeatedly turned down. But specific help is exactly what the experts would prescribe. When the bereaved individual eventually begins to exhibit feelings of acceptance and re-establishment, friends need to be there to encourage him or her to step out into life again.

Never hesitate to mention the name of the deceased to those who are still mourning for fear of causing more grief. This is a common reaction for friends and acquaintances; however, this misplaced concern has quite the opposite effect. When we refrain from mentioning the name of the one who has died, the mourners feel as though we aren't truly sharing their grief. Whereas, when we use the name of the deceased in our conversations or notes of sympathy, express our own feelings about the person who has died, and relate our fond memories or humorous incidents involving the loved one, these are real words of comfort to the bereaved.

HELPFUL BOOKS

For those who feel close to the bereaved and want to give them something other than flowers, consider one or more of the excellent books available to help them through their time of grief. Your post library or local bookstore will, no doubt, have some of these available. (The first two descriptions are extracted from a handout at an AWAG Conference by Chaplain (Col) Hiram "Doc" Jones, "Bibliography on Loss and Grief.")

Good Grief by Granger Westberg, and *Living When a Loved One Has Died* by Earl Grollman

These two books, one from the Jewish tradition [Grollman], the other from the Christian tradition [Westberg], are the resources to put into the hands of someone who has experienced a loss. The thing most feared by those who grieve is that they are losing their mind. The most helpful insight gained from these resources is that what they are experiencing is normal.

When Dinosaurs Die: A Guide to Understanding Death by Laurie Krasny Brown and Marc Brown.

This book is a comprehensive, sensitive guide for grieving children and families dealing with the loss of loved ones. *When Dinosaurs Die* helps readers understand what death means and how best to cope with their feelings.

Weird is Normal When Teenagers Grieve by Jenny Lee Wheeler.

Are you a teenager dealing with grief? You are not alone. Jenny shares her personal grief journey and reassures you that there are challenges of how to grieve in an adult world filled with "shoulds" and unrealistic expectations.

Surviving the Death of a Sibling: Living Through Grief When an Adult Brother or Sister Dies by T.J. Wray

Wray is a captivating storyteller who weaves stories of herself and many other sibling grievers to bring clarity and understanding to the complex process of sibling grief. Insightful, consoling, and filled with helpful, proactive steps designed to help surviving siblings cope with their devastating loss.

The Grief Recovery Handbook: A Step-by-Step Program for Moving Beyond Loss by John W. James and Frank Cherry

This is an excellent book that explains the different feelings encountered in everyday life after a loss. The authors help the readers to look at themselves honestly, evaluate where they are in the recovery process, and show how to move to the next step. They also deal with the "dumb" questions people ask, and how not to get angry with them.

The Courage to Grieve by Judy Tatelbaum
   This is a well-written source for creative living, recovery, and growth through the grief process. In it, Tatelbaum describes unsuccessful grief and offers guidelines for coping.

Getting to the Other Side of Grief: Overcoming the Loss of a Spouse by Susan K. Zonnebelt-Smeenge, R.N., Ed.D. and Robert C. De Vries, D.Min., Ph.D.
   Whether you're feeling alone, drowning under an ocean of emotions, or you've worked your way through the darkest nights of the soul and are now wondering how to get on with your life, you'll find comfort and guidance from Susan Zonnebelt-Smeenge, a clinical psychologist and Robert De Vries, a pastor and professor. Their empathy, valuable psychological insights, biblical observations, and male and female perspectives will help you experience grief in the healthiest, most complete way so you can move forward to embrace the new life that is waiting for you.

How To Go On Living When Someone You Love Dies by Therese A. Rando, Ph.D.
   Mourning the death of a loved one is a process all of us will go through at one time or another. But whether the death is sudden or anticipated, few of us are prepared for it or for the grief it brings. There is no right or wrong way to grieve; each person's response to loss will be different. This book will lead you gently through the painful but necessary process of grieving and help you find the best way for yourself.

When Bad Things Happen to Good People by Harold S. Kushner
   Rabbi Kushner's book is a moving personal account of the death of his son and is written for individuals of all faiths.

How to Survive the Loss of a Love by Melba Colgrove, Harold H. Bloomfield, and Peter McWilliams
   The first edition of this book sold nearly 2,000,000 copies. It's a must-have bedside companion for anyone who has suffered a loss of any kind—not just death. It has a companion workbook entitled *Surviving, Healing & Growing*.

No Time to Say Goodbye: Surviving the Suicide of a Loved One by Carla Fine.
   With this book, Carla Fine brings suicide survival from the darkness into the light, speaking frankly and with compassion about the overwhelming feeling of confusion, guilt, shame, anger, and loneliness that are shared by all survivors.

Drawing on her own experience and the experiences of the many other survivors with whom she has spoken, as well as on the knowledge of counselors and mental health professionals, she offers a strong helping hand and invaluable guidance through the various stages of the survival process.

None of us likes to think about death, but it eventually comes to everyone. Because of sheer numbers in the military, you will encounter death in your circle of acquaintances sooner or later during your spouse's career. If your spouse is just beginning a command assignment, accept the fact that sometime during the command time death will occur in the unit. Consider in advance how you can be of help in these difficult situations. This will permit you to act calmly and swiftly when you are needed.

HELPFUL RESOURCES
Survivor Outreach Services (SOS) Offices
Tragedy Assistance Program for Survivors (TAPS)
Veterans Administration Bereavement Center
Military OneSource
Military and Family Life Counselor Program (MFLC)
American Widow Project (AWP)
Family Life Chaplain
Snowball Express
Good Grief Camps
Gold Star Mothers of America, Inc.
Gold Star Wives of America, Inc.

* * * * * * * * * * * * *

The Army's tradition of "taking care of its own"
extends to the grave, and to those left behind.

Chapter 29

Military Courtesies

"The rules for military etiquette are founded on custom and tradition."
Oretha D. Swartz, Service Etiquette

From the first time you drive on to an Army post, your life will be affected by the customs of military courtesy. Each of these customs has evolved from the habitual practice of showing respect. They deal with respect—to the American flag; to older, more prominent, or more senior men and women; to other people in general; and sometimes even to one's own soldier. The courtesies described here are guides to help you understand these customs and, in so doing, encourage you to respect them.

FLAG ETIQUETTE
Flag etiquette, courtesy to the national flag, is not just a matter for military servicemembers; all Americans should know the proper ways to show respect for our nation's symbol.

At Retreat - The most frequent occasion you will have to show respect for the American flag on military posts is at Retreat. This ceremony honors our flag as it is being lowered from the flagpole at the end of each day (usually 5:00 p.m.). There is a comparable ceremony, Reveille, held each morning when the flag is raised, but few people are outdoors at that time to witness it.

Those who are outdoors on an Army post at the time of Retreat or Reveille are expected to stop what they are doing, face in the direction of the flagpole—or music if they cannot see the flag—and stand at attention with right hand over your heart until the flag is down and the music has ended. In addition, military in uniform salute. This courtesy is expected from everyone on post who is outside, not just those in sight of the flagpole. Even if you can't see the flag, you will know that it's Retreat when you see others standing at attention, or when you hear the music. (The music played on Army posts is "Retreat" and "To the Colors.") This music is usually played over loudspeakers throughout post so that everyone can hear it.

If you are driving on an Army post at the time of Retreat, stop your car (even if you're in the middle of the road), turn off the engine, get out, and stand beside the car. Any passengers should do the same. Of course, consider your safety and those of your passengers, as sometimes it is not conducive to stop on multi-lane post roads. (Passengers in a bus or truck may remain seated while the driver gets out and stands at attention.) It's important to note that Air Force and Navy courtesy differs from the Army with regard to what drivers and passengers are expected to do at Retreat. On an Air Force or Navy base, you are expected to stop your car until the flag is down and the music ends, but it's not necessary to get out and stand at attention.

Those who are outdoors in military housing areas are also expected to show courtesy to our flag at Retreat. Even though the housing area may be off post, those outside who hear the music should stop what they're doing and face in the direction of the flagpole or the sound of the music. Children should be taught to stop playing and stand still for these few moments, out of courtesy to the American flag.

To Display the Flag

- The flag is customarily displayed from sunrise to sunset. However, for a patriotic effect, the flag may be displayed twenty-four hours a day, if properly illuminated during the hours of darkness.
- The flag should not be displayed on days when the weather is inclement unless it is an all-weather flag.
- The flag should never touch anything beneath it, such as the ground, floor, or water.
- The flag should never be used as wearing apparel, bedding, or drapery. It should not be festooned, drawn back, or pinned up in folds; it should always be allowed to fall free. Bunting of blue, white, and red is appropriate to use for patriotic decorations, and it is always displayed with the blue above, white in the middle, and red below.
- A lapel flag pin is worn on the left side, near the heart.

Hand over Heart

- When the "National Anthem" is played outdoors and the flag is displayed, place your hand over your heart and face the flag. Indoors, protocol prescribes that for civilians, the hand should be over heart for indoor ceremonies and events as well.
- When the American flag passes in front of you at a parade or review, you should place your hand over your heart.

Pledge of Allegiance—This is a simple statement of loyalty to our nation's flag and all that it represents. Most of us learn to say this pledge by rote as small children in school and seldom take time to reflect on its full meaning and beauty. To say the pledge, we stand, hand over heart, and repeat the following words:

> "I pledge allegiance to the Flag of the United States of America, and to the Republic for which it stands, one Nation under God, indivisible, with liberty and justice for all."

It is appropriate to pause at the commas, for emphasis and clarity. Many people erroneously pause between "one Nation" and "under God," which gives a slight difference to the meaning of the words.

PROMOTION AND AWARD CEREMONIES

Promotion and award ceremonies are significant events in the lives of the Army members involved and their families. These are both joyous and somewhat solemn occasions to which guests and family members are invited. For informal ceremonies, often held indoors, chairs are usually provided and everyone is welcome to sit while waiting for the ceremony to begin. As the presiding officer enters the room and is *announced*, everyone present should stand (including their spouse). Later in the ceremony, as the award citation or promotion order is about to be read, the announcement of *"Attention to Orders"* will be made. All military members immediately stand at attention; civilians may remain seated, but, out of courtesy, should stand as well. After the orders have been read, the audience is seated. For a promotion ceremony, the spouse is usually invited to participate by pinning the new rank insignia on one shoulder (normally the left), while the presiding officer pins the new insignia on the other. Comments follow; then a line forms so that everyone can offer their congratulations.

At formal ceremonies, such as changes of command, the preamble "Attention to orders" is usually omitted; therefore, all observers (including the military in uniform) remain seated.

STAFF CAR SEATING

- The right rear seat of a staff car (or van) is considered the seat of honor and reserved for the senior person in the car. (This rule does not apply to women riding in a civilian car. Usually, in such a situation, the senior lady passenger sits in the front seat next to the door.)
- If you have occasion to ride in a staff car, the driver will probably open the rear door on the left side for you. When this is impractical because of heavy traffic, get in on the right side and slide across.
- Car seating normally conforms with that depicted in the following picture:

WHEN TO STAND
For a Senior Officer - A civilian woman is not normally expected to stand up for a man; however, in military settings there are a few occasions when *everyone* is expected to stand. When the senior officer walks into an official gathering (such as promotion or awards ceremony), or into an auditorium where he or she is to speak, *and is announced*, all those present rise. This includes not only military personnel, but family members, civilians, and even his/her spouse.

For Spouse of Senior Officer or NCO - Though the courtesy of standing as soon as the senior spouse enters the room is no longer practiced, military spouses still rise when the spouse of a very senior officer or NCO approaches to talk with them. In fact, true courtesy is to stand for *any adult* who walks up and remains standing to talk with you.

Although it's true that there is no rank among Army spouses, the spouses of senior officers and senior NCOs should be shown special respect because of their position in the community as the spouses of our senior leaders. There is little difference between this and the respect afforded the spouses of senior executives in the civilian corporate world. The only difference might be that most Army spouses in such positions have earned this respect in their own right—because of their experience, the responsibilities they shoulder, and the contributions they make to their communities. You can demonstrate your respect for these spouses not only by standing when they walk up to talk with you, but also in your actions and manner of speaking to them.

NO RANK AMONG SPOUSES
Army spouses can acquire poor reputations if they let their soldiers' rank go to their head. It's important to remember that spouses don't have any rank, unless they happen to be in the military themselves. These guidelines should help preclude problems of this nature:

- There is no rank among spouses, *only manners*. It's not appropriate to act or talk as though you are wearing your soldier's rank.
- To refer to your husband by his rank, as "Colonel Smith" or "the Colonel," is pretentious. Use either their first name or say, "my husband/wife."
- If your soldier is a commander, don't refer to the spouses in the coffee group as "my girls/guys"; that implies ownership.
- Avoid "name-dropping." Even if you are on a first-name basis with a senior person, use his or her proper title and last name in public and when speaking with others.

BLUF

It's all about human respect.

THE HONOR OF BEING FIRST

Since it is considered an honor to be first, the most senior person or couple present is accorded this honor. For instance, they should head up a reception line, or be invited to serve themselves first at a coffee, buffet, and potluck. For seated dinners, it is customary for the guest of honor to be served first and the host/hostess to be served last, even if they are more senior in rank to the guest of honor.

THE HONOR OF LEAVING FIRST

The honor of being the first person to leave a social function belongs to the most senior person or couple present, and then to any guests of honor (except at such free-flowing affairs as coffees, luncheons, receptions, and cocktail parties). If you are unable to wait for the senior person to depart, apologize to him or her—as well as any guests of honor—and explain the pressing reason for your early departure.

APPROACH SENIOR PEOPLE

At all social functions, the junior-ranking military members and their spouses should make an effort to talk at least briefly with those more senior, and any guests of honor. Don't wait for them to approach you first, and don't let their rank or position intimidate you. They will appreciate your thoughtfulness, good manners, and courtesy; and you will find them to be "real" people, just like everyone else.

CALL SENIOR OFFICERS BY NAME

Even though it seems natural to say "Sir/Ma'am" to someone to whom your soldier says "Sir/Ma'am," it's more appropriate for you, as a civilian, to call him or her by name. Unless you are in the military, you should not follow this military practice. Instead of "Good evening, Sir/Ma'am," say "Good evening, Colonel Smith." (If you can't remember their name, just say your greeting and smile.)

Many young spouses find it difficult to know when it's time to drop this courtesy they were taught in childhood, that of saying "Sir" and "Ma'am" to adults. Normally, it is dropped when a person reaches adulthood and replaced with the courtesy of using the other person's name. It's true that "Sir" and "Ma'am" can properly be used between adults when the person being addressed is significantly older. However, these titles of respect are usually reserved for use by children when speaking to adults, and for employees when speaking to their superiors, employers, or customers.

NEW YEAR'S RECEPTION—A COMMAND PERFORMANCE

Traditionally, Army commanders host receptions for their units' officers and NCOs and their spouses sometime during the New Year's season. Some hold their receptions in the home, others at the club. Unlike most social invitations, this one does not need to be repaid. However, it does have another obligation; it is considered a "command performance." That means that everyone's attendance is required. The only permissible reason for a military member to regret is illness or an out-of-town trip. Even a military member whose spouse is ill should go alone; however, if the military member is ill or away, the spouse should not attend.

Spouses of deployed soldiers who receive invitations are encouraged to attend and represent their soldier. Official Representational Funds may be used for some events, and those events will have limitations on whom may be invited.

BABIES AT ADULT FUNCTIONS

Babies are seldom welcome at adult functions in military society. When you're invited to a social function, do not presume that your baby is welcome. Unless a specific offer is made for you to bring your baby, it is expected that you will not do so. Most posts have childcare centers, as well as trained "childcare providers" who care for children in their homes. Hiring a babysitter or trading off with a neighbor who also has a baby are other commonly used options. If you can't find or afford childcare, or your baby is too young to be left with someone else, the polite course of action is to regret to the host/hostess and explain the problem. If they wish to offer for you to bring the baby, they will. Don't feel hurt that your baby isn't always welcome. Adult social functions are for adults; babies distract from everyone's socializing, and the hostess may not be equipped for babies or small children.

Similarly, older children should not be taken along to adult social events either. If they are wanted, they will be specifically invited, or the invitation will indicate that this is a "family" event.

CARRYING THE UMBRELLA

The Army finally came to its senses and now permits soldiers to carry umbrellas under most circumstances. The times when they are not allowed are when in formation or when wearing field or utility uniforms. Otherwise, the guidance is pretty standard—black, plain, no logos or design. Of course, umbrellas are not an issued item; you'll have to buy your own.

REVOLVING DOOR COURTESY

There is one male courtesy that seems to be unknown to almost all males and females, military and civilian. It is how a couple should go through a revolving door. While male courtesy is to allow a female to go through a door first, that isn't true when it comes to revolving doors. The reason is: A gentleman should go through a revolving door first in order to start the door revolving, which sometimes is hard to do. The lady follows in the next compartment while the gentleman is still *slowly* pushing the door. That means, after he is through the doorway, he doesn't just step out and move away; he should turn and ensure that the door continues to revolve for the lady, giving it a gentle push if need be. Simple, but oh so gentlemanly!

* * * * * * * * * * * * *

Learning military courtesy can prevent embarrassing moments.

SECTION EIGHT: MILITARY ROLES

CHAPTER 30

Commanders' Spouses

"Leadership is not bullying and leadership is not aggression. Leadership is the expectation that you can make the world a better place."
Sheryl Sandberg

At any level, selection for a command assignment is special. In fact, it should be viewed as an honor and a privilege. The message conveyed indicates that the Army has confidence and trust in the servicemember's ability to lead soldiers and civilians. While this assignment brings prestige and authority, it also means hard work and enormous responsibility and accountability. The family should rightfully feel proud but should also be aware of the impact. In particular, it often means less time together, but the personal sacrifice means that their loved one is making a wider impact upon the people who comprise the unit. It also means the spouse has to decide whether to take advantage of this opportunity and participate as a part of the command team.

BLUF

You have the power to create the leader you want to be.

IT'S YOUR CHOICE

Today, commanders' spouses, as well as all spouses, have more options available than ever before. Historically, many spouses would volunteer their time and talent to supporting their military spouse, family, and community, seldom working full-time outside their home. Commanders' spouses were expected to concentrate their efforts on the unit spouses' activities, even if it meant placing their families second. In the 1970s, all across America women moved into the workforce in ever-increasing numbers, both for economic reasons, as well as for professional gratification. Often times, this resulted in a smaller pool of volunteers, increased dissatisfaction with the military lifestyle as a result of frequent moves and discontinuity of employment opportunities, and a pronounced departure from the acceptance of traditional roles.

Wisely, the Army recognized the profound impact from the loss of a tremendously talented volunteer pool, as well as loss of the retention of career servicemembers. So, when spouses boldly and courageously made their suggestions at the Army Family Symposia in the early 1980s, the Army listened. Quickly, "quality of family life" became key watchwords. Emphasis was placed on improving housing, post facilities and services; employment assistance; volunteer recognition; and stabilizing tours of duty based on family concerns and requirements, such as school continuity. The value of a command team was recognized through the invitation for spouses to attend the pre-command course with a training program designed for their leadership. Then in 1987, DOD issued the Working Military Spouse Policy. It stated that "no commander or supervisor nor any other DOD official will, directly or indirectly, impede or otherwise interfere with the right of every military spouse to decide whether to pursue or hold a job, attend school or to serve on a voluntary basis." For the first time, commanders' spouses felt free to pursue their career or education without fear that it would harm their soldiers' careers.

With the passing of this policy, however, there was no intention to say leadership was not needed within the spouses of a unit. But it did provide some flexibility, even for those who were hesitant to assume that role. John Quincy Adams once said, 'If your actions inspire others to dream more, learn more, do more and become more, you are a leader." Leadership provides spouses with someone to coordinate and support their activities; a role model to learn from; an advisor to turn to; a "cheerleader" for their efforts; and the unifying spirit necessary for a productive, cooperative and informed group. If the commander is not married, if the spouse does not live at that location, or if the spouse chooses not to assist in this manner, it is important for another spouse in the unit to be asked to provide that leadership. It is strongly encouraged, however, that the commander's spouse at least takes the leadership training provided by the Army, then ask an appropriate and willing spouse in the unit—usually the executive officer or deputy's spouse—to take the leader's position, and pass along information gained from the Army's commanders' spouse training. So, the person who assumes these responsibilities is typically referred to as the "leading spouse."

It is significant to note that the majority of the commanders' spouses choose to accept this leadership role. While some may temporarily give up their jobs, other spouses work part-time, or telework during their command time. It's apparent that they recognize the importance of the contributions they can make and the personal gratification they will receive from being a part of the command team. In fact, often spouses find that the organizational and professional skills they use as leading spouses and readiness group leaders can be documented on their résumés when applying for future managerial positions. Whatever their motivation, commanders' spouses who become involved today do so by choice, even those who initially were reluctant due to a fear of assuming a leadership role. As Napoleon Bonaparte once famously said, "A leader is a dealer in hope."

PREPARING FOR THE ROLE

There are no "official duties," "job description," and no "prescribed role" for commanders' spouses. Nor is there a "typical" commander's spouse image that everyone should try to emulate or a yardstick by which their performance should be measured. So how can the commander's spouse, who chooses to take an active leadership role, prepare for such an undefined position? Here are a few suggestions:

• Recognize that the best approach is simply to be yourself. You will find that commanders' spouses defy description. Even at the same level of command, they can run the gamut in age, family composition, education, and experience—not to mention interests and personalities. Each brings to their leadership role something different. Build on your own personal talents and strengths. However, recognize that if you have a strong or domineering personality, it might require a shift in balance as you build consensus and cohesion among the spouses in your unit without interpretation of fear, threat, or intimidation.

- Discuss the implications of your involvement with the family. During this period, you will have more commitments, and, consequently, less time for family. Others in the family may have to help with household chores, particularly if you are going to balance work, children, home, and unit activities. It can be done, but it takes extra effort and flexibility from you, and probably from your family as well. Be sure they understand and support your efforts, and always try to save some quiet time for yourself. Depending upon age, prepare your children, too, for the possibility of comments they might hear regarding the actions of their soldier parent. Be sure to keep those communication lines open.

- Accept the fact that you will be much more visible and, consequently, a role model. Follow the common courtesies of responding to invitations within 24-48 hours, write thank-you notes, learn and use the first names of the other spouses, and take the lead in socializing within the unit. It may help to review the basic etiquette described in this handbook on such subjects as introductions, receiving lines, military etiquette, parades, seating arrangements, and table manners. Be sure to read Chapter 32, "General Officers' Spouses", as commanders' spouses and general officers' spouses have a great deal in common. People will observe how you engage with others, so always remember to gauge that your actions and words do not convey bias or favoritism.

- If this is your first experience as a commander's spouse, realize that leaders always feel isolated. Hence, you have probably heard the old saying, "It can be lonely at the top." You will want to be friendly and approachable to everyone in the unit, but do not confide to any of them. Your support will come from your spouse and your peer group. Seek friendships outside the unit among your contemporaries, some of whom may not even be local.

- Take advantage of any leadership training that is available to you. Accept the Army's invitation to attend the spouses' portion of the pre-command course, which is first offered at the battalion level. Collect and read books and pamphlets pertaining to relevant subjects—e.g., leadership, commanders' spouses, Family Readiness Groups, Army life, protocol and etiquette. Learn all that you can about the multitude of support services your Army community has to offer—e.g., ACS, JAG, Red Cross. You won't always have the answers to people's questions and concerns, but you can know where to find the answers or refer them.

- Consider establishing your personal "Gift Closet." As you see items that might serve as great hostess gifts or gifts to encourage people, begin purchasing and setting them aside for future use. Suggestions include such items as: candles; cocktail serving pieces; notecards; linen hand towel; picture frames; small pottery pieces or crystal pieces, just to name a few. Unit-specific items such as crest pins, crest lapel pins or tie tacks, crest pendants or charms, unit crest engraved or etched glass items; these are all token gifts with unit branding that can be kept within a modest budget. To complete your mini closet, it is also a good suggestion to purchase gift bags, tissue paper, and generic gift wrap/bows perhaps in unit colors so that you will always be prepared.

- Finally, consider the leadership principles, techniques, and guidelines discussed in the following section.

LEADERSHIP SKILLS

Principles of Leadership - Everyone has or will develop their own leadership style, as a result of personality and level of experience. No single style is the "ideal." However, there appear to be four principles of successful leadership:

1. Genuinely care about the people involved and treat all people with dignity and respect.
2. Provide the information that is important for this group to know.
3. Be reliable, someone the group can trust to do what you say you will do.
4. Set the example for others to follow, treating all fairly without the appearance of favoritism.

These principles can be applied to all leaders, regardless of which leadership techniques are employed.

Leadership Techniques - How you go about leading a group to accomplish particular objectives requires different approaches, depending on the nature of the group and the objective. Four leadership techniques have been described in *"The Commander's Link"*: teller, seller, participator, and delegator. The particulars of each technique are:

- Teller—When the group is very inexperienced or unwilling to make the decision for itself, the leader makes the decision and tells the group how to accomplish it. The entire project will require considerable guidance and supervision.

- Seller—For groups that are inexperienced but willing once they are convinced of the merits of the task, the leader makes the decision, but then explains the idea and "sells" it to the group. Once the majority are convinced, the leader gives direction and sustains the group's motivation by praising their decision to participate and their efforts.

- Participator—Experienced groups that are capable, but lack confidence or enthusiasm for the project, work best with a leader who lets the group talk the issues out and encourages a consensus. This technique may take longer to produce results, but the group gains confidence and cohesiveness in the process.

- Delegator—Groups comprised of very experienced, self-motivators only need a leader to delegate the responsibility for the various tasks involved, once the group decides on a project. Very little direction or supervision is required.

Leadership Guidelines—Regardless of the size, level, or experience of the spouse group you lead, there are leadership guidelines that can help you be more successful. The following fundamentals have been gleaned from the experiences of former commanders' spouses:

- *Be a friend to all—equally.* A leader must not APPEAR to like some people in the group more than others. If you are to be a leader to all, you should be a friend to all. This includes being consistent and fair to everyone—in other words, even-handed.

- *Always ask—never tell!* Even when using the "teller" leadership technique, a leader will achieve greater success by asking for volunteers or help on a project. Everyone belongs to the group voluntarily, and the leader has no authority to tell the others what to do.

- *Ask only of others what you are willing to do yourself.* This does not mean that you have to work with every committee and put your name on every list of volunteers, nor should you. Many times your role is best served when you only give guidance and supervision. However, you will want to "roll up your sleeves" occasionally and work like everyone else to demonstrate that you consider yourself a part of the group. Consider showing up during the course of an ongoing project, bringing a surprise of light refreshments to show your appreciation and encouragement, then working for a while before departing to let your group continue. A successful leader is not a micro-manager; when there are committees, let the committee chairperson take charge.

- *Always say thank you, both verbally as well as in writing.* Group members are the ones who do the work, and a good leader takes every opportunity to thank them and give them the credit. When you receive praise for what the group has accomplished, point out that it is the individuals in the group, not you, who deserve the accolade. Various examples include handwritten notes; asking the group to stand at an event for recognition; presenting certificates of appreciation; hosting a small coffee or dessert social in their honor, just to name a few. Make sure, however, that you do not minimize all who have helped, regardless of their level of assistance or participation.

- *Be a teacher.* Caring leaders see a part of their role as that of a model and mentor. Teach by simply providing a good example for others to follow, gentle guidance and advice where needed, and encouragement for others to excel.

- *Be aware.* Capture your own personal impressions and assessments of installation programs and community activities. There may be unique circumstances that affect how things have been done procedurally. This, too, will become a reference tool for you to potentially provide ideas or suggestions as it pertains to future projects.

- *Recognize your power and use it wisely.* Consider carefully the possible effects of what you say and do. Power comes with the role of leadership because your position gives you the ability to influence others.

LEADERSHIP ROLE

Although there are no official duties of a leading spouse, it is helpful to know what former commanders' spouses have seen as their "unofficial responsibilities." The following list is an adaptation from *Choices & Challenges*, which was written by the students' spouses at the Army War College in 1991. These unofficial responsibilities of leading spouses can be applied at all levels—platoon, company (battery/troop), battalion (squadron), brigade (regiment), and division.

Serve as leading spouse in the unit.
Develop and maintain good relationships with unit spouses.
Support the next higher-level leading spouse and keep them informed.
Identify needs and interests of the group.
Gather and pass information.
Plan activities.
Conduct meetings with other levels of leadership.
Assist less-experienced spouses.
Focus not only on unit but also community affairs.

At first glance, this list might seem intimidating to someone new to the role of a leading spouse. However, everything falls into place as you progress in the role, and you don't do everything at once.

To begin, a new leading spouse needs to get to know the senior spouses in your group. The easiest way to do this is to invite them to your home for a short get-together, perhaps a simple morning coffee or afternoon dessert. If you're still in the throes of unpacking, that's okay—use a packing crate for a table or simply sit on the floor! And if you aren't prepared to cook, store-bought goodies are fine. Alternately, you can meet at the club or a nearby coffee shop or restaurant. What's most important is to get together very soon after you take on this position of leadership and start forming those important bonds of friendly trust and cooperation. It's also an opportunity for you to learn from them about the spouse activities in the unit and community. Hopefully, the previous leading spouse met with you and passed along the important information you need to know. However, if that did not occur, now is the time and this is the group to ask about the workings of the coffee and Family Readiness Groups, the current representatives to different post activities, and the unit's responsibilities for various social and welfare projects. Additionally, the senior spouses of the unit will look upon this meeting as an opportunity to learn from you (either directly or by inference) what you hope to accomplish, how you want to work with them, and what you will be asking of them. It's best not to make any major changes immediately. Listen, look, and learn about the group before you decide on any changes you are contemplating.

Which senior spouses do you invite? Include the spouses of the following: the unit's executive officer (chief of staff at division level), subordinate commanders, and senior NCO (CSM or first sergeant). This should be the first of many such get-togethers with this group. At the brigade and battalion level, it is especially important to hold this type of informal meeting as often as possible and on a regular basis. These frequent get-togethers serve several important purposes. They strengthen your relationships and friendships with one another, making it easier for you to work together. They also serve as a group forum for the sharing of ideas and the decision-making that will affect the entire unit. By making decisions jointly with this group, you will probably find that the decisions are better because you pool the good ideas of the other leaders and jointly select the best. Additionally, when the other leaders are a part of the decision-making process, they support the decisions and willingly assist in their execution. Their "buy-in" enhances the chances for much greater success.

As times have changed, the value of a strong partnership between the officer's spouse and the NCO's spouse has become more apparent. Experience has proven that when the two leading spouses work together as a team, their mutual friendship and support help them both accomplish their goals for the good of the unit. They will hopefully become "battle buddies." Their teamwork also sets an example for the younger "spouse teams" to follow.

PRACTICAL ADVICE

After leading spouses understand the scope of their new role, there still remain issues they will confront that may be new to them. The following "lessons learned" by former leading spouses focus on such issues:

Advisory Positions

Spouses of commanders are often asked to serve in advisory positions. Depending on the size of your community and the level of your soldier's command, you may be invited to serve in an advisory or honorary capacity for the spouse club, community club, and other boards and councils. It is considered a privilege to be asked to serve as an advisor, however, you are under no obligation to accept. If you choose to decline, offer to assist in finding an appropriate representative. If you decide to accept, be sure to read the information on advisors in Chapter 22. However, as a general summary: advise but let the group decide, unless decisions border on legal or ethical troubles; listen and if necessary, ask leading questions that facilitate their making sound decisions; validate and encourage the opinions of all; read previous minutes in addition to the Constitution and By-Laws that govern the organization; and understand that you do not have to have knowledge of everything or express how things were done in the past.

In the past, advisory positions within a military community traditionally went to the leading spouses of particular units. That meant, if your predecessor served as advisor to a particular board, you probably would be expected to take over when your soldier assumed command. Today, a more flexible approach is prevalent. While the traditional structure may work for some, others may choose to ask someone else to serve in this capacity. The most important factor, however, is at least give the senior spouse the courtesy to choose or decline.

Public Speaking

Not all commanders' spouses and leading spouses are accomplished public speakers. Yet, they all find themselves in situations that require them to speak in public, often standing alone in front of a crowd with a microphone to amplify every quiver in their voice or a cell phone to capture this event. For those unaccustomed to speaking in public, the following suggestions will help:

1. If you know in advance that you will be speaking, plan what you want to say. Writing your remarks down forces you to clarify your thoughts and organize your presentation. If you are comfortable with adlibbing, it may only be necessary to make a topic outline or key bullets.

2. Rehearse your presentation—stand up and speak out loud, possibly recording what you say and how you say it. You may feel self-conscious at first, but that is exactly what this practice will overcome. Time yourself, if time is limited. After several rehearsals by yourself, rehearse at least once in front of someone, such as your spouse.

3. Visualize yourself giving your speech—you are calm, relaxed, confident; your presentation is smooth and flawless. Believe it or not, this visualization helps; sports figures use this technique all the time.

4. If your heart starts racing before you have to speak, you can slow it down. First, stop breathing; don't hold your breath, just stop breathing. Then, concentrate on your beating heart and on making it beat more slowly. When you absolutely have to breathe, do so, but repeat the process if your heart hasn't slowed down enough.

5. Remember that the audience is with you, not against you. Walk and talk slowly; keep your movements as natural as you can. If your voice quivers, just keep talking. If you make a mistake, just correct it and go on; everyone makes mistakes, and no one minds or will remember if you do.

6. If you must speak without time to prepare, concentrate totally on what you want to say. Don't let your mind worry about how you look or sound to the audience. Don't listen to your own voice as it comes out of the loudspeaker. Total concentration will prevent you from becoming distracted.

7. If you wish to actually read your intended message, find a poem or other writing that best summarizes your thoughts to share with your audience. This way of expression relieves pressure from forgetting what you might wish to say, yet gives an easy way to express a sentiment.

8. Finally, it is some consolation to know that public speaking does get easier with practice. Even so, a brief respite or a new audience may bring back the old anxieties. If this happens, it helps to remember that you are no different from most others; just go back to suggestion #1 and start again.

Your Leading Spouse

Every leading spouse has responsibilities that go up the "chain of concern" as well as down. You will want to support and assist your leading spouse, just as you hope the spouses in your group will support you. When a new commander takes over and, all at once, you have a new leading spouse, your loyalty belongs immediately to the new spouse. That does not mean you are no longer friends with the predecessor, but the current leader deserves your support. Recognize that all individuals have their own unique personality, leadership style, level of experience, interests, and talents. Do not tell your new leader how things used to be done, unless asked, or unless it is to share vital information that involves a program or policy that is unique and important. Do not tell the previous leader, or anyone else, how you think things are going since the new leading spouse arrived; keep your opinions to yourself. Simply stated, new leaders do not have to try to be their predecessors. If they do not champion the same causes or platforms, that is okay, too. Finally, remember that to you, they may appear self-confident and self-sufficient, but they have doubts and fears just like everyone else. A little praise for their efforts, without overdoing it, will bolster their confidence and ego, just as it would yours, reinforcing that they are on the right track.

Every leading spouse needs to get to know the groups that come within their sphere of influence. The best way this can be accomplished is for them to attend unit coffees and family readiness get-togethers occasionally. Other key members in the chain of concern should also be occasionally invited as special guests. For example, a battalion commander's spouse might invite the brigade commander's spouse, and occasionally the chief of staff's spouse, to special coffees or events. This gives many spouses in the group an opportunity to become acquainted with these senior spouses in a small-group setting and a more comfortable environment—something that they might otherwise never get to do. And by exposing spouses to other leaders, it can positively impact their impressions on how they would like to be as they continue to grow.

Before inviting special guests to a unit coffee or family readiness get-together, be sure to discuss their visit with the group the month before and remind everyone to take the opportunity to talk with the guests. Some spouses are reluctant to approach senior spouses, possibly based upon age, soldier's rank, etc. Simply reassure them that senior spouses have gone through similar experiences and can relate to their concerns about family and other similar topics. Remind them that it's an honor for your group to host your special guest, particularly considering the demands upon their time, and in some instances, the fact that they may not live locally. Everyone in the group should act as though they are a hostess and make the special guest feel welcome.

A Substitute Leading Spouse

When a unit commander does not have a spouse, or the spouse chooses not to take the role of leading spouse, usually it is the commander who finds a substitute to lead the spousal activities. If the commander's spouse is present but unavailable to serve in that capacity, the spouse may want to participate with the commander in the selection of their representative. Normally, the first person considered for this position is the spouse of the deputy commanding officer or executive officer. If this person doesn't accept the role, the spouse next in line usually follows the chain of command diagram. If these spouses are unavailable, other senior spouses who exhibit leadership qualities should be considered.

Ultimately, whoever decides to serve should, at a minimum, ask if the senior spouse wishes to be kept informed of what is happening within the unit, and follow up accordingly.

A commander who is seeking a substitute leading spouse will find it helpful to consider the following points:

1. Learn what community representative and spouse leadership positions need to be filled on behalf of the unit. This information is necessary so that the responsibilities can be clearly explained to the person asked to assume them. Likely sources for this information are the spouse of the unit's deputy commander or executive officer, and the next-higher leading spouse.

2. When asking someone to take on this role, it should be done in a business-like manner and setting—not at a social event. The considerate commander will invite the spouse to the office or call them on the phone, explain the need, and ask for assistance. Be careful to word the request so that the spouse feels free to decline, without prejudice to them or their soldier.

3. The commander should bear any related expenses incurred by the substitute leading spouse. For example, if they are asked to attend a monthly meeting in place of the unit commander's spouse, the commander should make it clear that they will pay any childcare, transportation, and food costs involved.

4. Single commanders are always responsible for their own social obligations. The substitute leading spouse should not take on or assist the commander with these. Even the little, time-consuming tasks, such as addressing invitation envelopes, are the commander's responsibility.

5. Remember to say thank you periodically for the time and effort the substitute leading spouse expends on behalf of the unit spouses and families. If there is a commander's spouse present, they, too, must be careful to give the substitute leading spouse the credit and praise they deserve.

Gifts

As the spouse of a commander, there will be occasions when you are involved with semi-official gifts. For example, the group you represent may want to give a gift to a departing spouse; if you accompany your soldier on an official visit, you may want to give a gift to the person who serves as your hostess; and, at the end of your soldier's command, others will want to farewell you and present you with a gift. Be careful in selecting the gifts you give and in receiving such semi-official gifts. They should not be so expensive or ostentatious as to appear in bad taste, nor should you ever request a specific gift. Remember, often it is simply the thought that counts. If you give departing gifts, remember to be consistent, and again, not display favoritism or bias.

The Army, as well as the judge advocate, recommends that everyone use good judgment and common sense in this matter. The intent is to avoid any—real or perceived—improprieties in the use of public office for private gain, the purchase of favors by subordinates, and command pressure to make involuntary contributions. If ever in doubt about the appropriateness or value of a gift, consult your local judge advocate.

It is especially important for the spouse of a commander to be circumspect in these matters, because you are frequently in the limelight and your activities are continuously observed. Consequently, you will want to reflect both propriety and good manners with respect to such gifts. Follow the Army standards that are designed to protect you, and acknowledge those who have taken the time to select something special for you with a written thank you note.

Entertaining

Good old-fashioned hospitality is very refreshing. If you enjoy people and entertaining, you will find that this is an aspect of command that can draw the unit together and add pleasure to everyone's busy lives. How, where, and the frequency with which you entertain are matters of personal choice for you and your spouse. However, if entertaining is something you dread, review Section Five, "Entertaining," to see if the suggested guidelines will help to ease your concerns.

There are certain groups that commanders and their spouses typically entertain. This is not just for the sake of tradition, but because experience has shown that it helps to build a stronger unit. Bringing the members of the unit and their spouses together for a few hours in a relaxed atmosphere gives them the opportunity to become closer and to become acquainted with the commander and the commander's spouse on a more personal level. The first step is to ask the adjutant for a copy of the social or FRG roster so you can become familiar with the names and be better prepared to greet the members of the unit at these wonderful social events. The Hail and Farewells are also a great opportunity to meet everyone and glean spouses' names for the coffee group.

The following list of suggestions for command entertaining at the battalion level is an adaptation from the U.S. Army War College's "Choices & Challenges" and has been updated to reflect our busier, working/command couples and families. Other appropriate modifications should be made for other levels of command.

Soon after the change of command:
- company/battalion/troop commanders, staff, CSM
- all officers and CSM (especially if there was no specific battalion event to welcome you)
- ideas—cocktails, buffet, ice-cream social, cook-out, heavy appetizers

Periodic get-together suggestions:
- CSM and XO and S-3 (new or annually)
- principal staff (annually, XO may entertain staff more frequently)
- company/battery/troop commanders (every six months; rotate houses, go out to dinner; informal BBQs or "happy hour" around a fire pit)
- farewells/welcomes (company/battery/troop commanders, XO, S-3, CSM)
- parties (all officers and CSM)—sit-down dinner, buffet, cook-out, brunch, luncheon, dessert, potluck, heavy appetizers

New Year's Reception:
- command performance (can be held at times during holiday season other than New Year's Day)
- all officers and CSM
- brigade/regimental commander, XO, CSM
- shifts by companies/batteries/troops (45 minutes each with 30 minutes between, or all at one time for 2 hours)
- traditionally Army Service Uniform (ASU with bow tie)/Sunday dress
- ideas—heavy appetizers, punch, desserts

Spouses only:
- monthly coffees or every other month (day or evening)
- get-togethers during deployments
- ideas—potluck, game night, desserts, food theme; dinner at a restaurant

Don't let the length of this entertainment list overwhelm you. Remember that these are only suggestions to be considered, and they are spread out over the entire command period. Additionally, many items listed can be very casual get-togethers; impromptu gatherings, or shared events, such as potlucks, rotating houses, or Dutch-treat dinners at a restaurant, even attending a sports event together during military appreciation. These shared events are especially popular today, even with everyone's overly busy schedule; they mean that no one person has to bear all the effort and expense. The New Year's Reception mentioned is not held at every level of command, usually only at battalion level and higher. In some instances, as you foster relationships with fellow (same level) command teams, you can assist each other in helping to set up and reset your receptions. Sometimes, senior commanders decide to forgo this tradition of asking subordinate units to help in order to leave time free for subordinate commanders to host receptions for their own units.

As you plan your entertaining, the following cautions are offered. Try not to overburden people with social engagements. Recognize that young Army couples tend to be very involved in working, family and community life, and value their free time. Avoid planning social activities on holidays or at times that would disrupt a normal family affair. This regard for family activity is one of the reasons the traditional New Year's Reception is seldom held on New Year's Day. Lastly, refrain from using executive officers and their spouses as "aides" for social entertaining. Occasional volunteer assistance is acceptable, but military members and their spouses should not be made to feel that this is an expected "additional duty." If they offer, it is a wonderful way to mentor. Instead, turn to friends or family to provide any additional assistance you may need.

WHEN THE COMMAND COMES TO AN END

By the time your soldier's command draws to an end, you will probably feel so much a part of the unit that it will be difficult to leave. But leave you must, and someone else gets the wonderful opportunity to command. You have one last opportunity to serve the unit and its people well—by making the transition from one commander to another as easy as possible for all concerned. The old Army tradition that held an incoming commander at arm's length—almost as though they didn't exist—until the moment the unit colors were passed, is out of date and not helpful in easing the unit through a change of commanders. The following suggestions are for the outgoing command team during this time of transition:

1. The unit usually ensures the incoming commander is given a sponsor as soon as possible. This is usually the executive officer, since the outgoing commander seldom has time to assume this responsibility.

2. You and your spouse should correspond separately with the incoming command team (commander and spouse). This opens the doors of communication and offers an opportunity for questions to be asked and answered. It says, "We care," and sets the tone for their welcome.

3. If the incoming commander's family will move into your quarters when you leave, try to move out early enough so that they can move in as soon as they arrive. This allows time for any maintenance or painting to be done prior to their arrival. If anyone has to be inconvenienced by living in temporary quarters, let it be the outgoing commander and their family. As the outgoing family, you have local friends and know the area well, two advantages that make your inconvenience less challenging than it would be to the newcomers. This also says, "We care about you and are trying to make your transition as easy as possible."

4. When the incoming commander and spouse arrive, welcome them warmly as new friends. Take them to dinner to introduce them to their new area. If you find positive things to say about the incoming commander and spouse, others in the unit are encouraged to look forward to meeting and working with them, and to plan a warm welcome.

5. Prepare concise notes concerning the important spouse functions, both unit and community, for the incoming commander's spouse. Include information on regularly scheduled meetings and social events, unit-volunteer commitments, coffee-group practices, and unit representatives to post activities. An up-to-date unit roster will also be helpful. A wonderful welcome gift could be an album containing a picture of each key family, possibly accompanied by their favorite recipe, as a way for the new spouse to begin associating who makes up their new unit family. Find time to sit down and discuss these notes with them, answering any questions they might have. Make sure not to discuss people, unless there are extenuating circumstances they should be aware of.

6. Once on post, the incoming commander and spouse should be invited to most unit social events. The exceptions would be the unit's farewell event(s) for the commander and the commander's spouse.

7. Consider that if the commander's spouse arrives during a regular meeting cycle, take the spouse to the meeting with you for the first time, particularly for community commitments.

8. Don't make key decisions to implement new changes on your way out. If you have noticed an area that might need to be tweaked, leave any decision to the incoming spouse, as she/he will be the one to live with the end results.

9. Review Chapter 26, "Parades, Changes of Command, and Retirement."

Remember how you felt as you and your soldier were just coming into the unit; you were the new ones and everything was so strange. Now, you are the old-timers and have grown to love the unit and its people, in addition to the community, and it is hard to say goodbye. But, it's your turn to leave—do it gracefully and do not reach back.

COMMAND REFLECTIONS

Reflecting upon what your soldier's command means to you is time well spent. Everyone acknowledges today that a commander's spouse does not automatically have to take on the role of leading spouse when their soldier is handed the unit colors. The role is theirs if they choose to accept it, but they are free to follow another path if they prefer. Nevertheless, there are four points worth contemplating.

1. Military service, in some ways, is similar to civilian employment. Typically, when someone reaches the executive level in civilian business, there are certain social and community responsibilities that may accompany the job. If they are married, their spouse may be expected to help handle those responsibilities. In some instances, if the spouse isn't a "team player," it may impact selection for jobs. This same philosophy also applies to other leadership positions—church ministers, those in high positions in state and federal government, and the diplomatic corps. If you feel that your soldier's command puts unwanted pressures on you, recognize that you very well might experience similar pressures in the civilian world. Positions of authority naturally bring privilege, and the companion to privilege is responsibility.

2. Someone is needed to serve as the leader for the unit's spouses. If the commander's spouse chooses not to take on the role of leading spouse, someone else should be asked to assume that position. Otherwise, the unit's spouses are a group without a leader, like a boat without a rudder. While it is certainly an honor to be asked to become a unit's substitute leading spouse, it also means a lot of time and effort. Additionally, anyone qualified to take on such a role is probably senior enough that they already have commitments of their own to the unit and community. Being asked to become the substitute leading spouse may be an honor, but it's also a commitment.

3. Consider all the benefits from being a part of the command team. It can add a new dimension to the partnership you already have with your soldier. You can develop new friendships as you work together with the other unit spouses, your contemporaries, and the senior leading spouses. You can expand your leadership and organizational skills, and you can have the satisfaction of knowing that your efforts contribute to the well-being of the families of your soldier's unit. To be part of a successful command time requires commitment. If you truly wish to achieve a great command time, establish a

great command climate, and make a difference, you should be prepared to stretch yourself. Most who have gone before you have seen this as a time of service. That's not to say that they didn't get involved when their spouses were not in command, but somehow, they worked harder at being of service as part of the command team. They tried to lead by example, and to make themselves available and responsive to the needs of others. They gave their time, talents, enthusiasm, and concern, whenever and wherever needed—to their own families first, the unit second, and the Army community third.

4. Recognize that occasionally you may encounter or have to deal with difficult people. Fortunately, this is the small minority, as most often you have the good fortune to work with and meet some incredibly positive, dedicated, hard-working and talented people. However, if this does occur, tread cautiously and carefully evaluate which course(s) of action you might have to take. If necessary, consult with a trusted confidant before deciding how you wish to handle the situation.

Reflect on what role you want to take during the time your spouse is a commander. You will be the "commander's spouse" no matter what you do, but you must decide whether or not you want to be the unit's "leading spouse." If you choose that path, you will have the opportunity to make a positive difference in many people's lives.

* * * * * * * * * * * * *

"Leadership is a privilege to better the lives of others.
It is not an opportunity to satisfy personal greed."
Mwai Kibaki

Chapter 31

NCOs' Spouses

"Army spouses are a benefit you don't have to budget for,
you just have to embrace."
Holly Dailey, Army spouse

Being the spouse of a noncommissioned officer is an honored position of which you should be *very* proud. Your soldier is being given more responsibility and authority through promotion, and is a part of that vital core of the Army that makes things happen! As the partner of your soldier, you will have the opportunity to share in the efforts, as well as the rewards, of making the Army way of life all that it can be for the soldiers and families in the unit. You will also have increased opportunities for enjoyable social interaction in your Army community. Unlike our Army of yesteryear, today's Army recognizes and promotes the importance of its Noncommissioned Officer Corps and their spouses.

LONG HOURS

The first thing you will probably notice as the spouse of an NCO is that your soldier works very long hours, leaving little time or energy for you and the family. The changing work schedule, frequent field exercises, and deployments also mean little free time for an active social life. Accepting this unique lifestyle you have chosen to be a part of will make for a positive and successful journey. Unlike the armies of most other nations, the U.S. Army works almost as hard in peacetime as in wartime. Our military leaders rely on sweat in peacetime to reduce bloodshed in wartime. Add to that the continuous mission to optimize total force readiness to perform warfighting, peacekeeping, and humanitarian tasks, and you have the makings for a very long day. The best approach any Army spouse can take is to accept the situation, to include periodically checking one's own expectations, and make the best of it.

BLUF

Bloom Where You are Planted.

- *Keep evenings and weekends as free as possible for relaxation and family activities. Try to accomplish your household tasks while your soldier is at work. Learn how to do simple household repairs and maintenance so there will be fewer* "honey-dos" and more time to relax together. Maintain family traditions in your household.

- *Keep your life interesting and active.* If you find yourself with too much free time on your hands, seek outside interests to fill the hours when your soldier is away. If you're a parent, consider scouting, youth sports programs, or school committees. If you enjoy working with adults, every Army community needs volunteers for the Family Readiness Group activities, Army Community Service (ACS), Red Cross, thrift shop, spouse club, and chapel. Another possibility is to develop one of your hobbies, talents, or interests into more than an occasional diversion. Broadening your education or career could be another option and will build upon your portfolio. Any of these activities can fill your time with gratifying results and have the added bonus of expanding your horizons.

- *Make the most of the time you have together.* Why waste time arguing with your soldier about the long hours they have to work? Learn to accept the limited time that a career-oriented NCO has for family life, and get on with sharing those precious hours in relaxing, enjoyable ways.

OPPORTUNITIES FOR INVOLVEMENT

As the spouse of an NCO, you will have added opportunities to become involved with the families of your soldier's unit and your military community. Whether or not you do so, and to what degree, are important issues for you and your soldier to discuss and decide upon. In fact, the degree of your involvement will probably vary with the different stages in your life, the growth of your soldier's professional responsibilities, and your current interests and family obligations. There may be periods in your life when you feel that you don't have time to devote to volunteer efforts—maybe you have a job, are a full-time student, or an active-duty member yourself; perhaps you have small children or elderly parents who need your attention. There may be other times when you welcome the opportunity to get involved in new interests, associate with adults, contribute to the well-being of the unit families, and work with others to improve your military community. What you decide to do for the unit and your community is totally voluntary; it should be based on what you and your soldier feel is right for your family's current circumstances.

If you have any time available, be sure to take advantage of the training that is available to you. Army posts abound with training for military spouses. Education courses and training—both in person and possibly online—will be offered to you when your soldier becomes a senior noncommissioned officer. Today's NCO spouses are fortunate that the Army has embraced the fact that they can be an important asset—to their soldiers, the units, and the Army. This is demonstrated by the First Sergeant Spouse Seminar as part of the Enlisted Spouse Training Series provided through Army Community Services and the excellent classes offered for spouses at the Army's Sergeants Major Academy. Also, the spouses of NCOs selected for senior CSM assignments are now invited to Fort Leavenworth for the spouse portion of the pre-command course. The Army recognizes that professional development for our Army spouses is an investment in our soldiers' and families' future, and therefore, continuously evaluates the quality of spouse professional development.

Whatever the degree of your involvement, it is important for you and your soldier to work as a team. In other words, discuss what needs to be done, how much time and energy you want to devote, and how best to proceed. Share your expectations with each other. This gives your soldier an opportunity to understand the challenges confronting you and your goals. Equally important, your soldier can offer suggestions, moral support, and perhaps on occasion assistance. When unit activities are involved, your efforts will complement your soldier's, and your combined efforts will demonstrate a unity that has a very positive influence on the atmosphere of the unit.

If you choose to become actively involved in unit activities, there's another unofficial team of which you will become a part—The Chain of Concern (refer to the communication chart later in this chapter). If you're the spouse of a junior NCO, you and the other active company (battery/troop) spouses will probably work together with the spouse of the first sergeant. In turn, the first sergeants' spouses work as a team with the spouse of the battalion (squadron) command sergeant major, and they will probably work with the spouse of the battalion commander. The cooperation, coordination, and friendship within your Chain of Concern team are important. This mutual support not only makes working together more enjoyable, but sets an example of teamwork and *esprit de corps* for others. Of course, it's always possible, for any number of reasons, that there is no spouse available in one of these key team positions; under those circumstances, someone should be asked or volunteer to fill the vacancy.

LEADERSHIP ROLES

If your soldier is promoted to a junior NCO or senior NCO leadership position, you should be extremely proud of them and the status they have achieved. Probably your first reaction to their promotion or new assignment is a mixture of pride, delight, and enthusiasm. However, your feelings may be tempered by apprehension about what this new position will mean for you and your family.

In order to decide how involved you want to become, it's essential to know what will be asked of you. If you have had some good role models, you already have an idea of what this new role can entail. However, for those who still have questions, a very helpful list was compiled in 1991 by a group of former commanders' wives at the Army War College and published there in a booklet entitled "Choices & Challenges." Following is an adaptation of their list of "unofficial responsibilities":

- leading spouse of NCO/enlisted spouses in unit
- maintains close relationship to unit NCO/enlisted spouses
- maintains close working relationship with commander's spouse
- link between commander's spouse and NCO/enlisted spouses
- supports commander's spouse and keeps her/him informed
- gathers and passes information
- organizes activities
- conducts meetings
- focuses not only on unit but also community affairs

The following discussion describes these spouse-leadership roles for different levels:

The Squad Leader's and Platoon Sergeant's Spouse

The squad leader's spouse is part of the smallest independent operating unit—"The Institutional Foundation of the Army!". The Squad Leader is the first leader in the soldier's Chain of Command. As your soldier's spouse in this first-line leader position, you have the most important and incredible opportunity to create a positive relationship among your spouses and to build trust in our Army Family. You will work closely with the other squad leader spouses in the platoon (as there is not an officer counterpart spouse at this level) and also with the platoon sergeant's spouse.

As the platoon sergeant's spouse, usually four squad leader spouses will look to you for coaching and mentoring, along with guidance on available resources and training. You will also have the chance to support and provide advice to the platoon leader's spouse and the first sergeant's spouse. Take advantage of this opportunity for self-growth and to enhance your leadership skills.

First Sergeants' Spouses

As a first sergeant's spouse, you are not only the most experienced spouse of the company, but also the leading lady or gentleman for the company (battery/troop) NCO/EM spouses. This puts you in a good position to get to know and make friends with these spouses and their families, especially if they are younger spouses new to the Army. By doing so, you can stay abreast of any potential problems, take action, provide resources, or give advice to help in difficult situations. It's particularly important to encourage spouses who don't work outside the home to participate in the unit and community activities, and to take advantage of the social and educational opportunities that are available to them. As a leader, you have a great opportunity to help others and to be a mentor.

Your "spouse team" is made up of the battalion (squadron) command sergeant major's spouse and the company (battery/troop) commander's spouse. Work closely with them and support them in any projects they undertake. Keep them informed of activities at your level and any potential problems that would affect them. Turn to them for advice and counsel on difficult problems. Share "Best Practices" that they could possibly pass onto sister companies in your battalion. They may each ask to meet with you periodically in order to make plans and share important unit and community news that needs to be passed to the company spouses. You may also want to take this opportunity to invite them to participate in your company's (battery/troop) activities.

> BLUF
>
> "Becoming informed is a proactive sport."
> *General David G. Perkins, 15th TRADOC Commander*

Take responsibility for staying informed yourself. While it's true that the details concerning unit and community happenings may be passed to you by the other members of your spouse team, don't sit back and wait for others to bring you the information. If everyone strives to find things out for oneself and then shares that information with others, we will all be better informed.

You can communicate with the company (battery/troop) spouses in a variety of ways—at unit functions, Family Readiness Group meetings, newsletters, bulletin boards, or the unit's legalized social media forum. Of course, you shouldn't have to do everything by yourself; suggest that the spouses of the platoon sergeants in the company also get involved. They can help you plan and put on company-level spouse get-togethers, and they can also meet separately with the spouses in their soldier's platoons. Experience shows that very young enlisted spouses usually respond better to getting acquainted on the platoon level first, before they feel comfortable at company (battery/troop) functions. Give them an opportunity to be part of the team, to take ownership, and to provide them with a sense of belonging.

Finally, in addition to being a part of the "Chain of Concern" for the families in your soldier's unit, you will want to stand ready to assist the battalion commander's spouse to welcome and farewell the command sergeants major spouses.

Command Sergeants Major Spouses

When your soldier becomes a command sergeant major, your role broadens greatly; if you choose to be involved, you are now the leading spouse of all of the unit's noncommissioned and enlisted spouses. Your new position puts you at a vantage point to see the unit as a whole. From this viewpoint, watch for problem areas that relate to families; listen to and evaluate the concerns they express to you. Share these family-related issues with your soldier who is in a unique position to help improve the situations. Effective leaders spend a significant amount of their time working to improve the quality of life for their subordinates and families. Command sergeants major are relied on heavily for their information concerning these issues. You can help in this effort by pointing out the family-related problem areas that come to your attention, and by monitoring the effectiveness of corrective actions.

As the leading spouse for the NCO/EM spouses of the unit, you are the one who gathers unit and community news for the first sergeants' spouses (at higher levels, for the CSM spouses of the subordinate units) to pass along to the junior spouses. Some of this information should come from the unit's commanding officer's spouse. That's one of the many reasons it's essential to keep the lines of communication open between the two of you. You both make up an important team, "Battle Buddies—a united front." It helps for you to meet frequently to compare views on issues, share information, plan your joint efforts, support each other, and become friends. For these same reasons, you will also want to maintain a close working relationship with the spouse of the next higher command sergeant major. You may also work with your sister brigades/battalions CSM spouses who will provide valuable support and relationships. All these relationships are critical as they "set the tone" in your unit.

You will also have an expanded opportunity to become involved with community affairs, such as serving as a unit representative on various boards. This works to everyone's advantage—you bring to the community the perspective of the NCO/EM families in your soldier's unit, and the community looks to you to pass information back to those families. If you are a spouse of a nominative command sergeant major, you will be exposed to a wider range and scope of the Army. This will give you the opportunity to make a broader impact.

Finally, as mentioned previously, the Army provides professional spouse training and courses to complement your soldier's position at different levels. It is important to become familiar with the various training available for you and for you to share with the other spouses. Therefore, Army spouses need to be aware of the opportunities available to them throughout their Army journey with their soldier. Mostly, it is essential that you, "the Army spouse," know that you are never alone on this Army journey!

LADDER OF COMMUNICATION AND SUPPORT

The relationships described in the preceding discussion may be more understandable in the following visual form shown on the next page. You will note that at every level, the spouse of the NCO and the spouse of that unit's commanding officer usually look to each other for support, information, and assistance—"Battle Buddies." Again, this relationship is important in contributing to the tone of your unit's atmosphere. In each case, if the appropriate NCO's or officer's spouse is not available or willing to take on this role, the spouse of the next-ranking person may be asked to assist. It is important to note that the relationships depicted on this ladder are not intended to limit or preclude other relationships from being formed.

Chain of Concern

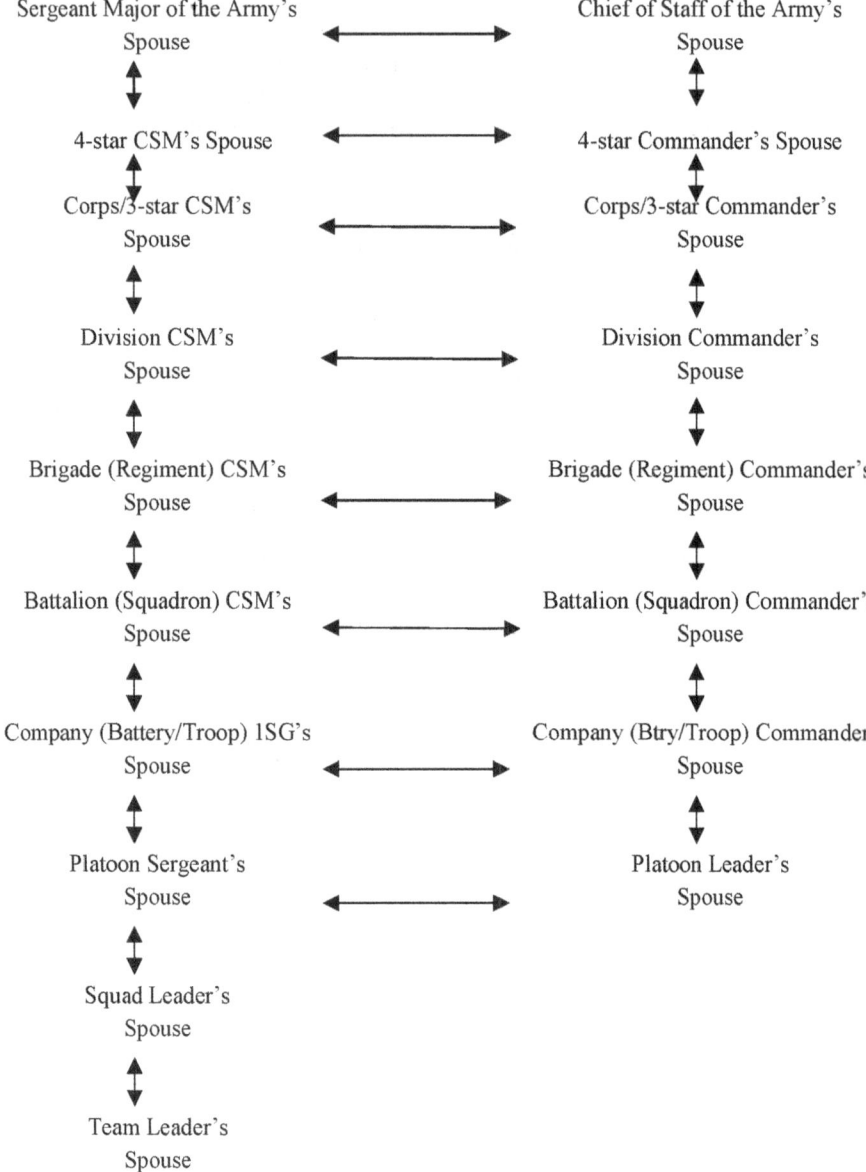

> **BLUF**
>
> We are leaders at all levels.

WHO'S IN CHARGE?
The person who assumes the leadership position at each level is in charge of the spouse functions that occur at that level. For example, for battalion (squadron) NCO/EM spouse functions, the spouse of the battalion command sergeant major is normally the person in charge, and depends on the assistance of the first sergeants' spouses and any other senior NCO spouses in the unit. For NCO/EM spouse functions at the company (battery/troop) level, the unit first sergeant's spouse, with the help of the other spouses in the unit, is usually in charge. The platoon sergeant's spouse can organize enlisted spouse get-togethers within the platoon. At each level, if the appropriate NCO's spouse or corresponding officer's spouse-team member is not available or willing to take on their leadership role, usually the spouse of the next ranking person should be asked if they are willing to do so.

Whether or not the officer's spouse-team member and/or the next higher-level NCO's spouse are invited to attend the NCO/EM spouse get-togethers on a regular or occasional basis is up to the NCO's spouse who's in charge. If they think the other spouses in the group would feel comfortable with them, then, by all means, invite one or both. There are many advantages to everyone getting to know one another. You, as the leader in charge at your level, may want to utilize these times to build relationship and mentorship opportunities—"lead by example." However, neither the senior officer's spouse nor the senior NCO's spouse should be expected to take charge. The responsibility for planning the agenda, running the meeting, and making the announcements belongs to the NCO's spouse who is the leading lady or gentleman of that group.

KEEPING THE PROPER PERSPECTIVE
One of the significant aspects of life as the spouse of a first sergeant or command sergeant major is your increased level of visibility. This is a natural result of your expanded sphere of influence and activity. As a result, you have a wonderful opportunity to serve as a role model for the younger spouses. It might help to review the military and social manners discussed in this handbook. This information can make you feel comfortable in most social situations, and assist you in setting a standard for others to emulate. Also, be sure to read the chapters entitled "Coffees," "Commanders' Spouses," and "General Officers' Spouses." A great deal in those discussions will be useful in your leadership role.

As you approach your new role, remember that no "extra points" are given for doing more than any other first sergeant's or command sergeant major's spouse. Everyone should do what they feel comfortable doing, and should not be judged by someone else's performance. Your capabilities will grow as you strive to become an effective leader and learn to balance responsibilities to family, unit, and community. As your capabilities expand, you will find that you can take on more responsibilities with ease. So, relax and enjoy this special time in your life, without turning it into a struggle or a contest with others. Be yourself!

> BLUF
>
> "It is 'Quality versus Quantity' time that counts!"
> *Ginger Perkins, Army spouse*

It's important to realize that the amount of effort you put forth will not necessarily have a direct correlation with the level of visible results, but don't let that deter you. Some efforts produce more noticeable results than others, simply by their very nature. For example, working with a committee to organize and coordinate a "welcome home" for the unit after a long field exercise or overseas deployment obviously results in a happy reunion that can be enjoyed by all. However, working to plan and put on a Family Readiness Group meeting that turns out to be poorly attended doesn't produce the visible results you had hoped for. If the latter situation occurs, remind yourself that it was a success for those who attended, and that continued efforts and a pleasant attitude will eventually pay dividends. You won't always know the good you accomplish because the effect is not always immediate.

As you work with commanders' spouses, you will find that the majority are caring spouses who bend over backwards to help you. However, if you encounter one who is hard to get along with, don't be tempted to generalize and label all officers' spouses that way. Look at each person as an individual, and form your opinion only after you get to know them. Everyone deserves respect!

Your attitude toward this role of leadership is all-important. If you have a cheerful, positive attitude, others will respond in the same way—smiles are infectious! If you genuinely care about people, they will sense your concern and appreciate what you are trying to accomplish. If your soldier sees that you are a willing teammate, he or she will be proud of your efforts and the results they bring. Your cooperative efforts can draw the two of you closer together and give you more in common to discuss and share. Remember that you both need to be trusted listeners for one another; you are both leaders, and that can be a lonely job. Respect each other. As you support your soldier and care for the families in the unit and your Army community, you will gain great personal satisfaction in knowing that your efforts are improving the quality of life for everyone. You have truly become a contributing member of the Army Family.

WORKING SPOUSES

Those who work outside the home, as many Army spouses do, can still be involved with the unit spouse activities. It may require a little extra help and understanding from your soldier, your family, and the spouses in the unit—but it can be done. For functions such as a unit bake sale, it isn't necessary for you to attend in order to support the unit. You can bake a delicious goody and drop it off on your way to work, or send it with your soldier or a neighbor. If comp-time is an option available to you at work, consider using that time for spouse activities. If you are a unit's leading lady or gentleman, you can schedule spouse get-togethers at noontime or in the evening. Evening functions are usually better attended anyway since the soldiers can babysit and working spouses are free to come. It's definitely harder to balance work and unit responsibilities, but it can be done! Ask yourself, "What is my passion?" "What are my skill sets?" "Where can I be effective?" "How can I make a difference?"

If, after careful consideration, an NCO's spouse who is in a leadership position decides that they do not have enough time or interest to get involved with the spouse activities, they should give their recommendation or input to: their soldier, officer-spouse counterpart, and commander. That way, someone else can be asked to take the lead. As a leader, this is a responsible action to take so that the leadership and mentorship in the enlisted-spouse chain are not broken and the commander and officer's spouse are supported. The first logical choice is the spouse of the next-ranking person in the unit. If that person cannot assist, then look for someone else in the unit who is enthusiastic, capable, and willing. Some spouses are natural leaders and do not need the support of their soldiers' rank to take on the mantle of leadership—they only need to be given the chance. Once a leader is selected and appointed, it is crucial for the commander to announce that leader and for the NCO's spouse to support her/him. Make this known to your unit families. This will give the appointed leader a fair chance and will build a positive atmosphere, especially if that NCO spouse attends a unit function later. It's important to realize that there is a need for leadership at these levels. The NCO and enlisted families in the unit deserve a committed leader.

HERITAGE OF SERVICE

NCOs' spouses, like commanders' spouses, have traditionally felt strong responsibilities to the unit. History tells us that our ancestral Army spouses who traveled west in covered wagons spent a great deal of their time helping younger spouses and soldiers. Though our means of transportation have improved, nothing has lessened the young families' need for support and guidance. If anything, we've learned that these needs have increased, due to the isolation of living farther from their own families and the frequent, lengthy separations caused by field exercises and deployments. We've also observed that helping the young spouses and families has a secondary benefit—not only are the young families better off, but the military unit is as well. Soldiers are more content and can concentrate on their jobs better when they know that their spouses and families are happy and secure. We are making future leaders for our Army and for our communities.

Any responsibilities you assume as the spouse of an NCO are voluntary. No one can force you to take them on, and no one will pay you for doing so. However, there's no doubt that the Army, the unit, your soldier's commander and spouse, your soldier, and the families of the soldiers serving under your soldier can benefit from your involvement. If you question whether or not you want to get involved, consider this:

Soldiers are in the Army because they want to be—they are volunteers; they obviously are ambitious and capable; they are advancing in rank and responsibility—and you are being given the opportunity to move up with your soldier. You will derive much personal satisfaction from helping where you're needed, make a lot of friends, and feel that you are a part of a team effort to improve the quality of life for the unit families. You will also be continuing the Army spouse heritage of service.

BLUF

"Army Families for Life!"

---

* * * * * * * * * * * * *

NCOs' spouses are the vital link in the "chain of concern" between the unit and the enlisted spouses.

CHAPTER 32

General Officers' Spouses

Watch your thoughts for they become words,
Watch your words for they become actions,
Watch your actions, for they become habits,
Watch your habits for they become your character,
Watch your character for it becomes your destiny.
--*Ralph Waldo Emerson*

General officers and their spouses are not born in heaven, nor does lightning strike the moment that proud spouse pins a brigadier's star on their soldier's shoulder to change them into brilliant, infallible creatures! Generals climb the ladder of rank and responsibility just like everyone else. Most general officers and their spouses realize that they are just ordinary people, and they don't let the promotion change them. They feel extremely fortunate to have been selected to join the general officer ranks, and are cognizant of the added responsibilities the rank brings.

INCREASED OPPORTUNITIES

A general officer's spouse today has the same choice as other military spouses. They can choose to become involved with family-related unit and community activities, or they can choose not to. However, based upon their soldier's rank and the elevated status accorded to both of them, they have greatly increased opportunities to make significant contributions for their military community. As a result, most general officers' spouses feel a sense of responsibility to use those opportunities and become actively involved.

What are the opportunities that call these spouses to service? First, general officers' spouses have the best vantage point available to spouses for a broad overview of the unit and community. They often sit on many community boards; participate in community discussions and seminars regarding major family issues; and hear the views of the military, civilian, and spousal senior leaders on important concerns. As a result, general officers' spouses usually have a better understanding of the "big picture" than many others; they often know why some things are done, or not done, in the community before others even see the problem. Second, due to the respect accorded generals and their spouses, they are in a position to express their opinions to those in authority about key issues—and to have their opinions count. They need to be aware that what they say and do can have far-reaching consequences, as their views could be assumed to be the general's views. Third, general officers' spouses are usually mature, seasoned, experienced spouses who possess a wealth of knowledge about Army-family issues and military communities. Others will ask them for their advice, and their wise counsel can save untold time and effort. Finally, generals' spouses have the opportunity to serve as role models for all the junior spouses with whom they come in contact. Their behavior and good manners can have a positive influence on those around them, as well as on the next generation. Such meaningful opportunities are available to very few, so tread wisely.

PITFALLS TO GUARD AGAINST

The path that those in authority must walk is always marked with pitfalls. This is especially true for general officers and their spouses, because they are instantly accorded their special status without having time to grow accustomed to it. Though these dangers may sound deceptively simple, they are worth remembering.

- Because people treat general officers and their spouses special, it makes them feel special; that's only natural. The danger comes if they start believing they really are special. Guard against letting the preferential treatment you are accorded go to your head. Remember some actions are done simply in accordance with protocol standards, and recognize there is an inherent level of respect given to you as a result of your soldier's rank. Always remain humble of this recognition and verbally appreciative of people's acts of kindness and respect.
- Because people look upon general officers and their spouses as being special, they tend to repeat and report their every comment and deed. They also tend to consider their "wishes" as "commands" or "directives." With the popularity of social media, this also extends to emails, texts, tweets, posts, etc. Remember to think carefully about possible consequences before you speak, act, or type. This also applies to the concept that your words could be considered an extension of your soldier's thoughts and opinions on a particular topic.
- Because people often look to general officers' spouses for answers and solutions, they may begin to believe they are or should be all-knowing. However, no one can have all the answers. If you don't know, say so; then try to find out, or direct people to the appropriate sources of information. Carefully weigh if you need to have input into a discussion. There are times when simply being silent shows depth and courage, emphasizing that you are not simply speaking out to impress.
- Because people potentially feel there is an inherent level of authority attached to you as a result of your soldier's position, carefully guard against developing a reputation for excessive control. Remember the tone and climate that you establish should not be one that could be labeled as toxic, nor should you establish a climate where it is perceived that only a few chosen ones are a part of your "inner circle."

The easiest way for you, as a new general officer's spouse, to avoid these pitfalls is to stay the same person you were before your soldier was promoted, UNLESS you have a strong personality that can be viewed as overly directive in nature. If so, you might wish to consider carefully as you are dealing with people. By taking on your new role without "taking on airs," people will truly respect you, not just pretend to.

BLUF

What do you want your legacy to be?

SETTING THE STANDARD

General officers' spouses are the subject of great interest and visibility in every military community. People watch them, listen to what they say, note how they dress and behave, and report to others their every move. In essence, they often represent many people's impressions of Army leadership, generating such thoughts as: are they inclusive to all; do they subscribe to cliques; are they approachable; do they embrace the uniqueness of each community. What a tremendous burden, but also a wonderful opportunity to set the standard for others to emulate! This can be especially effective in the area of social graces. By being gracious and always taking the time to be thoughtful of others, their good manners encourage those around them to act in the same manner. By responding to invitations promptly, arriving at functions on time, and writing thank-you notes, their example reminds others to do the same. New generals' spouses should not be tempted to let their busy schedules serve as an excuse for letting these social graces slide.

One of the first tests of a new general officer and their spouse's social graces comes as soon as their name appears on the promotion list to brigadier general. Dozens of congratulatory letters and notes flow in, mostly to the soldier, but some addressed to both. These letters of congratulations should be acknowledged, in the manner that they were received, to thank the well-wishers for thinking of them. The fact that probably the soldier is up to his or her neck with work and the spouse is preparing to move doesn't matter. Since there will be a large number of letters to answer, the soldier can lighten this task by writing one general response that will apply to everyone and have the administrative assistant type it. Letters to close friends can be personalized with an added handwritten message on the actual typed letter. It's a lot of work, but responding to letters of congratulations should not be overlooked. Additionally, if the spouse feels there are particular senior spouses who have provided encouragement, support, guidance or assistance, it would be a great idea to write a note of appreciation during this tremendous time in your military journey. Regardless, remind your soldier to ensure this task is completed.

An additional example of setting the standard is quickly getting settled and attending various installation programs and ceremonies. By simply lending your support and presence, not only will other spouses do the same, but the powerful message to those who have worked tirelessly to plan and organize the event is that you value their efforts—you care. Lastly, it is a great way to learn about your new community and the people who comprise it, even if it is an installation where you have been assigned previously. This establishes confidence and trust between you and your community and enhances the overall climate of the installation.

SOCIAL (OFFICIAL) CALENDAR

If you thought your social schedule was busy before your soldier became a general officer, as the saying goes, "You ain't seen nothin' yet." Invitations will arrive both at the office as well as at home—some for you, some for your soldier and some for both of you. A few will arrive months in advance, others only hours ahead of time; there will be conflicts, "command performances," and invitations that you "should accept." All of these must be juggled with your soldier's work schedule; your commitments; those of the family's if you have children at home or aging parents; as well as your small inner voice that occasionally pleads, "Save a little quiet time for me." If your soldier is deployed, make sure you still participate and support activities, events, and programs as your schedule permits.

Who juggles all of these invitations and decides which ones to accept and which to regret? The general officer and their spouse should, not the administrative assistant. It's true that the assistant knows the general's work calendar, when they will be out of town—sometimes even before the spouse—and probably recognizes the "should accept" invitations. It's certainly a good idea to ask the assistant's advice, but the final decisions should always be made by the general and the general's spouse.

There are several methods by which the administrative assistant can be of assistance with invitations. Following are the two most common:

- An invitation cover sheet can be used to aid the decision-making process. The assistant attaches one of these forms (see example) to each of the invitations that arrive at the office, and sends them home that evening for a decision. You and your spouse review the invitations and check the appropriate action to be taken, then add the information about the accepted invitations to your personal calendar. Note any of your concerns on the cover sheets for the assistant, such as, "Please find out which uniform the CG will be wearing to this function." or "Please remind me to send a note of congratulations, even though we cannot attend." All invitations, with their cover sheets attached, are then returned to the office. The assistant follows up on the action indicated, makes a copy of the "accepted" invitations for filing and returns all invitations with cover sheets to you or your soldier, including those that were regretted. Any answers to your questions are addressed on the bottom of the cover sheets. This closes the loop, and you know the desired action has been taken. Remember to keep your copies of all invitation and cover sheets/notes. (A foolproof method for keeping track of all accepted invitations is described in Chapter 9, "Responding to Invitations.")
- Calendars can be used as an aid in this process. Instead of using an invitation cover sheet, the assistant copies both the invitation and the computer-generated calendar for the week of the function. A similar response block to the invitation cover sheet is added at the bottom. By having the week's calendar included, you can see at a glance the other engagements that must be considered. After you've indicated your response, the assistant takes the appropriate action, retains the form on which you marked your response for their files, and sends home the original of all "accepted" invitations.

Invitations that arrive at home and involve your spouse can either be sent to the office to begin the process, or you can respond to them yourself. However, before responding personally to any of these invitations, check with the assistant to make certain your spouse's calendar is free. Any invitations or appointments that involve your soldier, which you personally accept, must be coordinated with the assistant so that the event can be placed on the calendar accordingly. These personal invitations or appointments should be indicated on the official calendar as "BLOCKED." Do not put personal information on official calendars that are shared with others. As a word of caution, be mindful that some official invitations may require a ruling from the legal team to prevent possible issues or conflicts of interest.

Maintaining the social calendar for a general officer and their spouse is a dual process. It takes an assistant at the office who keeps the social events listed on the general's calendar, and the spouse at home who keeps a calendar of their own individual activities as well as their joint activities. To keep these two calendars coordinated and up-to-date at all times requires cooperation and communication between the general officer's spouse and the assistant. It's also helpful for the assistant to send home copies of the general's short-range and long-range calendars with those events highlighted that the spouse is invited to attend. Of course, these calendars are frequently changed, but they can serve as reminders of upcoming events and aid in long-range planning.

INVITATION COVER SHEET

```
INVITATION TO:_____-_
DAY:_____DATE:_____TIME:_____
FROM:_____
_____

OCCASION:_____
_____

PLACE:_____
_____

ATTIRE:_____
COMMENTS:_____
_____

CONFLICTS WITH:_____
_____

BG DOE              ACCEPT_____REGRET_____
MRS. DOE            ACCEPT_____ REGRET_____
ACTION TAKEN:_____
_____

RECEIVED:_____
```

When the administrative assistant is asked to write invitations for official functions that the general officer and the spouse will host jointly, the assistant should be given the guest list with addresses, as well as a sample of how the invitations should be written. (See Chapter 7, "Extending Invitations," for examples.) Although this method is not used frequently, for written invitations ensure the assistant uses an ink pen that will result in a quality invitation. More often today, invitations are prepared utilizing computers. If the assistant has been working at this level previously, she/he may already have historical copies of invitations used in the past. Do not hesitate to seek their advice if that is the case. General officer invitations and stationery are appropriate for functions the general hosts alone or with their spouse; however, for those events not involving the general, it is not appropriate for any of the general officer stationery to be used, or for the assistant to be asked to prepare the invitations or accept the responses. Again, never should the stationery be used by the spouse. Remember the assistant works for the general officer, and not for the spouse.

AIDES

There are two types of aides for general officers: officer aides and enlisted aides. The officer aide position (Aide-de-Camp) was created to help general officers with minor details and tasks that, if performed personally, would be at the expense of the general's time spent on primary duties. Officer aides can be recognized by the gold aiguillette worn on the shoulder of their formal and informal uniforms. Enlisted aides are assigned based on the representational responsibilities and rank of the general officer. An enlisted aide is responsible for keeping the general officer's uniforms and military personal equipment ready for use, and the entertainment areas of their quarters orderly and clean. He or she also assists in the planning and preparation for official social functions and dinners, prepares and serves the food in the general officer's quarters for the general officer, provides security for the quarters, and serves as a point of contact for the quarters. Additionally, the enlisted aide maintains knowledge and oversight of any supplies that are purchased in connection with the general officer's entertainment requirements, such as china, glassware, etc. When the general officer is assigned to a position that is authorized an enlisted aide, the general's spouse becomes involved with the enlisted aide's activities, because they center around the home. However, the enlisted aide does not work for the spouse—they simply collaborate.

Not all general officers are authorized aides. They are authorized based upon the general's rank and assignment. If your soldier never served as an aide as a young officer, the duties and purpose of an aide may be new to you. While aides may seem "ready, willing, and able," some cautions are in order. When a general officer has an aide, it is important to remember that the aide works for the general, not the general's spouse. If the aide is married, their spouse isn't part of a "package deal." Any assistance that an aide and/or their spouse provide to you, as the general's spouse, should be offered by them, not solicited by you. And if it is offered, weigh your decision very carefully as you determine whether the intent and impression given are appropriate.

Officer and enlisted aides volunteer and interview for their jobs. While both positions require intellect and diplomacy, the enlisted aide position also requires culinary and serving skills. Some general officers are fortunate enough to find an enlisted aide who has had prior training and experience in this position. If that's not the case, the Army does periodically have training seminars available for enlisted aides. The general officer should work with the staff to ensure the aide is provided an opportunity to take advantage of this training. Additionally, the general officer and the general's spouse can be of great help in reinforcing the skills required. Helping an enlisted aide to master the needed skills is a boon to everyone—the aide, the general officer, the spouse, as well as to any others who have the good fortune of having this well-trained individual assigned as their enlisted aide in the future.

STEWARDSHIP OF QUARTERS

Most generals live in government quarters especially designed as general officers' quarters. These can vary greatly from one location to another, whether they are new homes or large old homes that have been maintained over the years. In fact, some are even listed on the National Register of Historic Places, and a few have been the homes of famous people in our nation's history. It is great to gain a working knowledge of the history surrounding your home as it is interesting to share with people you entertain there. The rooms of general officers' quarters are usually larger than those of other quarters, especially the entertaining area; the kitchen may have more equipment to accommodate the increased level of entertaining expected of most generals, and the entertaining areas are frequently carpeted for the benefit of the guests. To live in general officers' quarters is considered to be quite an honor. Consequently, be happy to share your quarters as a wonderful and comfortable setting for unit and community events.

Unlike other government quarters, the money spent to maintain each set of general officers' quarters is tabulated separately. Each general who lives in these designated quarters is allowed a budget for its maintenance and improvements. Some believe that they should be frugal with their budget and use only as much of the government's money as necessary to maintain the quarters. However, true stewardship requires that those who occupy general officers' quarters be concerned with maintaining those quarters for the future, as well as the present.

General officers need to concern themselves with infrastructure of their quarters and any remodeling that would improve its functionality. While the facility engineers routinely schedule major renovations of other quarters as they become necessary, this is not always the case with general officers' quarters. The budget allotted for each set of general officers' quarters is intended to cover all but the most extensive work. Unused funds from several calendar quarters can be retained and spent within that fiscal year for projects too costly for one calendar quarter's budget.

If you live in general officers' quarters, learn about the budget for your quarters, and how much is typically expended for your utilities and routine maintenance. Consider whether or not the quarters need more extensive work in order to be preserved and maintained for future generations of generals and their families. Remember this is one area where the general officer's enlisted aide should be informed of all work projects in order to assist with the coordination of workers at the quarters.

COPING SKILLS FOR NEW SPOUSES

Not all general officers' spouses are well acquainted with the life of an Army spouse; some just happen to fall in love with and marry a general officer! For these spouses, this military lifestyle may seem very strange indeed. The many traditions, taboos, expressions, acronyms, and protocol may all be new to them; yet they are expected to behave as though they have the experience of their peers. Because every person is unique and brings their own personality and skills to bear on each situation, there can be no hard and fast guidelines. However, the following explanations and suggestions may help new spouses feel more quickly at home among experienced Army spouses:

Army spouses are a friendly group; they extend a warm welcome to every newcomer and can truly be called friends within a few minutes of your first handshake. They don't remain private and hold back as they realize it is important to embrace those who are new to the journey. If their ways seem overly enthusiastic, don't be suspicious of their motives; their smiles and willing-to-be-helpful manner are genuine. In return, they expect you to be equally as open and ready to make friends.

While many Army customs have remained traditional and formal, Army society in general is a reflection of our more casual American society. For example, Army spouses today are normally on a first-name basis, something unheard of only a few decades ago. There's less of a chasm between spouses whose soldiers are of different ranks, especially between officers' spouses and NCO spouses, than there once was. Don't be misled into thinking that your interpersonal relationships need to be kept on a formal basis in order for others to respect your position as a general officer's spouse. The key is balance, and the best advice to help you weave your way through this labyrinth of interpersonal relationships is to just relax and be yourself.

Other Army spouses who know that you're recently married will understand that there's a great deal for you to become familiar with concerning the Army way of life—a learning curve. Most of them have had years to learn all of this, grow into all of this, and you're trying to do so overnight. Don't hesitate to ask questions; others will be pleased that you're interested and honored that you asked them. And take full advantage of any conferences, panels, or mentoring programs that are available to aid you in building your knowledge.

It's easy for an Army spouse to become over-committed, and that's especially true for a general officer's spouse. It may seem as though everyone wants you to come to their meetings or sit on their board. If you choose to become actively involved, that's great—but guard against spreading yourself out too thinly. A great alternative is to enlist the assistance of other general officer spouses and perhaps the nominative command sergeant major's spouse to serve with you on a particular board so that, if your schedule does not always allow attendance, the other spouse can do so and then provide any necessary feedback. If you must make a choice, it's a good idea to seek the counsel of other generals' spouses concerning which activities need your time and attention the most.

Because you are new to the Army and sometimes younger than your contemporaries and the spouses of your soldier's subordinates, you may feel ill at ease in a number of situations in which you find yourself—especially if you are thrust into a leadership position. You may not know as much about "the system" as the other members of the group do, but they will still look to you for direction and guidance. If you find yourself in such a position, the best approach is to get as much background information as you can, discuss the situation with the group, ask for their suggestions, and, whenever possible, let the group decide. A cooperative attitude on your part will be more warmly received than a take-charge attitude. It helps to try to view your situation from their perspective. Lastly, if you have friends/mentors who are not located with you, you can always seek their input or opinions in confidence.

Remember, at the end of the day, it is a delicate balancing act—conducting yourself in a manner that people will want to emulate your good qualities, yet feeling you are approachable as another soldier's spouse.

* * * * * * * * * * * * *

"Be kind to each other; take care of each other,
and show your love for each other."
*Laura Bush from "The Faith of America's First Ladies"
by Jane Hampton Cook*

SECTION NINE: MILITARY LIVING

Chapter 33

Community Life

"Home is where the Army sends you!"
Anonymous

Living in an Army community can be a wonderful experience that gives you a sense of belonging, security, order, and camaraderie. Yet, it also requires effort and an understanding of military life. There are regulations and military courtesies that everyone is expected to know and follow on each Army post. Add to that the diversity of each area—family housing, post facilities, support services, opportunities for professional growth and personal development. It's easy to see that Army community life is varied and complex for everyone, especially those new to the military. This chapter focuses on those features of community military living that everyone needs to know. It also contains information that will be helpful to those assigned to duty in that unique and complex city—our nation's capital.

BLUF

"I cast my lot in with a soldier,
and where he was, was home to me."
Martha Summerhayes, wife of LT Jack Summerhayes,
8th Infantry, Co A, American Civil War

LIVING ON POST/OFF POST/CAMP/STATION
When your spouse comes home with orders, probably one of your first questions is, "Where are we going to live?" If you have never lived on an Army or other service post/base/camp/station before, you need to know that it is different from living in a civilian community. While it offers many advantages, there are also advantages to living off post; you will have to look at the options and then make your decision.

- The first issue to consider is availability. The privatized housing office is the source for information concerning the availability of post housing. The privatized housing office maintains a list of the various types of quarters available, based on the number of bedrooms. All soldiers with family members are eligible for government quarters.

- The second issue is quality. Significant emphasis has been put into government quarters in the last decade. The Army's privatization program has improved government quarters substantially and significantly increased the number of government quarters for all ranks on all major Army installations.
- The third issue is quality of life. When you live in military family housing, it is much like living in a small town. You will have neighbors who share many common interests and experiences. Some housing areas elect "mayors"; some host "town meetings" so families can provide feedback and also express their concerns and requests, but most of these are now a function of the privatized housing system on your installation. Besides the commissary, post exchange, hospital and/or health center, clubs, churches, Army Community Services (ACS) and sometimes on-post schools, there are many Army Family and Morale, Welfare and Recreation (MWR) programs and services, Child and Youth Services, Sports and Fitness Programs, entertainment and dining, Travel Office, Army Lodging, Automotive Centers, and Better Opportunities for single Soldiers (B.O.S.S.) – resources and community support enabling readiness and resiliency. Some additional and very important advantages to on-post living are: the security of a closed post, proximity to work (usually within walking or bike-riding distance), utility bills as part of the BAH you pay each month where you are responsible for the amount of electricity you use over the average, and maintenance for the quarters and its equipment. However, those who live in quarters share a mutual responsibility with the Army regarding maintenance of their living quarters, possibly lawn care, depending on your privatized housing contracts, energy conservation, and clearance inspection.

Living "off post" also offers advantages. You can select your own neighborhood and your children's schools. You can become more involved in civilian community activities. And a most important advantage, you can accrue equity in property if you buy a house. Of course, selling it when you are ready to leave may present a problem. If you decide to rent rather than buy, be sure to protect yourself with an adequate military clause in your lease in the event of an unexpected PCS. It may also be a good idea to obtain renter's insurance whether you rent on or off the installation as privatized housing does not provide insurance for your household property and most rental contracts recommend you obtain renter's insurance.

ON-POST COURTESIES

Regardless of where you live, you will spend time on post and need to be aware of the following military courtesies:
- When driving on post, common courtesy should be extended to the gate guards. The guard will ask you to present your ID card, or your state or federal identification. (You should automatically assume when entering a military installation that you and/or guest will have to show state and/or federal identification and be subject to a search.) Most Army installations are going to ID card readers at the gate. You scan your ID card in the reader to gain installation access, and the

guards are there to assist if there is a problem. If guards are still verifying IDs for access, acknowledge the guard as you drive through the gate. A smile, wave, or nod, and thank-you (even if he/she can't hear your words) are small thanks for the service these guards perform. When driving onto post at night, courteously dim your headlights as you approach so the guard can see you, and have your ID card available. Also, every person in the vehicle will have to present a picture ID, regardless if they are in the service or not. A driver's license is acceptable for those entering the gate with someone who has military or dependent ID card.

- Caution is necessary when driving past units of marching or running soldiers. Drive *very* slowly (5-10 m.p.h.) or stop, if necessary, to ensure their safety. Watch for soldier "road guards" who may signal for you to drive on, or suddenly run into the middle of the street to stop traffic for a passing unit. Always give the right-of-way to pedestrian traffic.

- Be careful to observe speed limits in military housing areas. Children tend to feel more protected because of the slow speed limits on post and are frequently not as cautious in play areas as they should be.

- Respect the designated reserved parking places at the various post facilities. In addition to the spaces marked for handicapped, there are frequently spaces for the cars of general officers, certain commanders and senior NCOs, Gold Star Family parking, and volunteers of the month. This reserved parking is provided not only out of respect for rank or position, but also because these community leaders' time is important and limited. Observe this courtesy and leave these places free, unless you are entitled to use them.

- Flag courtesy must be observed when you are on an Army post at Reveille or Retreat. (See "Flag Etiquette" in Chapter 29.) During the ceremony of hoisting or lowering the flag, all people present or outside and close enough to hear the music should face the flag (or the direction of the flagpole or sound of the music). Soldiers salute and civilians stand at attention with the right hand over the heart. Civilians should remove their headgear with their right hand and hold it at the left shoulder, the hand being over the heart. On most posts, if driving in a vehicle during the playing of Reveille or Retreat, the operator must come to a complete stop until the playing is complete and the flag is secured. The soldier is directed to get out of the car and civilians may stand outside of his/her vehicle to render appropriate honors. Children who are playing outside should be taught to stop and stand still for these few moments out of courtesy to the American flag.

> **BLUF**
>
> Be a part of the Army Family;
> it's only as good as your involvement!

BEING A GOOD NEIGHBOR
Wherever you live, you will want your relationship with your neighbors to be pleasant. Quarters on post often mean a high density of people and require special courtesies.

Welcome Courtesies - Welcoming newcomers is always a good first step. Any simple gesture of friendship will bring a smile to both of your faces, and get your relationship off to a good start. Don't be hesitant about calling too soon on a new arrival. During the morning, a pot of coffee or a jar of "instant" and some cups, or simply baked goods (not necessarily homemade), can be a good start to a lasting friendship. If it's lunch or dinner time and they're in the midst of unpacking, you can offer to run to the local "fast food" restaurant for burgers, especially if there are small children in the family. If you work all day, you might stop and pick up a dessert surprise for them on your way home. Also consider a simple "welcome" card with your contact information or a "Community Packet" with local area maps, brochures, or website addresses; these are a great addition or alternative, as food allergies may be considered when giving consumable gifts. Although much of what you can offer may not actually be needed, especially if the sponsorship program is working well, it is the gesture of friendship that's important. It's also important never to *assume* that someone else is looking after your new neighbors.

Children - Your children's behavior in the area where you live is an important part of being good neighbors. Simple courtesies are particularly important for everyone's harmony and safety in high-density areas, especially common in military housing areas. Children should be taught to play in their own yard, their friend's yards, or the designated play areas—not to play in or walk through other people's yards without permission. They should put their play equipment away when they're finished playing—not leave it on driveways, sidewalks, or in stairwells. All trash, such as candy wrappers, should be properly disposed of, and soft drink cans put in recycle bins—not left to clutter the area or to be picked up by someone else. It might be helpful to point out to your children that these are not special burdens you are imposing on them, but the way all good neighbors behave toward one another.

Pets - Well-loved pets are an important dimension of family life. Before you move, find out about any necessary shots or quarantine requirements at your new location. If you decide you are unable to take your pet with you, be sure to find it a good home. When living on post, you will need to register your pet with the post veterinary office and obtain the proper tags. Well-behaved, well-controlled pets are appreciated in any neighborhood; if you cannot guarantee your dog's behavior, be sure to keep it on a leash. Any damage your pet does to a neighbor's property should be compensated for immediately. Always observe designated pet areas; this is important not only for appearance, but for health reasons. High-density living in apartments makes walking your dog a must; cleaning up after your pet is also a must. Most installations have dog parks for your dog to run free and socialize with other dogs. You must ensure that your dog plays well with other dogs and you must maintain a presence at the dog park. The park is not a place to drop your dog off and leave alone. Also, you must clean up after your dog.

Simple good manners practiced by everyone in your neighborhood can make a housing area a real community.

BEING A GOOD SPONSOR

Whoever dreamed up the idea of assigning each incoming family a sponsor deserves a medal! Sponsorship assists soldiers and families during the reassignment process. We only need to devise a way to ensure that no family gets overlooked, and every sponsor takes this additional duty seriously. Of course, that's a tall order with no simple solution; as with every other program in the Army, the sponsorship program isn't perfect. However, everyone can do his or her part to make the program work.

When your soldier is assigned as a sponsor, you can assist by helping to welcome the incoming soldier and family to the unit and community. Sponsors make a lasting impression, and you only get one chance to make a positive first impression. You should both take the responsibility to heart. Start by helping the family prepare for their move to your area by providing all of the information you think they will need, as well as the answers to their questions—"pre-arrival assistance." Welcome them when they arrive, and assist them during their adjustment period. In other words, *treat them just as you would want to be treated.* These are some of the things you can do, in more detail:

1. As soon as you are given the family's name and address, contact them to provide information on housing, school, childcare, pets, climate, vehicle registration, and anything else pertinent to the area you think they should know. Sponsorship (Relocation) packets are usually available from the S1/G1/J1 at your command. Ensure you make contact with the family you are sponsoring as soon as possible (before they move). Provide them with the installation's home page and link for the Relocation Assistance Program at ACS. The availability of privatized government housing will need to be researched in order to provide accurate, up-to-date specifics. When you write, offer to provide answers to any remaining questions. If the family has special needs—e.g., child with a disability, large pet, unusual medical needs—stand ready to find out about any needed facilities or services in your area. The recommended way to provide sponsorship after initial coordination is to have the military-to-military member contact and the spouse-to-spouse contact as each has different expectations and questions regarding the move. Also, you may need to provide information for spousal employment, so ask if the dependent spouse will be looking for a job when

they relocate. The other information to obtain is whether the spouse will be looking in the private sector or transferring from a government-to-government position. The Military Spouse Preference Program is available to provide assistance and is located within each installation's Civilian Personnel Advisory Center.

2. If government housing will not be available as soon as the family arrives, assist with reservations for temporary living arrangements for them—then assist in finding other accommodations to tide them over until privatized housing is available.

3. Whether the temporary living accommodations are at a military facility or civilian hotel, many Family Readiness Groups or soldier foundations can help with items that can ease the transition to a new duty assignment. You can check with them ahead of time. Also, refer to the local ACS for temporary items the family may need. Your unit or their new neighborhood may provide a welcome wagon with a few extra amenities as a welcome to the family—fresh fruit basket, sodas, beer or a bottle of wine, and snacks for the children. Remember what you do for one family, please remember to do for all. Pay it forward. You don't have to spend much to let them know you care.

4. If the family must fly to this new assignment, meet them at the airport, or ensure that they have transportation to the temporary living accommodations you have arranged. By this time, you should know the size of the family and be able to arrange transportation appropriate for that number of people—plus their luggage and any pets. Before the family leaves their old location, be sure they know what transportation will be available for them, and how to find it.

5. The evening the family arrives, ensure they have their basic needs met for the first 24 hours, such as: where to find economical meals, economical means to entertain children, local grocery stores, etc. You may invite them to your home for dinner or provide a meal at their temporary accommodations. If the family doesn't have a car, you can offer to drive them, or tell them what local transportation is available.

6. During the first weeks, the family may need your help in learning where facilities are located and occasional transportation. You aren't "adopting" them, but you will want to help until they learn their way around and can arrange for their own transportation. Check with ACS, as they may offer a tour of the post especially for families new to the area.

7. Don't expect any more than a big thank-you for your efforts. Your reward comes in knowing that you've helped another Army family by making their move a bit easier. Hopefully, they will pass the favor on when they are assigned as a sponsor. If everyone does that, when your orders come, some terrific sponsor at your next assignment will be waiting to welcome you in the same warm manner.

The sponsorship program is generally managed by the gaining unit. This means that the unit at the new post to which the military member will be assigned is responsible for formally designating a sponsor. If you are overlooked, and no sponsor contacts you in a reasonable time after your soldier receives PCS orders, your spouse may request a sponsor. Officers may contact the adjutant of the new unit; NCO/EM may contact the CSM. It's important to know that this option is available to you.

The ultimate goal for the sponsorship program is to reinforce to our incoming soldiers and families that we are their "Army family." Sponsorship can provide the sense that they are never alone and helps to create a sense of belonging. It gives them the information they need to make sound decisions that are best for their family so they can settle quickly. A good sponsor experience will affect the whole family's well-being, retention, and resiliency. Check with your local ACS for the availability of "Sponsorship Training for Spouses." The Army wants the sponsorship program to work.

MILITARY TIME

Military time, like time in most of the world, is measured on a 24-hour clock. Consequently, operating hours for most on-post activities are listed using this method of telling time. It may be confusing at first, but once you are accustomed to it, the 24-hour clock is actually quite simple. Its principal feature is that it eliminates the need to differentiate between a.m. and p.m.

24-Hour Clock

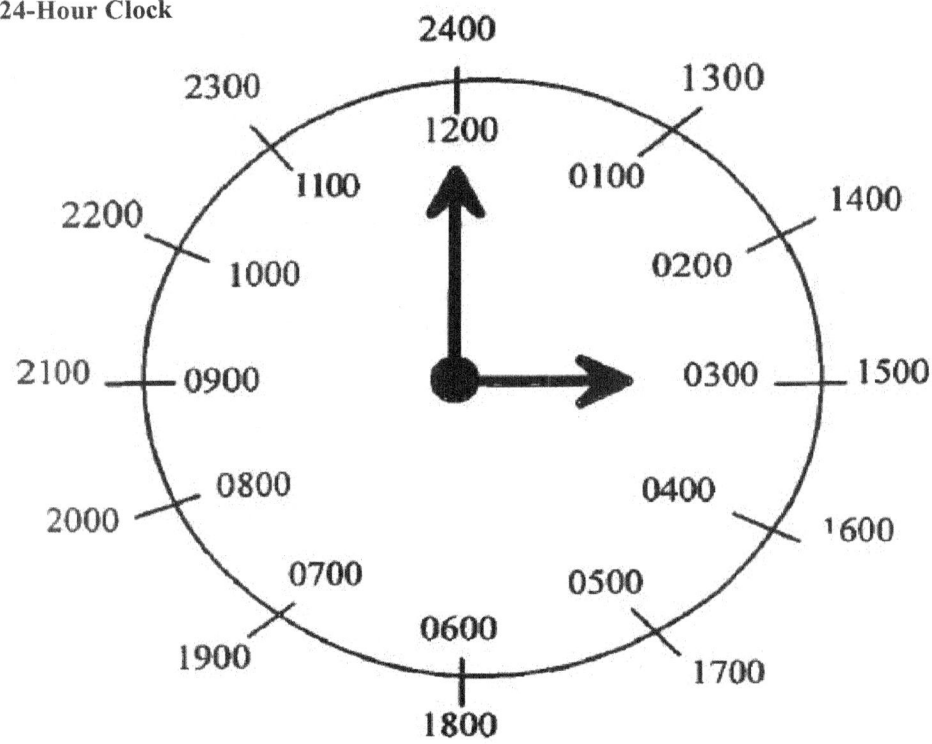

Unlike the 12-hour method of telling time, the 24-hour clock does not use a colon to divide the hour from the minutes. The following simple explanation will help you understand the 24-hour clock:

- The *first twelve hours of the day* look familiar to us because they correspond to the 12-hour clock. The only exception is that time, under the 24-hour system, always contains four (4) numbers. Therefore, for any hour before 10:00 a.m., a zero (0) is added before the hour. Thus, eight o'clock in the morning becomes "zero eight hundred," or 0800. From 10:00 a.m. to 12:00 a.m., these hours are referred to as "ten hundred," "eleven hundred" and "twelve hundred;" they are written as 1000, 1100, 1200.

- For the *remaining twelve hours of the day* (any hour between noon and midnight), simply add twelve hours to the time as it would appear on the 12-hour clock. For example, 2:00 p.m. becomes "fourteen hundred," written 1400 in military time. However, more frequently you will need to do your mental calculations the other way, by converting military time to the 12-hour clock. In that case, simply reverse the process; mentally add the colon, and then subtract twelve hours from the military time, to end up with "real time." Thus, a social function that begins at 1830, starts at 6:30 p.m. (1830 becomes 18:30, minus 12 becomes 6:30).

BLUF

As Jimmy Buffet says, "It's 1700 somewhere!"

SUPPORT SERVICES AND RESOURCES

Army spouses have access to many special programs and resources not equaled in most civilian communities: recreational facilities; arts and crafts centers; automotive hobby shops; physical fitness facilities; educational services that offer college-degree programs, vocational and technical certificates, as well as high school diplomas; legal assistance; youth centers and child development centers; chaplain support; Red Cross; Army Community Services; and, starting in 1995, Army Family Team Building.

Army Community Service (ACS) - It's a good idea to visit the local ACS office soon after you arrive at a new post. This military-sponsored agency, founded in 1965, staffed by both civilian professionals and dedicated volunteers, has grown over the years in the variety of services it offers to the military community. If you didn't receive a sponsorship packet before you arrived, you can pick one up at ACS. Packets are also available online to expedite learning about the post and local area. Some of the services they provide, regardless of your geographic location, are Deployment or Mobilization and Stability Support Operations Readiness to include Family Readiness Groups, Soldier and Family Readiness (emergency Family Assistance, Relocation Readiness, Employment Readiness, Financial Readiness) and Outreach service. They have information available on most Army posts; so when your soldier receives the next PCS orders, ACS can help you find out about where you're going. Other ways they can help: If, for example, your household goods haven't arrived yet, ask to look at the ACS loan closet where you will probably find the essential household items (such as dishes, pots and pans, iron, ironing board, etc.) that you can borrow temporarily. Be sure to find out what programs the local ACS is currently offering—typically such classes as financial planning, parenting, stress management, couples' communication, and training for childcare providers. Other worthwhile projects that some ACS offices sponsor are the layette program (basic clothing necessities provided for newborns to junior enlisted Army families) and assistance for spouses who need help in preparing to enter the job market. For soldiers, ACS is probably most important because it hosts the program known as Army Emergency Relief (AER). This important aid has assisted soldiers and families with interest-free loans, grants, and scholarships that can help you in areas such as emergency travel, funeral expenses, rent or mortgages, minor home repairs, and much more. Not only is ACS a good place to visit in order to find out about your community and what ACS can do for you and your family, but it's also a very worthwhile place to do volunteer work in your community. ACS serves as the information and referral service on many posts for volunteers. It also coordinates a post-wide volunteer recognition program.

Volunteerism - The volunteer has always been critical to Army community life, and Army spouses have traditionally responded to the need. Even with today's increased number of working spouses, many seek a balance between their salaried and volunteer time, thus satisfying both professional goals and community responsibilities. If you decide to volunteer, you will find that you are most effective in areas where you have a personal interest—working with the PTA or on a school advisory council if you have school-age children, financial counseling if you are a trained financial management expert, volunteering at the thrift shop if you have a retail-sales background. However, there are volunteer opportunities for all Army spouses, and no one needs prior experience to volunteer. In fact, many a military spouse has received training through a volunteer position that provided the background for them to later obtain a paid position in that field.

The Volunteer Management Information System (VMIS) provides many tools to manage volunteer activities for Army volunteers. The VMIS tools are public tools, volunteer tools, and volunteer supervisor tools. Public tools are available for all users on the site, including non-registered users of MyArmyOneSource.com website. Volunteer tools provide approved volunteers the ability to track their hours. Volunteer Supervisor Tools provide Army Volunteer Corps Coordinators the ability to manage and view volunteer program activities within their areas of responsibility.

Volunteers receive training and support through a variety of sources. For example:

- *VOLUNTEER - The National Center* is a civilian organization that offers important training for volunteers and volunteer managers at an annual conference in June. Since 1986, the Department of Defense (DOD) Office of Family Policy has joined the conference with DOD panel presentations that focus on volunteer issues

in the armed forces. Each of the services sends both professional volunteer managers and volunteers to the conference. The message is clear for volunteer Army spouses—the Army cares about supporting its volunteers!
- *Army Community Service* - ACS serves as the information and referral service on many posts for volunteers. It also coordinates a post-wide volunteer recognition program.

Spouse Employment – Spouse Employment is very critical to today's army spouses as many families rely on dual income, but spouse employment also provides self-worth, a sense of belonging, and skills sets. The army spouse is an asset in our communities; therefore, telling your stories on the challenges and best practices of spouse employment improves the journey, especially during PCS times. The following initiatives assist Army spouses in finding employment promptly before, during, and after each relocation:
- *Employment Readiness Program (ERP)* - Accessed through Army Community Centers, this program is designed to help Army spouses find employment and plan careers to fit their mobile lifestyle. The program assists with obtaining skills, networks, and resources. ERP is the local connection to our communities and creates partnerships for Employment Fairs (i.e., with the US Chamber of Commerce) and referrals for spouse employment opportunities.
- *Military OneSource* - This initiative enhances the Employment Readiness Program as it reaches beyond the local communities and provides a worldwide connection 24 hours a day. Military OneSource and ERP work hand-in-hand to provide the best possible support to our Army spouses. Military OneSource offers career exploration under Spouse Education and Career Opportunities (SECO), career connections, education, training, licensing, employment readiness, entrepreneurship, and My Career Advancement Account Scholarship (MyCAA). The Military Spouse Employment Program (MSEP) connects spouses with vetted employers, therefore providing a trusted "one-stop shopping" for spouse employment.
- *The Military Spouse Preference Program* - In accordance with DOD Policy, spouses of active-duty military members can, under certain circumstances, receive preference in hiring for civil service positions in the DOD. Preference means priority in consideration for a vacancy, and applies when a spouse accompanies the military member to a new duty station. You can get specific information at the Civilian Personnel Advisory Center (CPAC). Our Army, DOD, US Department of Labor, and the current administration, along with many of our community partners, are working together to improve spouse employment.
- *Spouse Education and Training* - This effort is to prepare and support Army families and to build and sustain family readiness with a positive impact.
- *Self-employment in military housing* - In 1985, the Army authorized the operation of home-based businesses in family military housing. However, any business you plan to operate from your quarters must be approved first by the installation commander. To obtain approval, prepare a written request describing your proposed

business and requesting permission to operate the business from your quarters. The installation commander will approve your request, so long as it's suitable for a military housing area and does not adversely impact on utilities and traffic.
- *State Licensing and Career Credentials Initiative* - Whether you're already a licensed or certified professional or are planning to become one, you won't have to dread going through the credentialing process every time you PCS to a new state. So keep making those career plans—soon it'll be easier than ever to do, or keep doing, what you love. Visit the CareerOneStop Licensed Occupations page to find out which occupations and states require a license or certification. The page describes the profession, provides the name and address of the licensing agency, and lists the types of licenses required. Some of the many occupations requiring licensing or certification include health care, education, real estate, childcare, and food services.

ARMY FAMILY TEAM BUILDING (AFTB)

The Army has taken another giant step forward in its effort to help Army families cope with this unique military lifestyle. In 1993, the Chief of Staff of the Army signed the Army Family Team Building Action Plan. This broad education program, which is being designed for both military members and their appropriate family members, is intended to ensure that all Army families understand Army life in general, what services are available to them, and how the various organizations on post can assist them. Additionally, training offered to spouses will help them improve their leadership and management skills, as they progress in their volunteer and/or command-team roles. Both military-member and family-member programs are designed to include progressive, sequential, and standardized training that is available throughout the Army.

The education program for family members is a modular training program that is organized into three levels of training. Level I (Military Knowledge) is for all spouses and appropriate family members who are new to the Army life. One will discover military acronyms, the structure and purpose of the chain of command, Family Readiness Groups, and resources to name a few. Level II (Personal Growth) is appropriate for those spouses who are about to take on positions of responsibility. Spouses have the opportunity to build upon and enhance personal growth skills in the areas of communication, stress management, time management, resiliency, and the development of healthy relationships. This is all in conjunction with the FRG, community agencies, or other military and civilian organizations. Level III (Leadership Development) is for spouses who have assumed positions of responsibility and are now confronted with planning, organizing, and managing. This training offers the development of advanced leadership and mentorship skills. It's important to point out that Army spouses have been very involved in developing this program.

This family-member training program is available to spouses of all ranks and positions; the level at which a spouse begins this training or when she/he progresses to the next level is determined by what's appropriate for her/him. A spouse can enter the system at any level, and progress up, or even go back, as their needs change. Spouse training is voluntary, but it is hoped that everyone will want to take advantage of this education. Keep in mind that the primary goal is to make Army families self-sufficient and self-reliant, especially during times of family separation. Another important consideration is that the leadership and management and resiliency skills taught are not only useful in volunteer roles, but also can be beneficial in paid-employment positions as well.

AFTB builds a great foundation for Army spouse training. To build upon it, the Army also offers many courses for Army spouses (many of which are accessible through your ACS or online). Formal training is offered to spouses throughout their soldier's career progression. The Army Family Team Building program demonstrates the Army's concern for all family members and its recognition of the important role they play in unit readiness.

As an Army family, you may want to familiarize yourselves with many of the non-profit and private organizations in your communities that are not only advocates and resources for military families but opportunities to volunteer and obtain gainful employment. Examples include Army Emergency Relief (AER), Association of the United States Army (AUSA), National Military Family Association (NMFA), United Service Organizations (USO), Red Cross, Easter Seals, Tragedy Assistance Program for Survivors (TAPS), Armed Services YMCA, United Through Reading (UTR), Hiring Our Heroes, Blue Star Families, to name a few. Check with the ACS Army Volunteer Corps and Military OneSource when interested in participating with an organization, as they can ease the overwhelming sensation on where to start and whom to trust. They are your network connection.

DUTY IN WASHINGTON, D.C.

If you are headed with your soldier for a tour of "duty" in our nation's capital, this section is written for you. Life in the Washington, D.C. area is guaranteed to be exhilarating, challenging, frustrating, and expensive! Happily, there are many compensating factors that will contribute to making your time in D.C. rank among the more memorable of your soldier's service assignments. You will be able to picnic in the shadow of the Washington monument while enjoying special concerts on the Mall, view exhibits at the Smithsonian and National Gallery, and, of course, visit the Pentagon, the Capitol and the White House. You will treasure memories such as taking visitors to the Vietnam Memorial for the first time and seeing the perfection of the Army's Twilight Tattoo or the "Evening Parade" at the Marine Barracks, long after you have forgotten the frustration of finding a place to live, the right schools for your children, a job/position suited to your own needs, the endless traffic, and the long lines at the commissary or hospital pharmacy. Here is some practical information to enhance living as an Army spouse in Washington, D.C.

Before You Arrive - Unless you are one of the lucky ones who gets quarters on one of the posts in the D.C. area, your first challenge will be to find a place to live. It's a good idea to contact your sponsor for the ACS-arrival packet and other crucial information as soon as possible. In addition, you could ask friends, or someone where your soldier (or you) will work, to suggest a reliable real estate agent. Expect to be bombarded with calls and packets when the "word gets out" that you are inbound. The packets that agents send may contain useful information about the different areas and schools, a description of available transportation, and maps of the area. Do some sleuthing on your own; your post library or bookstore will have helpful information. Of special note are Perry Smith's book, *Assignment Pentagon: The Insider's Guide to the Potomac Puzzle Palace,* and the USO's *The Guide to Washington.* Smith's personal view of life in the Pentagon is a helpful guide for all the family. Included are tips on house hunting, family and cultural opportunities, as well as how the "puzzle palace" really works. Teenage children will especially enjoy the list of "pentagonese," such as "green door" and "Potomac fever." The USO's guide is regarded by many as the most useful book available for those who live in the area. Another great source of information on the D.C. area can be found at the Armed Forces Hostess Association (AFHA) that has an office at the Metro Entrance of the Pentagon, (703) 614-0350. (You may request a packet to be mailed to you.) They maintain "Welcome Packets" or "Sponsor Packets" with a myriad of information for the Washington Area. The packets include an up-to-date Washington D.C. Area Information Handbook, maps of all the major military installations in the area, lists of businesses/hotels/restaurants giving military discounts, transportation options (Metrorail Pocket Guide), and information on the Pentagon, Thrift Shop, the National Mall, and memorial parks.

Employment - Employment opportunities in the Washington area are generally good. Fortunately, in the last decade an increased emphasis by the Department of Defense on military spouse employment assistance ensures that you will find a strong network in place. There are eight military-sponsored centers in the D.C. area that cooperate to provide employment assistance to military spouses. The services they provide range from job-search counseling to current job-bank data. The locations of these employment assistance centers are as follows:

Anacostia Naval Station, D.C.   Fort Myer, VA
Andrews AFB, MD   Henderson Hall, Arlington, VA
Bolling AFB, D.C.   Quantico, VA
Fort Belvoir, VA   Walter Reed Army Medical Center

Many of these resources may be linked to your present post through ACS. For those who have a home-based business that they plan to transfer to the D.C. area, if living on post you must receive permission from the installation commanding officer. Other resources and guidance for entrepreneurship and self-employment are Military OneSource, Department of Labor, and Small Business Administration.

Education - Educational opportunities in the area are limitless. There are study programs available from the GED to the Ph.D. level. Community colleges offer associate-degree programs in many areas and, depending on whether you live in Virginia or Maryland, each has excellent state universities (George Mason University in Fairfax and the University of Maryland in College Park). Georgetown University,

Howard University, American University, and George Washington University are centrally located in the District and have superb reputations. If you are currently enrolled in a program on post, check with your education-service officer to see if a comparable program is provided in D.C. Several colleges offer degree programs for military members and their families on posts worldwide. Your chances are very good that the same program will be available at one of the many military installations in the Washington area.

In addition to formal degree programs, there are numerous informal educational opportunities. For example, the Smithsonian Museum issues a catalog of courses (including digital archives) in a variety of fields for its members and members' children. The annual membership fee is nominal and includes museum publications, event calendars, and invitations to special events available only to "in residence" members. The county in which you live may also offer a wide variety of interesting and educational courses through its adult education program.

Volunteerism - Outstanding volunteer activities are available in the area. In addition to the normal ones, there are volunteer programs unique to the Washington, D.C. area. A few, such as the Arlington Lady or the Armed Forces Hostess Association, reflect the long history of service by military wives/spouses. The Arlington Lady volunteers are described in Chapter 28, "Military Funerals"; to volunteer, contact any D.C.-area spouse club. The Arlington Ladies constitution and by-laws only allow ladies to join. The Armed Forces Hostess Association has its office in the Pentagon. Their volunteers provide answers to the queries from military personnel and their families regarding any U.S. military facility in the world. The information comes from the extensive files that the Association maintains on all branches of service. The Association's only requirement is that you're a military spouse, so males are eligible to join.

Living in the D.C. area offers other unique opportunities to do volunteer work at the national level. For example, volunteers are always welcome at the headquarters of the National Military Family Association, located in Alexandria, Virginia; and you can check if volunteers are needed at the White House. During an election year, there are thousands of inaugural invitations to be hand-addressed by volunteers. Another opportunity not to be overlooked is the possibility to train as a docent for one of the national museums, or historic federal or state sites. Docents serve as guides, assist in the gift shop and tea room, and extend the professional staff—a wonderful opportunity to learn about our history and feel as though you are a small part of it.

Cultural - The pace of life in the Washington area is hectic. You will value your free time, but get ready—friends and relatives are sure to arrive and seek your services as a tour guide. You can, of course, send them on a tour or give them directions and let them go sightseeing on their own; however, when time permits, it is a great pleasure to guide them yourself. You'll find that you never tire of seeing our nation's capital. Many of the finest sights are free of charge or have only nominal admission fees, and public transportation is excellent. It's a good idea to keep a guidebook in your car (or saved on your phone); good used copies are often available at the thrift shop. A detailed map is *essential*.

Don't take long to venture out and discover the important cultural events that our nation's capital has to offer, because you can never have enough time to enjoy them all. The easiest way to get around and not struggle with finding a place to park is the Metro. You can purchase a metro card in various ways, from a daily to a monthly card. And here are some of the important sites to visit. The Kennedy Center, home for the National Symphony, as well as plays, ballets, and a variety of entertainment, has events to suit everyone. There are excellent theaters offering the latest from Broadway, including the famous Lincoln Theater. The finest Shakespeare can be seen at the Shakespeare Theater. In the summer, you can enjoy lawn seating and a picnic at Wolf Trap, an outdoor amphitheater located between Fairfax and Dulles Airport that features a wide variety of nationally and internationally known entertainers. There are also many opportunities for free musical entertainment in D.C. For example, each of the service bands provides marvelous summer band concerts at the various memorials and on the Mall. Not to be missed is The Old Guard's weekly Twilight Tattoo at Fort Myer in the summer. Other events, including the fabulous, nationally televised Fourth of July celebration, feature famous guest artists. Whether you're a fan of country and western, pop, march music, or opera, you can find it on the Mall.

Social and Recreational - Social life in Washington, D.C. has many variables that are determined by the nature of your soldier's assignment and your own personal circumstances. Since privatized government housing is limited and distances are great, most Army families live in civilian communities spread all around the Beltway. Personal entertaining tends to be with colleagues from work, friends from former assignments also stationed in the area, and neighbors. Yet, military units and staff directorates do gather periodically to welcome newcomers and honor those leaving or retiring.

Military Clubs - There are many clubs for military members in the area, and they are a great value. The larger clubs usually have swimming pools which can be a welcome benefit for the club members during those hot D.C. summers, and those with restaurants frequently offer a bountiful Sunday brunch, in addition to the regular club amenities. All-ranks clubs are located at most of the military installations in the area. Many occupy historic buildings with panoramic views of the city or the Potomac River. The private Army Navy Country Club located in Arlington, Virginia, is well known for its fine golf course and other recreational facilities.

Spouse Clubs - The Army Officers' Spouse Club of the Greater Washington Area is one of the oldest spouse clubs in the Army. It was reorganized in 1947 when Mary Bradley (Mrs. Omar Bradley) and Mamie Eisenhower (Mrs. Dwight Eisenhower) invited a group of women to lunch at Quarters One at Fort Myer to discuss activities and projects to encourage "young wives." Since that era, it has been a vital part of the Washington Army community. It has evolved into a non-profit organization that strives to provide and promote socio-economic and educational outreach to the National Capitol Region Army community and to organizations benefiting soldiers. In addition, the inclusive participation offered to all spouses within the Army ranks promotes encouragement, support, and interaction among its diverse membership. The Army Spouse Club is extremely proud of its heritage, including the creation and exclusive ownership of the distinctive Army Wives' Club Seal. Today the club is a vital part of the Army family, enabling its membership to easily connect with old and new friends through an array of activities, like monthly luncheons, tours (including Quarters One), volunteer and fundraising opportunities, as well as other social events in the D.C. area like the Joint Armed Forces of Washington Luncheon.

Capital Area Military Spouses is an all ranks club in the National Capital Area open to all spouses of the Armed Forces, both active and retired. CAMS was formed in January 2016 as a new all-service organization looking to embrace and gain from the diversity that members from each of the services bring. This allows "Many Minds" to come together and reflect the "One Heart" of our military-spouse community.

Recreational facilities - Washington's civilian and military communities abound in many types of recreational facilities. Every Friday, *The Washington Post* publishes an entire section describing the events that will take place during the weekend and upcoming week. Many events are free, and the variety is phenomenal. They also publish an online "Going Out Guide" for the D.C. area nightlife, events, and dining. A trip to the wonderful D.C. zoo is not to be missed. Those who are interested in professional and college sports will have no difficulty filling their free time. And the military offers not only the normal on-post recreational facilities, but also rental cottages at nearby coastal areas to provide an inexpensive get-away. Stop by the travel and ticket office on post for additional information and possible discounted tickets.

Historic sights - Washington, D.C. is filled with historic sights. Its famous buildings, monuments, and documents attract tourists from all over the world every day. You have the privilege of living within easy access and will want to see them all. Your heritage as an American citizen and the history of the country you serve will come alive as you view the Declaration of Independence at the National Archives, read Thomas Jefferson's words on the walls of his memorial, and tour the White House. The reality of today's issues will unfold before you when you sit in the gallery of the United States Congress. These are opportunities not to be missed!

Washington, D.C. is one of the most beautiful cities in the world, but not everyone is eager to live there. Army spouses may find that they look forward to living and working in the Washington area far more than the military member. The long hours, travel time to and from work, and lack of immediate, visible results from their efforts usually combine to make most military members dread an assignment to D.C. Nevertheless, for the family, it can be one of the most memorable assignments, if you use your time there to enjoy the wonderful experiences Washington, D.C. has to offer.

KEEPING INFORMED

> **BLUF**
>
> "Becoming informed is a proactive sport."
>
> *General David G. Perkins,*
> *the 15th TRADOC Commander*

Living gracefully in any Army community is considerably easier and more enjoyable if you are well informed. Not only is it important to keep up with what's happening in your own military community, but it's also important to keep abreast of the significant issues being discussed and decided upon in Washington that can have an impact upon your life.

There are a number of periodicals available that can serve to keep you informed. For local news, look to your own post or division weekly newspaper. For Army-wide news, there are a variety of weekly, monthly, and quarterly publications. The best known is probably *The Army Times*, published weekly and available either by subscription, through AAFES, or online. Besides reporting on topics of current interest to the Army, this popular newspaper also publishes all promotion and command lists. Other publications of interest, available by subscription and at your local post library, are *Armed Forces Journal International*, *Army*, *Military Review*, and *Soldiers*. Each provides insights about the Army from different prospectives.

* * * * * * * * * * * * *

Generations of military spouses have helped to shape, and benefited from, the community life of America's armed forces.

Chapter 34

Overseas Assignments

> ... for whither thou goest, I will go; ...
> *Ruth 1:16*

As the spouse of a member of the U.S. Army, you may have the opportunity to accompany your soldier on an overseas assignment. The United States has maintained a military presence abroad since the late nineteenth century and the U.S. Army has been a critical part of that presence. The nature of the Army's peacetime mission after World War II and for much of this century has meant that Army families are a vital extension of the Army overseas. Clearly, the foreign and domestic changes occurring in the late twentieth century and the subsequent global restructuring are reducing this presence, and force reductions abroad mean fewer families living overseas. Yet, because of the continued global commitment of the Army, you and your family may still have the experience of living outside the United States sometime during your soldier's career. In preparation for living, or even traveling, abroad, it's important to recognize that social customs and traditions vary in this world's multitude of diverse cultures.

OVERSEAS LIVING

Duty in a foreign country will provide you with many exciting opportunities, but your time overseas will be far more enjoyable if you take time to learn about the country and its culture before you go. Research is the key. Your local ACS office has information on most Army locations worldwide. When PCSing to a foreign country, some stateside ACS locations offer a class to help prepare you for your overseas move. After arrival to your new overseas location, check in with your local ACS to sign up for any familiarization courses they offer. Many offer free, foreign language classes, local tours, and tips to help you understand the customs and culture of your new location. It is important to note that a lot of internet resources are available. All posts have web pages and probably Facebook pages (official and unofficial) with valuable information. Ask questions regarding children, pets, and special considerations. There is one thing that is predictable about living overseas—it is going to be different! The next section may help you deal with some of those differences.

HOST NATION COURTESY AND CUSTOMS

Living or visiting overseas means that you and your family are guests in a foreign country. You also serve as a representative for America. What you say and do and wear are important. When a host countryman sees you in his country, he believes you are typical of all Americans. In order not to offend, it is important to learn what is considered appropriate and inappropriate behavior in that country.

There are many publications that can provide information about foreign nations.

For example, the Department of State publishes area handbooks that should be available at your post library. More current might be the "culturegrams" available from the David M. Kennedy Center for International Studies at Brigham Young University. These brief publications are available for over one hundred countries and are updated annually.

Your actions and the courtesy you show the members of the host nation are vital in cultivating cordial relations. Here are a few helpful tips arranged by region.

East and Southeast Asia - Asians have an unusually strong sense of politeness. You need to be aware of this and treat them in an equally courteous manner. Start by always referring to them as Asian, rather than Oriental.

- In Japan, a small bow before you shake hands is appropriate. Women, who seldom shake hands, simply bow slightly. In all of Asia, touching, as with a hug or pat on the arm or back, is not acceptable behavior.

- Use an individual's title and family name. Never presume that you are welcome to speak to, or refer to, someone by his or her first name.

- In most of the Far East, broad arm gestures and pointing with an extended finger are not considered polite. Neither is beckoning for someone to come to you with the index finger, palm up.

- Practice keeping your voice lowered when out in public or visiting, so as not to be offensive. Also, when conversing with a Japanese, it is not polite to look directly into the eyes; looking at the necktie knot is considered more respectful.

- If you are invited to a person's home, consider this a great honor. Normally, Asian homes are never used for entertaining, and certainly not for casual acquaintances. Because space is at a premium, the home is not usually considered appropriate for entertaining.

- Whether you are entertained in the home or, more likely, at a club or restaurant, it is proper to take a small hostess gift. When presenting a gift at a restaurant, it's best to wait until the end of the dinner to make the presentation. Gifts should always be nicely wrapped and are seldom opened in the presence of the giver. In Japan, the quality of the gift wrapping is considered as important as the gift itself. In South Korea, there is a strong sense of generosity, your host may also have a gift to present to you; let him present his gift first.

- In Japan, the number "4" equates to death. In fact, the Japanese word for four means death. Therefore, buildings do not have 4th floors, goods are not packaged in fours, and that number will never be seen in advertising. Remember that when giving a gift; it should not be in quantities of four.

- In Japan and Korea, one should remove one's shoes before entering a home. Even though your host may invite you to leave your shoes on, don't do so if the host family removes theirs. Be prepared to take your shoes off by either wearing socks or by keeping a pair of socks handy. Note that this custom is also observed at some restaurants. It helps if one wears shoes that can easily be removed, so as not to delay the event or the guests waiting to enter.

- In Korea, do not use red ink for your correspondence (even for signing Christmas cards), as this signifies death.
- Learn and practice the use of chopsticks, as a sign of your interest in the host country's customs.

Europe - Europeans continue to be more formal and private than Americans. Many have traveled extensively in the United States in recent years and have first-hand knowledge of our country and customs. However, as hosts, they will appreciate your observance of their customs.

- The handshake is standard greeting between both sexes. The French and Italians embrace with a "kiss" on each cheek; the Belgians, Danish, and Russians use three "air kisses" – not actual kisses on the cheek, starting with the right cheek.
- First names are rarely used, even among colleagues and long-term acquaintances. Those with academic titles and degrees expect them to be used, i.e., Frau Doctor or Herr Professor.
- Avoid personal questions, especially references to jobs and professional positions.
- Speak quietly when out in public or dining in a restaurant. Americans' typical loud manner of speaking is considered to be in very bad taste.
- Being entertained in someone's home is a great honor and will normally be a lengthy affair. Most Europeans are gracious hosts and will offer you special dishes and take great care in their preparation and presentation. Friendship, once offered, is considered lifelong.
- Whether in a private or public setting, accept your first drink but never take a sip until everyone has received their glass. A toast is usually offered before anyone in the group begins to drink.
- A small gift to your hostess is customary. If you give flowers, they must be cut. (They probably will be artistically arranged at the florists and enclosed in clear wrap.) Do not bring a potted plant, as you might do for an American. Avoid chrysanthemums that are associated with death, and red roses that are only for wives and sweethearts. Other gifts should be wrapped, with a gift card enclosed or attached. The gifts may or may not be opened in your presence.
- Even if you prefer to continue using the American-style of eating, do not put your hand in your lap while at the dining table. When you are not using your hands to eat, rest your wrists or upper arms (never your elbows) on the table edge, thus keeping your hands above the table at all times.
- In Spain and Italy, it is customary to dine late. If you arrive at a restaurant in Spain

before ten or eleven o'clock, you will probably dine alone; in Italy, dining starts around eight o'clock at night.

The Middle East - Although we tend to lump all countries in the "Arab world" together, accepted protocol may vary from one to another. Be alert for these differences. The most significant variation from our culture that you will find in Islamic countries is the way women are treated and regarded. They tend to be relegated to less-than-equal status, and have lifestyles very different from those of Western women. Even though you may not agree with it, you are expected to conform to the dress code prescribed for women by the host nation's religion, when you go out in public. Before you arrive, make sure you have clothing that will comply with local customs.

- Women are seldom included in official entertaining. It is not appropriate to ask about a man's wife.
- Though men may shake one another's hands, they do not shake a woman's hand; to touch a woman is considered bad manners. A typical Arab handshake is limp and lingering, not at all like the quick, hearty handshake of most Americans.
- Muslims do not drink alcohol, and dietary rules based upon their religion are very important to them.
- Accept all of the cups of coffee or tea offered, so as not to offend your host.
- Use only the right hand for greeting, eating, drinking, or when offering or receiving anything. The left hand is considered unclean because, in their tradition, it is only used for sanitation purposes; therefore, to use it for any other purpose is considered a gross insult.
- The "space bubble" of someone from the Middle East is smaller than yours. When they stand very close to converse, do not back away, as this would be impolite.
- Women should keep their arms and shoulders covered and avoid crossing their legs. In addition, in some countries in the Middle East such as Saudi Arabia, women must keep their legs covered.
- Keep both feet on the floor in order to avoid exposing the soles of your shoes or feet. To do so implies that you consider the viewer "under your feet," or equal to dirt.

Frequently, the AAFES bookstore will carry well-written booklets about the host nation's customs. Also, the post protocol office should be able to give you additional information. Since the proper etiquette concerning public behavior and entertaining in many countries is very different from our own, it is better to ask questions beforehand and save possible embarrassment.

LANGUAGE TRAINING

When you live overseas, you have an opportunity to be an informal ambassador of the United States within your host country, as well as reap the benefits derived from living in another culture. One of the best ways to accomplish both is to learn as much of the local language as you can. Learning at least a few words of the local language is possible for everyone. Polite greetings, please and thank you, and other short phrases go a long way toward dispelling the impression of an American who expects everyone to speak English. By simply making the effort, the sincerity of your goodwill is eloquently communicated. Remember that the behavior and attitude of Americans living abroad deeply influence the host nation's impression of all Americans.

Classes in the language of the host nation are offered on almost all overseas Army installations. Check with the education center and ACS for class schedules and times. The more you learn to communicate in the local language, the more you will enjoy your tour and establish lasting friendships. You will also have the added benefit of learning another language among native speakers, something few Americans have the opportunity to do. (Also see "Attaché Duty" in Chapter 35.)

HOME LINK

All local media can be found at the local post/base exchange. *USA Today* and the *New York Times* are now available worldwide and provide a broad view of current events; likewise, the *International Herald Tribune*, one of the best newspapers in the world, is also available. The American Armed Forces Radio and Television Network broadcasts on radio (AM and FM) and television for the military audience overseas. It offers a wide variety of stateside television programs—news, sports events, and many of the popular shows being seen by viewers in the United States. AFN has multiple channels for movies, sports, children's programs, etc., and cable is now available overseas.

Some television programs are broadcast live via satellite, such as the news and some sports events; however, most are broadcast from tapes, which are especially useful in overseas areas because of the time difference. Also provided are news segments on local events, both from the military and civilian communities. Another source of news to those living or traveling overseas may also be the U.S.-based Cable News Network (CNN). Just as in America, this continuous news broadcast can now be seen in many countries. The material presented is more international in content, a fact that makes it even more interesting to watch, should you have the opportunity.

* * * * * * * * * * * * *

Learning about and respecting cultures different
from one's own are international courtesies.

Chapter 35

Joint Assignments: Service Traditions

"First to fight for the right...."
"Anchors aweigh my boys...."
"Off we go into the wild blue yonder...."
"From the Halls of Montezuma...."
— *Service Songs*

The various United States Armed Forces must coordinate their efforts in order to protect America and its interests abroad. To facilitate this coordination, a limited number of members from each service are routinely assigned to a tour of duty with a unified command or specified command or another service. As the spouse of a soldier, you may have the opportunity to accompany your soldier on such an assignment, and find yourself living on a post or base where many of the service professionals are not in the Army. These joint-service assignments offer a welcome opportunity to make new friends among the Army's sister services; yet, they also offer new challenges. While we all belong to the family of American Armed Forces and have many similarities, each service provides a different flavor of social activity.

DUTY WITH A UNIFIED COMBATANT COMMAND

Duty with a unified combatant command is one type of joint-service assignment. A unified combatant command includes components from two or more services and has a broad and continuing mission. Examples are the United States European Command, with its headquarters in Stuttgart, Germany; and the United States Pacific Command, with its headquarters at Camp Smith, Hawaii. Members serving in such joint-service assignments refer to themselves as "purple suiters." This term is a reference to their mythical uniform that is neither green, white, blue, or olive—but referred to as "purple," symbolizing a mixture of all the services.

Unified combatant commands present an environment and social life similar to that found at an Army major command, with the added opportunity to interact with members and families from sister services. You will quickly observe that each service has developed its own special traditions, customs, and ceremonies. In a unified combatant command, the members of each service tend to retain their own individual service traditions when in a service-specific forum (both socially and formalized ceremonies). However, spouses of different services blend those traditions and customs to create good harmony and esprit among all the spouses, regardless of service. Spouses in the "purple environment" try to put aside their own service customs to make sure every spouse feels welcome and part of the Team. Each service has programs and spouse events that are very important but, to create an atmosphere where all spouses feel equal, sometimes those programs and traditions are set aside or blended for the betterment of the Command.

> **BLUF**
>
> "Embrace new traditions and practices while sharing Army customs."
> *Brad Combs*

DUTY WITH ANOTHER SERVICE

Duty with another service is the second type of joint-service assignment. Because there are only two services working together in this type of duty, it is an especially exciting and challenging time. Under these circumstances, both you and your spouse serve as representatives of the Army and learn firsthand about the practices and traditions of the other service. Typically, Air Force members are assigned to duty with the Army because of the Army's need for airlift and close air support. Since the Navy and Marines seldom provide these functions for the Army, joint assignments with those services are more limited. Army members are usually assigned to the Air Force for the purpose of ground liaison.

SISTER SERVICES

In order to understand and appreciate the differences between each service, it is important to know a bit about their history, structure, and mission.

The U.S. Air Force - This is the youngest member of America's armed forces. It traces its lineage back to the Air Service that was formally organized in 1920, after a fledgling start in 1907. The Army Air Corps built a strong base of public support after it was created by the 1926 Air Corps Act. Commanded by Gen. H. H. "Hap" Arnold during World War II, the corps became the Army Air Force, until the National Security Act established a separate Department of the Air Force in 1947. The mission for the redefined force was to provide the country with the air power needed to defend it at home and to meet the nation's commitments abroad. The Air Force is divided into ten major air commands, under which there are numbered air force units, wings (the basic unit), groups, and squadrons. Air Force servicemembers, male and female, are called airmen. "The Air Force Song," beginning with the familiar words, "Off we go into the wild blue yonder," is this service's official song. Its colors are ultramarine blue and golden yellow.

The U.S. Navy - The Navy is the nation's second senior service after the Army, founded in October 1775, by General Washington and authorized by the Continental Congress. The primary mission of the Navy is to protect the United States by the effective prosecution of war at sea, including the seizure or defense of advanced naval bases; to support the forces of the other military departments; and to maintain freedom of the seas. Navy servicemembers, male and female, are called sailors. Like the Army, the Navy has its own aviators, most trained for service on aircraft carriers. The Navy song is "Anchors Aweigh." Blue and gold are the official colors.

The U.S. Marines - The proud tradition of the U.S. Marine Corps began in November 1775, when the Continental Congress established two battalions of "American Marines." The Marine Corps is a separate service within the Department of the Navy, organized, trained, equipped, and ready to go almost anywhere in the world. The Marine Corps includes Operating Forces, which emphasize air-ground operations both ashore and afloat. Marines also provide the security detachments to protect our embassies around the world. The Corps is especially proud of the U.S. Marine Band, known as "The President's Own," headquartered at the Marine Barracks in Washington, D.C., and led at one time by John Philip Sousa. The Marine Corps Hymn is better known as "From the Halls of Montezuma." The official colors are red and gold.

The U.S. Coast Guard - The Coast Guard, created in 1790, is actually the oldest of America's seagoing armed forces when measured by continuous service. In peacetime, the Coast Guard is responsible for maritime safety and law enforcement, and operates under the Department of Transportation, not the Department of Defense (DOD). In wartime, the Coast Guard can come under the jurisdiction of the Navy and DOD. The Coast Guard motto is *Semper Paratus*—always ready; this is also the name of its famous marching song. Blue and white are the Coast Guard colors.

The Reserves - The Reserve forces of each of the five services and the National Guard serve as a trained and committed part of the "total force," ready to ensure national and international security. The Reserve forces belong to the federal government. The Army National Guard and Air National Guard belong to the individual states. The National Guard has a dual status: as state troops under the state governors that can be called upon in times of local emergencies; and as reserve components of the U.S. Army and Air Force that can be called to active federal service as needed.

SERVICE CUSTOMS

One of the most interesting aspects of assignment with another service, or serving in a unified combatant command, is the opportunity to observe the customs of the other services. Though many are the same as those of the Army, a few are interestingly different. You will be able to experience and perhaps participate in the ceremonies unique to their way of life. Here are guidelines to help you understand some of the special customs and traditions of each service.

Air Force—Because the origin of the Air Force is the Army's Air Service, many Air Force traditions are adaptations of Army traditions. However, the Air Force has developed its own unique social style.

- In Air Force receiving lines, the sponsor precedes his or her spouse/date through the receiving line. In the event of a dual-military couple, the servicemember whose unit is sponsoring the event goes first.

- At most Air Force "dining-outs," the first glass of wine is poured with great ceremony—the decanters are passed from members hand-to-hand until all the glasses are filled, with care taken that the decanters never touch the table. Toasting includes, after the President, toasts to important individuals in the Air Force, rather than groups of individuals as in traditional Army toasts. The final toast is "To our honored guests," to which the appropriate response is, "Hear, hear." Later in the evening, any "infraction of the rules" usually requires a trip to the grog bowl for a drink of its ominous concoction as penalty for the error.

- Official Air Force functions may include a toast with water, "To our comrades, killed in action, missing in action, or prisoners of war."

- When driving on an Air Force base when the flag is lowered at Retreat (usually 4:30 p.m.), stop the car, turn off the engine, and remain in the car until the music ends. It is not necessary to get out and stand by the car.

Navy—Navy traditions may be the most unique. Many of the social customs are distinctly Navy.

- "Wetting downs" are Navy promotion parties. The custom is to pour a drink over the new stripe, thus wetting it down. This is supposed to make it blend with the older and presumably more-tarnished stripes.

- In Navy receiving lines, the military member precedes his or her spouse (or date).

- "All-hands" parties imply unit events that include everyone in the unit.

- "Mister" and "Miss" are the proper forms of social introductions for Navy officers below the rank of Commander.

- Shipboard etiquette: "Captain" is the courtesy title for the commanding officer of all naval vessels, regardless of actual rank.

- When going on board a Navy ship for a ceremonial occasion, such as a change of command, if there are "side boys" rendering honors to your soldier (side boys are sailors who participate in honors ceremonies on the quarter-deck), precede your soldier up the walkway (called a brow) and then move quickly out of the way when you hear the boatswain's (pronounced bosun's) pipe. Your soldier will return the salute as he steps on board and the piping ends.

- Low-heeled shoes are recommended when you are invited to tour a ship, for you will be doing a lot of walking and going up and down ladders. Avoid slim-fitting skirts. Slacks and rubber-soled shoes are advisable when visiting a submarine.

- When dining in the wardroom, the Captain is the first to begin eating.

Marine Corps—The *esprit de corps* of the U.S. Marine Corps is known worldwide. Because of its close association with the Navy, many of the social customs are similar, with some important additions.

- The birthday of the Marine Corps, November 10, 1775, is celebrated by Marines all over the world. The form of the party may vary, depending on the number of Marines, but it *will* be celebrated. Traditionally, there is a birthday cake-cutting ceremony with the youngest and oldest Marine participating. Whether a formal ball, dinner, or reception, it is a significant honor for non-Marines to be invited. A guest's expenses are normally paid for by the hosting Marine; it is appropriate to repay this courtesy with a comparable invitation.

- In the Marine Corps, only officers are invited to Mess Night. Spouses and dates are invited to Marine Corps dining-outs.

- In Marine receiving lines, the Marine precedes the spouse/date.

- If you visit Washington, D.C. during the summer months, you will want to attend "Evening Parade" at the Marine Barracks. The United States Marine Band and part of the Ceremonial Guard perform, following the special Retreat ceremony. Ask any Marine stationed in the area for ticket information.

Service Songs—One of the nicest traditions when serving with other services is the opportunity to stand when "your" service song is played at various social functions. The traditional end to a musical program will be a medley of service songs, from the youngest service to the oldest. Everyone associated with a particular branch stands as their service song is played.

Selected Special Dates—

June 14, 1775	U.S. Army anniversary
August 4, 1790	U.S. Coast Guard anniversary
September 18, 1947	U.S. Air Force anniversary
October 13, 1775	U.S. Navy anniversary
November 10, 1775	U.S. Marine Corps anniversary

ATTACHÉ DUTY

Another important type of joint-service assignment is attaché duty. Attachés represent the American military on the U.S. diplomatic staffs in foreign capitals. This duty is similar to the Foreign Service, in that the servicemembers and their families play a significant part in reflecting our country's image abroad. The wives and teenage children of the Army members (both officers and enlisted) assigned as attachés, or as members of the attaché staff, have an opportunity to receive training and a good overview of what to expect at their overseas station. They are encouraged to retain their own varied interests and involvements, while representing Americans in a positive way to the rest of the world.

Training for attaché spouses is offered concurrently with the servicemembers' course at the Defense Intelligence College, Department of Attaché Training, Bolling AFB. Approximately 85% of the attaché training program is open to the spouses. Additionally, a Young Diplomats program is offered to the teenage children. In this program, they have the opportunity to meet and talk with other teens who have had the experience of living in an overseas diplomatic environment. The Young Diplomats program also includes "Image Management: Dressing to Represent the United States Abroad," and other etiquette and protocol topics.

Language training for the overseas assignment is also considered an important part of the preparation for attaché duty. It may be offered at the Foreign Service Institute in Washington, D.C., or at the Defense Language Institute at Monterey, California. Those who are offered the opportunity to study a foreign language will want to accept. It is a very concentrated period of study; however, the language skills acquired are invaluable in helping to make the most of the challenging world of attaché duty.

RANK INSIGNIA CHART

The Army rank structure and insignia may be confusing at first. The fact that, within officer ranks, silver denotes a higher rank than gold insignia of the same design adds to the confusion. But ours is an army of strong tradition, and most of the insignia (as well as its coloration) predates the American military and the founding of our nation.

The following rank insignia chart provides a useful guide to the insignia of all services:

Enlisted Personnel

Pay Grade	Army and Marine Corps	Navy and Coast Guard	Air Force
E-1	Private	Seaman Recruit	Airman Basic
E-2	Army: Private (PV2) Marines: Private First Class (PFC)	Seaman Apprentice (SA)	Airman (Amn)
E-3	Army: Private First Class (PFC) Marines: Lance Corporal (LCpl)	Seaman (SN)	Airman First Class (A1C)
E-4	Army: Specialist (SPC) Corporal (CPL) Marines: Corporal (Cpl)	Petty Officer, Third Class (PO3)	Senior Airman (SrA)
E-5	Sergeant Army: (SGT) Marines: (Sgt)	Petty Officer, Second Class (PO2)	Staff Sergeant (SSgt)

Grade	Army and Marine Corps		Navy and Coast Guard		Air Force
E-6		Staff Sergeant **Army:** (SSG) **Marines:** (SSgt)		Petty Officer, First Class (PO1)	Technical Sergeant (TSgt)
E-7		**Army:** Sergeant First Class (SFC) **Marines:** Gunnery Sergeant (GySgt)		Chief Petty Officer (CPO)	Master Sergeant (MSgt) First Sergeant (—)
E-8		Master Sergeant **Army:** (MSG) **Marines:** (MSgt) First Sergeant **Army:** (1SG) **Marines:** (—)		Senior Chief Petty Officer (SCPO)	Senior Master Sergeant (SMSgt) First Sergeant (—)

Grade	Army and Marine Corps		Navy and Coast Guard		Air Force	
E-9		**Army:** Sergeant Major (SGM) Command Sergeant Major (CSM) **Marines:** Sergeant Major (SgtMaj) Master Gunnery Sergeant (MGySgt)		Master Chief Petty Officer (MCPO) Fleet/Command Master Chief Petty Officer (—)		Chief Master Sergeant (CMSgt) First Sergeant (—) Command Chief Master Sergeant (—)
	Special Grades					
E-9		Sergeant Major of the Army (SMA) Sergeant Major of the Marine Corps (SgtMajMC)		Master Chief Petty Officer of the Navy (MCPON) Master Chief Petty Officer of the Coast Guard (MCPOCG)		Chief Master Sergeant of the Air Force (CMSAF)

Grade	Warrant Officers			
	Army and Marine Corps		Navy and Coast Guard	Air Force
W-1		Warrant Officer 1 **Army:** (WO1) **Marines:** (WO)	DISCONTINUED 1975	NO WARRANT
W-2		Chief Warrant Officer 2 **Army:** (CW2) **Marines:** (CWO2)		Chief Warrant Officer 2 (CWO2)	NO WARRANT
W-3		Chief Warrant Officer 3 **Army:** (CW3) **Marines:** (CWO3)		Chief Warrant Officer 3 (CWO3)	NO WARRANT
W-4		Chief Warrant Officer 4 **Army:** (CW4) **Marines:** (CWO4)		Chief Warrant Officer 4 (CWO4)	NO WARRANT
W-5		Chief Warrant Officer 5 **Army:** (CW5) **Marines:** (CWO5)		Chief Warrant Officer 5 (CWO5)	NO WARRANT

Commissioned Officers			
Grade	Army and Marine Corps	Navy and Coast Guard	Air Force
O-1	Second Lieutenant **Army:** (2LT) **Marines:** (2ndLt)	Ensign (ENS)	Second Lieutenant (2d Lt)
O-2	First Lieutenant **Army:** (1LT) **Marines:** (1stLt)	Lieutenant Junior Grade (LTJG)	First Lieutenant 1st Lt
O-3	Captain **Army:** (CPT) **Marines:** (Capt)	Lieutenant (LT)	Captain (Capt)
O-4	Major **Army:** (MAJ) **Marines:** (Maj)	Lieutenant Commander (LCDR)	Major (Maj)
O-5	Lieutenant Colonel **Army:** (LTC) **Marines:** (LtCol)	Commander (CDR)	Lieutenant Colonel (Lt Col)

Grade	Army and Marine Corps		Navy and Coast Guard		Air Force	
O-6		Colonel **Army:** (COL) **Marines:** (Col)		Captain (CAPT)		Colonel (Col)
O-7		Brigadier General **Army:** (BG) **Marines:** (BGen)		Rear Admiral Lower Half (RDML)		Brigadier General (Brig Gen)
O-8		Major General **Army:** (MG) **Marines:** (MajGen)		Rear Admiral (RADM)		Major General (Maj Gen)
O-9		Lieutenant General **Army:** (LTG) **Marines:** (LtGen)		Vice Admiral (VADM)		Lieutenant General (Lt Gen)
O-10		Chief of Staff of the Army Commandant of the Marine Corps General **Army:** (GEN) **Marines:** (Gen)		Chief of Naval Operations Commandant of the Coast Guard Admiral (ADM)		Chief of Staff of the Air Force General (Gen)
	Army and Marine Corps		Navy and Coast Guard		Air Force	

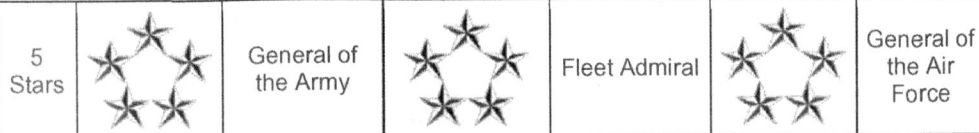

The rank of 5 Stars is solely reserved for wartime, and as such there are currently no 5-star Generals or Admirals.

* * * * * * * * * * *

Chapter 36

The United States Army

March along, sing our song, with the Army of the free.
Count the brave, count the true, who have fought to victory.
We're the Army and proud of our name,
We're the Army and proudly proclaim.

First to fight for the right, and to build the nation's might,
And the Army goes rolling along.
Proud of all we have done, fighting till the battle's won,
And the Army goes rolling along.

Then it's Hi! Hi! Hey! The Army's on its way.
Count off the cadence loud and strong (Two! Three!).
For where're we go, you will always know
That the Army goes rolling along.

"The Army Song"

    The United States Army, established in 1775, is the oldest of America's Armed Forces. Its birthplace was on the battlefields of the American Revolution and, because of the dedication and sacrifice of those patriots, our nation was born. Still today, the Army focuses on providing our primary ground fighting force, while other services have been added to our nation's armed forces to guard the skies, oceans, and shores. The dual purpose of all the services is to defend our nation and its interests abroad. Like those soldiers of our original Army, the men and women of today's Army must be dedicated to their work and, when necessary, willing to sacrifice all for their nation. In return, within the bounds set by Congress and its allotted budget, the Army strives not only to be combat ready, but to provide a good quality of life for its members and their families.

    Being a part of this vast, vital, powerful organization—even as a family member—can be wonderfully exciting and rewarding. However, the first step toward enjoying being a part of the "Army Family" is simply to understand it. The sheer size of the Army is staggering. The complexities of the chain of command, the various branches, unit organization, rank structure, insignia, and the all-too-frequent use of acronyms and abbreviations create a foreign environment for a new Army spouse. This chapter should make him or her feel more "at home" with their new family.

> **BLUF**
>
> "The greatest military force in the world; know who we are and how we operate."
>
> *Brad Combs*

CHAIN OF COMMAND

The chain of command is a straight line that runs from the President of the United States, who serves as the Commander-in-Chief of the Armed Forces, down to the newest, lowest-ranking soldier. Every member of the Army is responsible for his or her actions to the person above. Each soldier, no matter how senior, receives orders and guidance from above, and is rated on his or her performance. In addition, every soldier is responsible for providing leadership and loyalty down the chain of command—to our hard-working, brave American soldiers.

ARMY STRUCTURE

The Army is divided by function into various branches (e.g., Infantry, Armor, Field Artillery). All soldiers, up to the general officer ranks, belong to a specific branch. General officers are supposed to be "generalists," able to function in all branches; therefore, they do not belong to any specific branch.

For command purposes, the entire Army is divided into major commands, each composed of all of the Army units within its geographic area or sphere of influence. The major troop commands are:

U.S. Army Forces Command (FORSCOM), Ft. Bragg, NC
U.S. Army Training & Doctrine Command (TRADOC), Ft. Eustis, VA
U.S. Army Materiel Command (AMC), Redstone Arsenal, AL
U.S. Army Futures Command (AFC), Austin, TX
U.S. Army Central Command (USARCENT), Shaw AFB, SC
U.S. Army North Command (USARNORTH), Ft. Sam Houston, TX
U.S. Army South Command (USARSO), Ft. Sam Houston, TX
U.S. Army Europe & Seventh Army (USAREUR), Wiesbaden, Germany
U.S. Army Pacific Command (USARPAC), Ft. Shafter, HI
U.S. Army Africa Command (AFRICOM), Stuttgart, Germany
First U.S. Army, Rock Island, IL
Eighth U.S. Army (EUSA), Pyeongtaek, Korea
U.S. Special Operations Command (SOCOM), Ft. Bragg, N.C.
U.S. Army Space and Missile Defense Command/Army Strategic Command (USASMDC/ARSTRAT), Redstone Arsenal, AL
U.S. Army Cyber Command (USCYBERCOM), Ft. Gordon, GA

Within these major troop commands, the units can be broken down into progressively smaller units—each with its own command structure—until you reach the smallest squad or section. Large units (corps, divisions) are made up of soldiers from many branches. Small units (squads, platoons) may include soldiers from a single branch. The chart on the following page describes those units in a generic manner; there are numerous exceptions.

Common unit abbreviations are as follows: Div for division, Bde for brigade, Regt for regiment, Bn for battalion, Co for company, Btry for battery, Plt for platoon, Sq for squad, Sec for section.

Unit	Composition	Unit Commander
corps	headquarters 2-5 divisions fires brigade engineer brigade battlefield surveillance brigade military police brigade sustainment brigade	lieutenant general (command sergeant major)
division	headquarters 3 brigades combat aviation brigade sustainment brigade	major general (command sergeant major)
brigade or regiment	headquarters 3 or more battalions or squadrons	colonel (command sergeant major)
battalion or squadron	headquarters 3-5 companies or batteries or troops	lieutenant colonel (command sergeant major)
company or battery or troop	headquarters 3-4 platoons	captain (first sergeant)
platoon	3-4 squads or sections	lieutenant (platoon sergeant)
squad or section	4-10 soldiers	staff sergeant

Thus, PFC Jones might be a member of the 1st Squad, of the 1st Platoon, in A Company, 3d Battalion, 41st Infantry, which is a unit of 1st Brigade, 2d Armored Division of the III (U.S.) Corps. When an Army spouse is asked what unit their soldier is in, it's understandable that it may be difficult to answer. However, if the inquirer is someone outside the unit, they usually want to know the battalion designation.

UNIT STRUCTURE

Each unit is organized like a pyramid, with the unit commander at the top and the entire unit branching downward beneath. A simplified battalion unit organization looks like the diagram below:

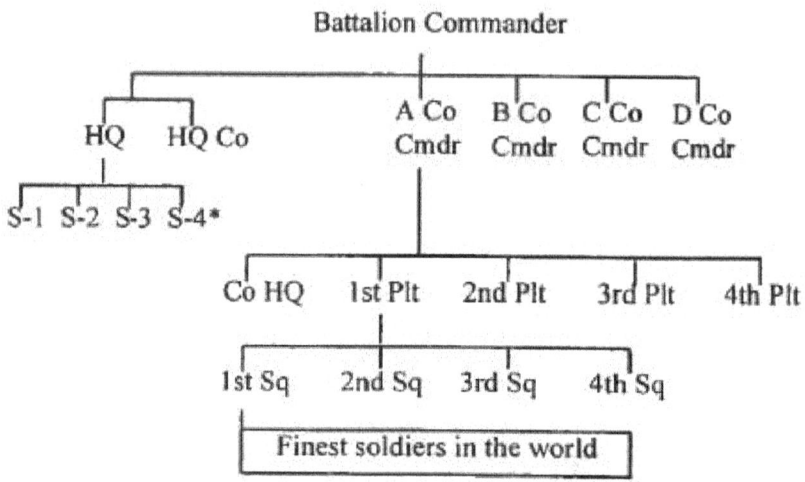

* An easy way to remember what each of the principal staff is responsible for is to remember the word "PITS." Each letter stands for the first letter of the primary staff jobs in numerical order—(PITS) S-1 is personnel, S-2 is intelligence, S-3 is training, S-4 is supply.

RANK STRUCTURE

The following list is provided to show the sequence of officer and enlisted ranks, their abbreviations, pay grades, and the appropriate titles used when speaking to servicemembers. Titles of respect, such as "Sir" for men and "Ma'am" for women, without adding the last name, are often used when a military member addresses an officer who is senior. The term "field grade officer" is used to refer to majors, lieutenant colonels, and colonels, but is never used to address someone. Pictures of the rank insignia for all the branches of the service, including the Army, appear in the previous chapter.

Title (abbreviation, pay grade)	Address Orally as ....
Officer Ranks:	
General of the Army (GEN, Special)	"General"
General (GEN, O10)	"
Lieutenant General (LTG, O9)	"
Major General (MG, O8)	"
Brigadier General (BG, O7)	"
Colonel (COL, O6)	"Colonel"
Lieutenant Colonel (LTC, O5)	"
Major (MAJ, O4)	"Major"
Captain (CPT, O3)	"Captain"
First Lieutenant (1LT, O2)	"Lieutenant"
Second Lieutenant (2LT, O1)	"
Warrant Officer Ranks: (See history below.)	
Chief Warrant Officer (CW5, W5)	"Chief" "Ms.," "Miss," or "Mrs."
Chief Warrant Officer (CW4, W4)	"
Chief Warrant Officer (CW3, W3)	"
Chief Warrant Officer (CW2, W2)	"
Warrant Officer (WO1, W1)	"
Enlisted Ranks	
Sergeant Major of the Army (SMA, E9)	"Sergeant Major"

Command Sergeant Major (CSM, E9)	"Sergeant Major"
Sergeant Major (SGM, E9)	"
First Sergeant (1SG, E8)	"First Sergeant"
Master Sergeant (MSG, E8)	"Master Sergeant:
Sergeant First Class (SFC, E7)	"Sergeant"
Staff Sergeant (SSG, E6)	"
Sergeant (SGT, E5)	"
Corporal (CPL, E4)	"Corporal"
Specialist 4 (SPC, E4)	"Specialist"
Private First Class (PFC, E3)	"Private"
Private E-2 (PV2 or PVT, E2)	"
Private E-1 (PV1 or PVT, E1)	"

Warrant Officer History

The Warrant Officer rank has a long and interesting history that goes all the way back to the Royal British Navy prior to Christopher Columbus. The following background is from the Warrant Officer Historical Foundation. The British Navy would often rely on the technical expertise and cooperation of a senior sailor to help operate the ship and cannons. These sailors became indispensable to less experienced officers and were subsequently rewarded with a Royal Warrant, a special designation to set them apart from other sailors but not violate the British Navy class system of that time. Napoleon is said to have used Warrant Officers as communication links between his officers and the rank and file soldiers.

The United States Congress established the Army Mine Planter Service as part of the Coast Artillery and the Army Warrant Officer Corps in July 1918. The Warrant Officer Corps, at that time, provided officers and engineers for mine planting ships, a principal armament of coastal defense works.

In today's Army, Warrant Officers are Commissioned Officers who play a similar role as their predecessors, serving as technical experts in their respective fields. Commanders rely on Warrant Officers for their leadership, deep technical expertise, and vast experience. While Warrant Officers are authorized to command units, they normally serve in staff roles. Commanders treat Warrant Officer leaders as trusted advisors and staff officers.

The senior Chief Warrant Officer of a unit and their spouse can be an integral part of an Army Unit. The spouse can, if they choose, help with functions, host events and provide counsel to their group. The senior spouse may choose to include the CWO spouse to round out the command team.

MILITARY EDUCATION

Professional military education is available to all officers and noncommissioned officers to help improve their performance and potential for promotion. Some military education courses may be taken by correspondence, while others must be accomplished in residence. One important point for Army spouses to note is that in-residence military education frequently offers an optional program for spouses.

Noncommissioned officer education (NCOE) begins for E4s and E5s with the Warrior Leader Course (WLC) taught at the division level. Next up the ladder of education for NCOs is the Advanced Leader Course (ALC) and the Senior Leader Course (SLC), which are taught at either the division NCO Academies or branch schools. The First Sergeant Course is taught at the Army's Sergeants Major Academy at Fort Bliss, Texas. In addition, that is where the Army's sergeants major and command sergeants major attend the senior enlisted courses.

There are five levels of professional military education for officers. The Basic Officer Leader Course (BOLC), offered at each branch, trains new 2nd lieutenants in the basic skills of their branch. The Captain Career Course at each branch prepares captains for company and battery command. The Command and General Staff College (CGSC) at Fort Leavenworth, Kansas, is designed to prepare new field grade officers for the next 10 years of service. It produces field grade officers with a warrior ethos and joint, expeditionary mindset who are grounded in warfighting. Alternatively, some officers are sent to the equivalent training offered by the sister services or the Armed Forces Staff College at Norfolk, Virginia, which is a joint service school. Senior service schooling for select LTCs and COLs is provided by: the War College at Carlisle, Pennsylvania, the equivalent schools of sister services, or the National Defense University at Ft. McNair, Washington, D.C.

Both noncommissioned officers and officers who are selected for battalion and brigade command positions attend a short pre-command course at Fort Leavenworth. Their spouses are also invited to attend and participate in the spouses' course.

ACRONYMS AND ABBREVIATIONS

Soldiers and civilians associated with the Army frequently develop acronyms or abbreviations to shorten the titles of names they use often. Acronyms are words formed by using the first initials of a series of words, a form of oral shorthand. This military jargon becomes so much a part of "Army talk" that we often forget that those who are new to, or unfamiliar with, the Army don't understand it. Hopefully, this list of commonly used acronyms and abbreviations will help you learn this "foreign language." (Note: When the acronym or abbreviation is pronounced as a word, the pronunciation is enclosed in parenthesis; otherwise, the letters are pronounced individually to create the acronym.)

AAFES (a-fus)—Army and Air Force Exchange System; they run the PX.
AAM—Army Achievement Medal
ACOM—Army Command, i.e., FORSCOM, TRADOC, AMC, Futures
ACS—Army Community Service
ACU—Army Combat Uniform
AER—Army Emergency Relief
AFTB—Army Family Team Building
AG—Adjutant General
AGI—annual general inspection
AIT—Advanced Individual Training
ALC—Advanced Leader Course
APO—Army Post Office
AR—Army regulation
ARCOM (ar-com)—Army Commendation Medal
ARTEP (r-tep)—Army Readiness Training and Evaluation Program; assessment of unit training
ASAP (a-sap)—as soon as possible
ASCC—Army Service Component Command, i.e., ARCENT, USAREUR, USARPAC
ASU—Army Service Uniform
AUSA—Association of the U.S. Army
AVN—aviation
AWAG—Americans Working Around the Globe, formerly known as American Women's Activities Germany
AWOL (a-wall)—absent without leave
BAQ—basic allowance for quarters
BEQ—bachelor enlisted quarters (for NCOs)
BOLC—Basic Officers Leader Course
BOQ—bachelor officers' quarters
CCC—Captain Career Course
CG—commanding general
CGSC—Command and General Staff College
CJCS—Chairman Joint Chiefs of Staff
CO—commanding officer, sometimes referred to as "the Old Man"
COB—close of business
COLA (cola)—cost of living allowance
CONUS (co-nus)—continental United States
CP—command post
CPO—Civilian Personnel Office
CPX—command post exercise
CQ—charge of quarters; NCO representing the unit commander in the barracks after duty hours
CSM—command sergeant major
CWOC—Catholic Women of the Chapel

DA—Department of the Army
DAC (dac)—Department of Army civilian
DACOWITS (dac-o-wits)—Defense Advisory Committee on Women in the Service
D.C.SLOG (des-log)—deputy chief of staff for logistics
D.C.SOPS (des-ops)—deputy chief of staff for operations
D.C.SPER (des-per)—deputy chief of staff for personnel
DEERS (dears)—Defense Enrollment Eligibility Reporting System
DEH—Director of Engineering and Housing
DENTAC (den-tac)—U.S. dental activity
DEROS (d-ros)—date of estimated rotation from overseas station
DF—disposition form; a memo best described as a "buck slip"
DFAS—Defense Finance Accounting System
DISCOM (dis-com)—Division Support Command
DIVARTY (div-arty)—Division Artillery; FA units of a division
DOD—Department of Defense
DODDS—Department of Defense Dependent Schools
DPCA—Director of Personnel and Community Activities
DRU—Direct Reporting Unit
EO—equal opportunity
ETA/ETD—estimated time of arrival/departure
ETS—expiration of term of service
EUCOM—U.S. European Command
FA—Field Artillery; Redleg
FLO (flow)—Family Liaison Office
FM—field manual
FOD—Field Officer of the Day (MAJ and LTC duty officers)
FORSCOM (force-com)—Forces Command
FSG—family support group
FTX—field training exercise
FYI—for your information
GED—general education diploma (equivalent of high school diploma)
GI—government issue, slang for a soldier
GO—general officer
HHC—headquarters and headquarters company
HQ—headquarters
HRAP (h-rap)—health risk appraisal program
HRC—Human Resources Command
HUMMER (hummer)—highly mobile, multipurpose wheeled vehicle
ICAF (i-caf)—Industrial College of the Armed Forces
IG—Inspector General
ILE—Intermediate Level Education
JAG (jag)—judge advocate general
JCS—Joint Chiefs of Staff

KD—Key and Developmental (jobs for each rank; i.e., company commander for CPT; bn S3/XO for field grade)
KP—kitchen police; soldiers detailed to work as cook's helpers
LES—leave and earnings statement (monthly pay slip)
LOI—letter of instruction
MAC—Military Airlift Command
MACOM (may-com)—major Army command
MEDAC (med-ack)—U.S. medical activity
MI—Military Intelligence
MILCOM (mill-com)—military community
MILES (miles)—multiple integrated laser engagement system
MOS—military occupational specialty
MP—Military Police
MREs—meals, ready to eat
MSM—Meritorious Service Medal
MWR—morale, welfare, and recreation
NAF (naf)—nonappropriated funds
NATO (nay-toe)—North Atlantic Treaty Organization
NCO—noncommissioned officer, E4-E9
NCOER—NCO Efficiency Report
NCOIC—noncommissioned officer in charge
NEO (nee-o)—noncombatant evacuation operation/order
NLT—not later than
NTC—National Training Center, Ft. Irwin, CA
OCONUS—Outside Continental United States
OCS—Officer Candidate School
OD—officer of the day (LT and CPT duty officer)
OER—Officer Efficiency Report
OIC—officer in charge
OJT—on-the-job training
PAC (pack)—Personnel Administration Center
PAO—public affairs officer
PCC—pre-command course
PCS—permanent change of station
POC—point of contact
POV—privately owned vehicle
PT—physical training
PWOC—Protestant Women of the Chapel
PX—post exchange
RA—regular Army
RBI—reply by endorsement
RIF (riff)—reduction in force ("down size")
RON—remain overnight, as in TDY
ROTC—Reserve Officer Training Corps

SATCOM (sat-com)—satellite communications
SBP—Survivor Benefit Plan
SECDEF (sec-def)—secretary of defense
SHAPE (shape)—Supreme Headquarters Allied Powers Europe
SITREP (sit-rep)—situation report
SLC—Senior Leader Course
SOP—standing operational procedure
SQT—skill qualification test
SSN—social security number (usually the sponsor's number is requested)
SUPCOM (sup-com)—Support Command
TA-50 (t-a-fifty)—regulation field equipment issued to soldiers
TDY—temporary duty
TLA—temporary living allowance
TLE—temporary lodging entitlement
TMP—transportation motor pool
TO&E—Table of Organization and Equipment; official authorizations for personnel and equipment in a unit
TOC (tock)—tactical operational center
TRADOC (tray-doc)—Training and Doctrine Command
TRICARE—formally CHAMPUS (Civilian Health and Medical Program of the Uniformed Services)
UCMJ—Uniform Code of Military Justice
USAFE (u-safe)—U.S. Air Force Europe
USAREUR (use-a-reur)—U.S. Army Europe
USARPAC—U.S. Army Pacific
WLC—Warrior Leader Course
XO—executive officer
YAC (yack)—Youth Activities Center

ARMY FAMILY

The Army is a large, complex, and diverse organization. In spite of its complexities, the collective term traditionally used to describe the institution, its members, and their immediate families is "Army Family." As an Army spouse, you are a part of this family.

The benefits associated with being a part of such a large family group are too numerous to mention. However, the most important benefit is that the Army does its best to take care of you and to help you take care of yourself, just as any caring family would do for its members. Simply being an Army spouse makes you eligible to be a part of this giant family and offers you many opportunities—to live in different places and parts of the world, to interact with and learn about a variety of diverse communities, to be a newcomer yet never an outsider, and to form wonderful friendships that will last a lifetime. When the time comes for your spouse to leave the Army, you will look back on the time you have spent as a member of the Army Family with fondness and pride. You, too, are a part of the legacy.

* * * * * * * * * * * *

Be all you can be!

Chapter 37

Lessons Learned

"Never look at what you have lost—look at what you have left."
Robert Schuller

Whenever and wherever Army spouses and their friends gather, you will often hear lively discussions about their experiences of coping with the military lifestyle. Each willingly shares from the "baggage" they have collected along their journey. Over time, many of these "lessons learned" achieve the status of conventional wisdom that is passed along. Thus, the lessons of the past help to shape the future. The following "lessons learned" are shared with you, the Army spouses of today and tomorrow, to ease your "journey."

BLUF

"You can't grow and improve unless you review the past."
Brad Combs

FRIENDS ARE ESSENTIAL
Friends are like mortar; they help to hold your life together. Cultivate them, cherish them, and keep them. Frequent moves may make that difficult, but they also provide the opportunity to develop friendships all over the world. Make the most of these opportunities and make friends wherever you go. When it's time to move on, remember that it is more difficult to be the friend left behind than the friend who leaves. The one who remains must fill a void, while the other gains a new environment, new experiences, and new friends. But don't let distance come between you and good friends. Keeping in touch only takes a few letters or phone calls a year. Your circle of friends will grow ever larger and enrich your life beyond measure. Friends share your joy when you're happy, bolster your strength when you're sick or sad, encourage you when you need to be strong, congratulate you when you succeed, and sustain you when you are alone. You, in turn, do the same for them. Friends are truly an essential part of your life and, as a military spouse, you need them more than your civilian friends ever will.

CHILDREN ALSO SERVE
   Being an Army "brat" is a challenging role. "Brat" is not a derisive term, but an affectionate one applied to all children of military members. These youngsters who spend their childhood years moving with their families from state to state, and country to country, also serve our nation through the sacrifices they make. The servicemember's "life" moves with him—in terms of status, job security, and position; it is, in many ways, simply a continuation in a new location. However, for family members, this is far less true. They must redesign their lives with each move, and the children often find that adjustment more difficult than the adults. For example, children who are slow to make friends often have just found that special friend when it's time to move again. Youngsters in sports programs must repeatedly re-establish their position on a team, and frequently miss out completely—either because of a mid-season move or because of a coach's preference for "local" students. Moving high-school teens is frequently listed as one of the major "dissatisfiers" of service life.
   There are ways to lessen the burden of frequent moves on our children. The primary way is to help them have a positive attitude when it comes time to move. The parents need to show excitement about the new location and talk about all of its positive aspects. Big smiles say a lot. Emphasize the pleasures of seeing new places, making new friends, and living in a new house. Reassure them that it's natural to feel bad about leaving old friends—that you hate to leave yours too—but that they can write, text, email, call and perhaps even visit their friends in the near future. Remind them that they are not losing their friends, but will keep those and make more!
   Other ways to generate enthusiasm for the move involve focusing on the interesting sights and activities that your travel and the new location can offer. Internet and Google searches; a trip to the local library, bookstore, and ACS for information about the next duty station, the state and country; possible side trips to enjoy during the move—these are great ways to start building that enthusiasm. New games, books, or ideas for quiet activities to help pass the time while traveling are always welcome. Don't forget family discussions about the opportunities the new location holds and the family trips you can take once there.
   If your youngsters are teens or pre-teens, they are probably more interested in what activities will be available for them at the new location. Ask your sponsor to tell you about the youth activities and sports programs offered, or contact the school yourself to find out what clubs and activities are available. You might even be able to find a peer of your youngster who has recently moved from that location for your child to talk with. Generating interest about the new area is the key to easing the departure from the old.
   There is also a growing number of families who prefer to homeschool their children, and many communities now offer support for this option. Reaching out to the school counselors at the new location before you arrive would be helpful. As for your teenagers, in preparation for future jobs or school applications, it would be a good idea for them to collect letters of recommendation *before* they move to a new school and new area.
   Even though it's not easy to live the life of an Army "brat," there are some definite

benefits. Our children have seen the world, not just read about it. Because of their experiences with adapting, they tend to be more self-reliant than civilian youngsters. When it comes time to move away from home, it is usually less traumatic for our youngsters than it is for those from civilian families. In time, Army brats learn to deal with new situations and make friends more easily than others, though these traits may not be apparent until they reach adulthood. Finally, to your great surprise, your children may grow to love this vagabond way of life so much that they join the military service themselves, or marry someone in the Army. Focus on the benefits while you help your children adapt to this lifestyle, and remember that they are serving their nation through the efforts and the sacrifices they make.

The Interstate Compact on Educational Opportunity for Military Children is a vital tool for transitioning families. It aims to reduce the educational and emotional issues encountered when our military children are required to transfer from school in one state to another. The compact has been signed by all states, but as spouses, we must be aware of the limitations.

Serving with a Special Needs Child—Over the recent years, more and more military families are faced with finding services for their special needs children. This includes physical, emotional and mental health needs. Transitioning a special needs child can make a hill seem like a mountain. Finding the right schools and services can take a PCS from being stressful to traumatic. Military parents of special needs Brats might find the burden eased with thorough research and adequate time to make the transition. Most families in this category are familiar with the EFMP (Exceptional Military Families Program); if you are not, it is a good place to start your research. Making a list of services your child will need both academically and medically can give you a blueprint of what you need to obtain at the next duty station. There is quite a bit of support for families of these dynamic Brats. Military Education Coalition and National Military Family Association are good resources, as well as Military OneSource. Don't forget about supporting yourself through your own self-care. Sometimes reaching out to those who know your struggle can help tremendously. There are several support groups on social media for Military Families with Special Needs. We all feel better when we know that we are not alone and, as Army spouses, we continue to be the best resource for each other.

AVOID FAMILY SEPARATIONS WHEN POSSIBLE

Army families experience unavoidable separations for a wide variety of reasons and for different lengths of time. In addition to wartime, separations routinely occur because of field exercises, TDY, deployments, and remote assignments where families are not permitted to accompany their servicemember. Separations can also occur when young enlisted families are not command sponsored in an overseas area, and the military member is unable to afford the transportation costs for his family and living expenses on the economy. The Army has responded to the needs of those left behind by establishing the commander-sponsored Family Readiness Groups, and they do help to prepare spouses and families to cope better during these times of separation. As for the families themselves, there's little they can do to lessen the strain except strive to keep the lines of communication between the servicemember and family as open and all-encompassing as possible.

It's the lengthy separations that are *avoidable* that we can do something about. These are situations that are all too familiar, such as when a husband receives orders for a distant or overseas assignment, but the spouse has a great job or an educational commitment and hates to leave and begin again. Or, the servicemember is reassigned just before a child's last year of high school, and the non-serving spouse feels stability of school is important and the child needs to stay and finish school with all their friends. Possibly the assignment is to an area where children have to attend boarding school because there is no acceptable local school. These and similar scenarios occur frequently throughout the Army and merit serious family discussion.

If you find yourself in such a situation, sit down with your spouse and the entire family, and review all the options. In such situations, it's important for everyone, including the children, to express their feelings freely. It might be helpful, if you know a family that has had to make a similar choice, to ask them to share what they learned. If you are thinking about "staying behind" for an extended period, you may want to visit your chaplain or someone from the ACS before making a final decision. Research shows that, while families may deal effectively with one such tour, the second often brings very serious problems. The lesson to be passed along is this: Though involuntary separations are a reality, it's important to avoid extended voluntary separations whenever possible and practical, if you value your marriage.

PROMOTION LISTS REQUIRE KINDNESS

Promotion lists are a source of joy—and pain. Be sensitive to these feelings for all concerned. Before a promotion list is published, it is usually preceded by rumors of who's "on the list." These rumors are often incorrect and, if passed on, may cause great heartbreak. So, if you hear a rumor, keep it to yourself. Once the list is released, it's only natural to feel tremendous joy if your spouse or friends have been selected. However, in the excitement of the moment, don't forget about friends who had hoped to be on the list but weren't, and don't let your embarrassment over not knowing what to say to them cause you to ignore their silent pain. Conversely, if you had hoped your spouse would be selected and wasn't, don't let your disappointment keep you from offering congratulations to those who were.

Today's use of social media has provided another opportunity to both celebrate and cause pain. Many people excitedly post the list on their social media as soon as it is released. This is not the way for a spouse to learn of it, as they may see the list before their servicemember tells them. Social media is appropriate, but only after a period of time, especially in a situation such as the senior service colleges.

Remember that the disappointment of non-selection is rather like going through a period of mourning. Close friends can best help by sympathizing over the non-selection, adding a few words of encouragement, but then going on to other topics. Acknowledge it, but don't dwell on it.

WHEN A COMMANDER IS RELIEVED

Commanders are seldom relieved in the Army, but occasionally it does happen. Not every person picked for command is right for the job. Regardless of why a soldier is relieved of command, it is related to business—it has nothing to do with the spouse. So, if it happens, friends should not hesitate to call and sympathize, and help in any way they can as the family prepares to move.

It's important to understand what the spouse's feelings and needs are, if you're going to be of help. The spouse probably has feelings of anger and embarrassment, and needs an opportunity to verbalize these feelings to their closest friends. The spouse also needs the support of contemporaries. The spouse needs to hear from them that what has happened doesn't change their friendship or respect, and that what the spouse has worked to achieve in the unit and community has been worthwhile and appreciated. A small, informal farewell given by close friends in the unit, perhaps a small lunch, would be appropriate and probably welcome. However, be sure to consult with the spouse before planning begins. Whatever else you do and say, don't forget to tell them that you'll miss them.

If the senior commander's spouse wants to be of help, she or he can call or drop by briefly, expressing sadness about how things turned out. They should be sympathetic, but should not try to defend their spouse's actions.

DIVORCE HAPPENS

When there's a separation or divorce in your unit or community, gossip makes matters worse. It's not helpful for friends to take sides or make the couple's private lives a topic of conversation. That doesn't mean that you can't or shouldn't discuss the matter with those involved and be sympathetic.

The pain and sorrow that accompany divorce can be worse than experiencing the death of a loved one. If the spouse still loves the servicemember, the shock of rejection, coupled with the loss of the soulmate and home, can be devastating. If you are a close friend, be there and let them talk out their feelings. Encourage them in their plans for the future. But remember that there are two sides to every divorce, and one seldom hears both. (Even if we could, it's not for us to judge.) If you were friends with the couple before the divorce, continue to be friends with both of them, throughout and after the divorce.

During times of divorce, you may want to reach out and offer support, but are unsure of what to say. A written note from anyone, but especially from the spouse of a senior officer or NCO (depending on the situation), is a courtesy that will be valued. You don't need to mention the divorce; you can thank the person for their help, say you're sorry about their leaving and that they will be missed, and wish them luck.

If the separation or divorce is between a commander and spouse, the spouse's departure is not noted by those in the unit as it would be if they were leaving under different circumstances. Even if the spouse has given a great deal of time to the unit and community, she or he will probably not receive formal acknowledgment of those contributions. Friends may want to have a farewell and say thank you in private, but this should not be done publicly. Divorces and separations are no cause for celebration.

THE ARMY IS NOT A DEMOCRACY

Soldiers know that the Army is not a democracy, but some Army spouses have difficulty realizing that this truth has application in their lives as well. Unfortunately, it can become all too apparent, and sometimes quite painful, when a young Army spouse comes into conflict with the commander or the spouse. This doesn't happen often, but once is too often for those personally involved.

If you have a disagreement with the commander or their spouse and you can't, with reasoned logic, convince him or her of your point of view, then give up. You will not win the disagreement, no matter how right you may be or how many supporters you have. The commander and spouse have the prerogative of their positions to make decisions without accepting your advice or taking into account your views. You may not like it; you may feel that it's wrong, but that's the way it is because the Army isn't a democracy.

NURTURE YOUR SPIRITUAL HEALTH

Leaders of soldiers must be concerned with the moral strength of those who serve under them. This moral strength is the very soul of our Army, and the bedrock of that strength is based on discipline and religious principles. To encourage religious conviction, troop leaders can set the right example by attending church where their soldiers worship—the unit or post chapel.

The Army recognizes the importance of the spiritual health of its members and their families. It provides chaplains—ministers, priests, and rabbis—to tend to the religious needs of their Army "flock." They provide pastoral counseling for families, as well as individual servicemembers. In addition to regular chapel services, post chapels offer religious education classes, retreats/seminars/workshops for spiritual and personal growth, and nursery facilities for infant care during chapel programs. There are also special spouse programs, such as Protestant or Catholic Women of the Chapel. If you avail yourself of the chapel services provided for your spouse's unit or post, they can fill your spirit with joy and address the special needs of your military family. Whatever your religious background or past practices, take the time to nurture your spiritual health in peacetime, and it will be there to support and sustain you in times of crisis.

BLUF

Faith in the foxhole, hope on the homefront.

DEALING WITH THE MEDIA

Occasionally, an Army spouse is interviewed for a television newscast, especially when an unusual circumstance has occurred in their soldier's unit or to the service-member personally. Should that happen to you, here are guidelines to help you handle media inquiries, and tips to help you prepare for the interview.

- When you are asked to speak to the news media, be sure to get the reporter's name, employer, and telephone number.

- The unit public affairs officer (PAO) can be helpful to you in a variety of ways if you have prior knowledge of the interview. The PAO has a working relationship with the media and can provide you with crucial advice on how to be most effective. They can help you clarify the points you want to make, be there to assist you during the interview, and help protect your privacy.

- Remember that you're not in the Army, and cannot express its position on any matter. Accordingly, it's not helpful to speculate about the Army view, even in casual conversation, with those representing the media. Realize that you are always "on the record," not just when the camera or recorder is rolling.

- Before you go on camera—whether for a stand-up interview, news event, or talk show—remember:

    * Your audience will remember less of what you say than they will of the way you say it. About 7% of your statement is in your words, 38% in your voice and vocal-quality delivery, and 55% of your meaning is communicated in purely non-verbal ways.

    * Know the two or three points you want to make, and keep your answers short and concise. Be warm and friendly, but watch out for the "loaded" questions.

    * Set the pace yourself and, if you need time to think, ask the interviewer to repeat the question.

    * Unless you're being televised "live," take all the time you need to compose your answers; "dead time" will be cut out.

    * Stay positive, and don't say any more than you intend.

    * Prepare yourself, stay confident, and maintain control.

MOVING SMART

Every post Transportation Office provides valuable guidance on preparing your personal effects for shipment. However, the information doesn't include the "lessons learned" from experienced Army spouses. The following information and suggestions are intended to help you "move smart."

- A door-to-door or "direct" move within the U.S.A. is the best type of move. Your household goods are handled less and, because they are out of your possession the shortest period of time, less subject to theft. If there's any way to avoid having your shipment off-loaded and put into temporary storage, do it.

- Prepare for the packers by clearly separating items that are to go into different shipments. If physically separating each shipment isn't possible, then do so within each room—and label the items or sections of the room clearly. Labeling can be done with directions written on masking tape applied across a cabinet door or group of items. Additionally, each room can be cataloged on a sheet of paper that's taped to the door (e.g., "All items in closet stay.").

- The difference in shipping time for hold baggage (unaccompanied baggage) and household goods has decreased to the point that it may be better to combine both into one shipment, or make two household goods shipments at different times (with no hold baggage shipment) before you leave. A household goods shipment has a number of advantages over hold baggage: It is packed better and can contain any items of furniture desired, whereas hold baggage is limited to small, unbreakable items. Hold baggage also has a restricted weight limit, and cannot be stored at the destination. Discuss your options with the Transportation Office.

- A clothes steamer can save a great deal of time and/or money, especially after a move. The post exchange sells an inexpensive 110/220 volt steamer that is easy to use and especially handy for wrinkled wool clothing, sweaters, jackets, and overcoats. Even wrinkled upholstered furniture and drapes can be renewed with a steamer.

- Moving overseas in the winter is especially difficult because large winter uniforms and winter clothing are bulky and heavy to pack. Footlockers used to fit within the size limitations for baggage checked on international flights, but no longer. The only suggestion that comes to mind with regard to winter clothing is perhaps to ship a few packages by air freight, mail, or UPS. If you use air freight, the packages may arrive when you do. Otherwise, perhaps you can send them to the unit you are going to, or to your sponsor, in time for the uniforms to be there when you arrive.

- Keep a current inventory of your household goods, along with the original receipts of major items. For electric items, note: brand name, model number, serial number. Your inventory will help to determine how much personal property insurance to carry, and provide the information needed in the event of loss or damage. Claims require original cost and date of purchase. If you have valuable antiques, collections, or art, appraisals are recommended.

- Take videos and photos of your personal property, and keep these in a fireproof box or safety deposit box. It's best to photograph each room, with closets and cabinet doors open. Carry these pictures on a disk or thumb drive with you when you move; do not let them get packed in your shipment. An electric engraver can be used to mark appropriate items with your name.

- If you use one of the popular time-management "organizer" books, it's a good place to record all of your pre-moving and post-moving notes and instructions. These will serve as a good checklist the next time you move.

- Finally, remind yourself each time you move that "life begins now!" Unpack the boxes, hang your curtains and pictures, plant some flowers, get out, and get involved. Don't postpone your life looking forward to the next assignment, a nicer house, a different country, or retirement. Enjoy what you have!

TRAVEL SMART

Space-available travel is a wonderful benefit for military families that provides inexpensive vacations to places you might otherwise not be able to afford. Especially if you are living near an Air Force base that offers Space-A flights, consider letting Uncle Sam's planes take you on your travels. Following are some suggestions for space-available travel:

- Be sure you understand the rules about signing up, both to go and to return. Keep your travel plans flexible, if at all possible, so that if you are shut out of travel to one location, you can take advantage of a flight to another. Always remember that this travel is offered only on a space-available basis; be prepared to pay your way home on a commercial flight, if that becomes necessary. An excellent resource for Space-A travel is on the internet at military.com.

- When considering space-available travel that won't get you all the way to your final destination, be sure to compare the cost of the commercial travel from your point of origin with that from the space-available destination. You may find that a short commercial flight (from Space-A destination to final destination) with no advance reservation is almost as expensive as the airfare from your original departure point when the reservation can be booked in advance. In such cases, the uncertainty of the Space-A flights and the extra time that must be allowed on both ends of your trip may not be worth the small amount of money saved.

- Make use of the USO courtesy booths or waiting rooms at many major airports for information and assistance. They can also serve as a good meeting point.

DECORATING TIPS

The mobile lifestyle of an Army spouse creates an opportunity for innovative and imaginative decorating. Whether furnishing government quarters on post, a house of your own, or a rental apartment, you can make it uniquely yours with a few personal touches. There is a wonderful story about a five-year-old little girl that reflects this positive attitude. She was keeping her younger brother occupied in the front lobby of the post housing office, while her parents talked to the housing representative, when another customer asked her, "Are you looking for a home too?" The little girl immediately replies, "Oh no, we have a home. We're just looking for a house to put it in."

Government Housing

Rank determines not only eligibility but also amenities (appliances, carpets, drapes) available for on-post housing. Normally, furnishings are minimal; but privatized housing offers different amenities depending on which company has your privatized housing contract. Some offer you choices in colors to paint the house, updated appliances, garages, and other items.

Window treatment - Experienced Army spouses know drapes never fit twice, but there are some tricks to help you cope with that problem. Simple window treatments are in fashion today. With a single wooden dowel, supporting brackets, and a length of your favorite fabric, you can trim a window in short order. Simply swag the fabric across the top, loop it over the ends of the rod, and let it cascade down each side. It may take a little time to drape the swag attractively, but a few attempts should result in something you like. If the fabric is long enough, let it "puddle" onto the floor and you won't even have to hem it.

Another idea is to use decorator sheets; they have always been a favorite with Army spouses for curtains. Not only do they come in pretty colors and patterns, but they are wider than the fabric you can buy. Use them just as they come, or open up each end of a hem and run a curtain rod through, or cut them as you would regular fabric. If pleated drapes are more to your liking, you can do what many spouses before you have done: Either buy or make a number of long, identical, pleated drapes—all from the same, neutral-colored fabric. If you make them, don't sew the panels together. Then, no matter how wide or narrow your windows, you simply accommodate them by hanging more or fewer panels together. With this type of drape, you'll probably have to let the hems up or down with every move, so plan ahead and make deep hems. Other sources for good ideas and easy-to-use fixtures are fabric stores, home remodeling centers, home-making magazines, and the internet.

Floor coverings - Frontier Army spouses tacked their carpets down to cover bare wood and often dirt floors. You can be assured your flooring won't be dirt, but it may be black vinyl or some other surface you want to cover immediately. This can be costly, especially since wall-to-wall carpets, like drapes, never fit a second time. However, if you've invested in such carpets, they can often be recut and pieced to fit. Most Army families purchase standard-size rugs, none too large, and just let the floor show around the edges. An economy measure is to buy carpet remnants and bind the edges for use as area rugs. If your tastes include Oriental rugs and your budget allows, don't overlook the opportunity to acquire them at a substantial savings during some of your overseas assignments or trips.

Walls - Most post housing allows you to put fabric or wallpaper on the walls, so long as you leave the walls clean when you move out. The easiest is fabric, held in place either by staples or fabric starch. To use staples, start across the top and fasten the fabric edge next to the ceiling with staples or a narrow wood strip tacked into place; then sparingly use staples or small dots of glue from a glue gun to hold the edges and bottom in place. To use fabric starch, precut the fabric to the correct size, wet the pieces in undiluted fabric starch, and smooth it onto the walls. To remove fabric from the walls after the starch dries, simply peel it off. It can be washed and used again later. Using wall covering works especially well in small areas, like the bathroom, or on only one wall of a room. As for pictures, you can save considerable money by taking a framing course at the post craft shop. Then you can frame inexpensive but interesting calendar prints, family photos, or pieces from the local "starving artists" fair. When you leave, a little spackle will fill the nail holes, and any starch remaining on the wall can be easily removed with a wet sponge.

Furnishings - If you are a new Army spouse, you can look forward to the fun of acquiring home furnishings in a great variety of interesting, faraway places. Until then, there are many economical ways to find the furnishings you need. First, check your post and community thrift shops, and watch the paper for local auctions or estate sales. These are great places to find furniture and appliances, many in excellent condition and at bargain prices. Another possibility is the Defense Reutilization and Marketing Office (DRMO) auction. The Army periodically disposes of unneeded or damaged items at these auctions. Advance advertisements list the categories that will be available for sale, and when they can be seen. Although many will not be of interest to you, frequently such useful items as furniture, lawn equipment, and bicycles are sold. Once you have the essentials, it's time to think about your long-term goals. Rather than always looking for the most economical buys, decide what you would really like to have and start saving for it. You will find that good quality furniture can take the wear and tear of frequent moves better than cheap furniture, and you will have the pleasure of using and enjoying nice things—even if you can only buy one piece at a time.

Your Own House

The same guidelines as above apply with several caveats. Consider investing in quality window treatments (such as shutters, vertical blinds, or custom-made drapes) and floor coverings. If you plan to rent your home when you move, these features will make it easier to rent. When you get ready to sell, they will increase the value of your home.

ARMY SPOUSES CARE IN WAYS THAT COUNT
The following letter to the editor appeared in *The Stars and Stripes*, May 8, 1988. The writer had a powerful message for all Army spouses, whether they're living overseas or in the United States; it captured what caring is all about in a few, well-chosen words. Even though the letter was written during a different era, the sentiments still ring true. Therefore, the letter is presented here in its entirety.

I went to a company coffee last night. There was the usual griping about the Army, griping about the weather, and some not so usual attacks on the commanding officer's wife because she couldn't attend. Under the sniping was the attitude that nobody cares.

Well, ladies, I've got some good news and some bad news. The bad news is that the commander's wife, the sergeant major's wife, the first sergeant's wife, and others have houses to keep clean, jobs, volunteer work, children to tend to, husbands to baby, college to attend, bills to pay, and their share of homesickness, boredom and despair over living overseas. We don't have time to coddle you about your boredom, loneliness or non-existent social life.

The good news is, we do care. Call the same commander's wife, too busy to make it to a coffee, and tell her your car broke down and you need to get the baby to the doctor, and she'll be there to give you a ride. If your husband is in the field and the kids are making you crazy, call the first sergeant's wife, and she'll be more than willing to listen, maybe even baby-sit so you can get away for awhile. Need a job? Call the platoon sergeant's wife who works at CPO and find out how willing she is to show you how to fill out the maze of paper work.

We care because we're all in this together. But you're grown women, and your first responsibility is to care about yourselves. You have to reach out, and let people know you have problems. You have to take the steps to ensure your happiness.

Homesick but can't afford to call Mom? Write letters, or send cassette tapes and post cards. Can't afford to travel? Travel through the post library, or discover the city or village where you live by foot or bus.

Don't know anybody and you're lonely? Reach out. Invite possible friends for coffee. Encourage your husband to bring single soldiers home for dinner; macaroni and cheese in a homey atmosphere beats steak in the mess hall any day. Be the nice lady who bakes a birthday cake for the single men, or sews stripes on their uniforms.

Bored and can't find a job? Volunteer. Being needed a few hours a day is a terrific remedy for a sagging self-esteem. Are your kids making you crazy, but you can't afford a baby sitter? Find someone else in the same situation and time-share the child care. No night life where you are stationed? A bottle of wine, a deck of cards, and a few friends can be a lot of fun on a Saturday night. Husband in the field? Have a slumber party or a potluck dinner.

Make yourself useful, do favors for people, develop your skills and talents. Start caring about yourself and you'll be pleasantly surprised to find out how many people out there care about you.

Another great principle to adopt is "Do what you can and don't sweat the rest," as everyone has competing priorities and your family comes first. Your attending every coffee or family readiness group won't get your spouse promoted. Other spouses want you to attend and help when you can, but they want you to do so because you have a desire to help, make long-lasting friendships, and support the Army way of life.

LIFE AFTER THE ARMY

When your spouse leaves the Army, for whatever reason, a transition occurs that affects the entire family. Whether the career is twenty years or more, they will still be at a comparatively young age in terms of normal life expectancy. If you have internalized the values and traditions of Army life, you also will feel the various aspects of transition. The possibilities of establishing a second, satisfying career; fulfilling a dream or two for you and your spouse; building your own stable career at last; buying or building a house and settling down—these are some of the hopeful and positive aspects of the transition. Yet, you will both leave with some sadness, recognizing that you are leaving a way of life that has influenced your lives for a number of years. The future may be uncertain, and you should expect some symptoms of adjustment as you seek to find a new community and define your social interaction within it.

Fortunately, the Army helps to prepare members for this transition with programs specially tailored to facilitate the process. Preparation is the key, and the great challenge is for you to have access to all the necessary information. ACS is a good place to start. Try to attend all classes offered to spouses regarding transition.

For you, more challenges and options lie ahead. One of those options is to retain ties to the Army community nearest you. Many spouses of retired military members have found that continued association with the volunteer and social aspects of military life remains important to them after their spouses retire. It helps them to keep that feeling of belonging, the friendship of other Army families, and the sense of usefulness that is crucial for everyone—no matter their "age or stage." Spouses of retired Army members offer this wise counsel for everyone preparing to leave the Army: If you still have children living at home, prepare them as well for "getting out"; resolve family and marital problems before you approach the time for retirement or separation; develop outside interests, such as returning to school or acquiring new job-related skills; and believe the always-resounding affirmation, "There is life after the Army!"

LETTING GO GRACEFULLY

Ellen Goodman wisely writes, "There's a trick to the Graceful Exit. It begins with the vision to know when a life stage, a job, a relationship is over—and to let go."

If you are leaving a leadership position at any level, there are things you can do to help in the transition for the one replacing you. You might think of it as "How to Pave the Way for the New Spouse." Some suggestions are:

- Make sure that the new spouse is welcomed, and that the other spouses are ready to provide their support during the transition.

- You might prepare some type of resource or continuity guide to help in the transition.

- It is important for the other spouses to allow the new leader room to grow roots of their own. He or she will bring their own leadership style, experience, talents, and interests. Encourage the other spouses to immediately look to the new leader for guidance, and not to compare it with the way things used to be done. In this way, you set the tone for the transition, and that is the best way to honor the new leader and what you are leaving behind.

As an Army spouse, you will make many such exits; the trick is to exit gracefully. Before a move, change of command, separation from the service, or retirement, wrap up the loose ends. Write your thank-you notes, complete the circle of friendship by entertaining those who have entertained you, turn in the final report for the committees on which you served, say farewell to colleagues at work or in your volunteer activities, and obtain those needed letters of recommendation. Stop and visit briefly with members of the community who have provided the services upon which you and your family depended—the library, fitness center, arts and craft center. Give the plants you have nurtured to friends who will care for them until it's time to pass them on again. Empty your "bag of community responsibilities" so that you can fill it with memories. You will then be able to let go gracefully, as you prepare for the next stage.

THE MILITARY SPOUSE—A NOBLE CREATION

Someone, unknown to us, truly understands the military spouse. Read the following story that captures the imaginary moment of our creation, and see if you don't agree. Could it possibly be that the reason the author remains unknown is that she was the angel who was there?

The Military Spouse

The good Lord was creating a model for military spouses and was into his sixth day of overtime when an angel appeared. The angel said, "Lord, you seem to be having a lot of trouble with this one. What's wrong with the standard model?"

The Lord replied, "Have you seen the specs on this order? The spouse has to be completely independent, possess the qualities of both father and mother, be a perfect host/hostess to four or 40 with an hour's notice, run on black coffee, handle every emergency imaginable without a manual, be able to carry on cheerfully, even if dealing with a medical issue and has the flu, and must be willing to move to a new location 10 times in 17 years. And oh, yes, must have six pairs of hands."

The angel shook her head. "Six pairs of hands? No way."

The Lord continued, "Don't worry, we will make other military spouses to help. And we will give them an unusually strong heart so it can swell with pride in the servicemember's achievements, sustain the pain of separations, beat soundly when it is overworked and tired and be large enough to say, 'I understand,' when nothing makes sense, and say, 'I love you,' regardless."

"Lord," said the angel, touching his arm gently, "go to bed and get some rest. You can finish this tomorrow."

"I can't stop now," said the Lord. "I am so close to creating something unique. Already this model heals itself when sick, can put up six unexpected guests for the weekend, wave goodbye to their servicemember from a pier, a runway or a depot, and understand why it's important that they leave."

Finally, the angel bent over and ran her finger across the cheek of the Lord's creation. "There's a leak," she announced. "Something is wrong with the construction. I am not surprised that it has cracked. You are trying to put too much into this model."

The Lord appeared offended at the angel's lack of confidence. "What you see is not a leak," he said. "It's a tear."

"A tear? What is it there for?" asked the angel.

The Lord replied, "It's for joy, sadness, pain, disappointment, loneliness, pride and a dedication to all the values as a couple they hold dear."

"You are a genius!" exclaimed the angel.

The Lord looked puzzled and replied, "I didn't put it there."

Author Unknown

Chapter 38

Suggestions for Further Reading

"And now ... for the rest of the story."
Paul Harvey

The United States Army is built on the heritage of the brave military men and women who came before us. The same can be said about Army wives who formed the genesis of the support to our soldiers and military families. Yet nothing stays the same and, as the Army has transformed over the years, so have the Army wives. They have shown us the transformation to the current spousal role, as more women have assumed the role of soldier, while many of their male spouses have joined the ranks of the support network.

This chapter focuses on the history and resources for military spouses. You will see that most of the history written by and about military spouses refers only to them as "wives." That's because there were no men taking the role of military spouse until recently. While we as an Army have progressed to the term "spouses," with the male "spouse" playing a more prominent role, it helps to understand our history of spouses and how we arrived at this point.

You might find it useful to know about some of the many books available on each of the subjects discussed in this handbook. There is an increasing number of books in print today on these topics: military wives' history, manners and etiquette, surviving in the military lifestyle, entertaining, service etiquette, and many official publications that define and explain the Army institution. The following bibliographic essay will guide you to these more specialized readings.

MILITARY WIVES' HISTORY

There are numerous diaries, collections of letters, and books by frontier Army wives that are of great importance in understanding the evolution of military society. Many have been reprinted recently, an indication of the great interest in our predecessors' views, lifestyles, and contributions. The most famous and best-selling were the writings of Elizabeth Custer. Her books have each been reprinted and make interesting reading. The most popular is *Boots and Saddles: or Life in Dakota with General Custer* (University of Oklahoma Press, 1961); see also *Following the Guidon* (University of Oklahoma Press, 1966). Elizabeth Custer survived her husband by 57 years and dedicated that time to perpetuating his heroic memory, as related by Michael Tate in "The Girl He Left Behind: Elizabeth Custer and the Making of a Legend," (*Red River Valley Historical Review*, Vol. 5, Winter, 1980). After Elizabeth Custer's death, "Libbie" and "Autie's" personal correspondence was edited and published by her friend and literary executor, Marguerite Merington in *The Custer Story: The Life and Intimate Letters of General George A. Custer and his Wife Elizabeth* (University of Nebraska

Press, 1987). Another interesting account of frontier social life during that same period is included in Katherine Gibson Fougera's *With Custer's Cavalry* (University of Nebraska Press, 1986).

Four other classics written by Army wives in the nineteenth century include: Frances M. Roe, *Army Letters from an Officer's Wife* (University of Nebraska Press, 1987); Mrs. Orsemus Boyd, *Cavalry Life in Tent and Field* (University of Nebraska Press, 1982); Lydia S. Lane, *I Married a Soldier* (Horn and Wallace, 1964); and Martha Summerhayes, *Vanished Arizona: Recollections of the Army Life of a New England Woman* (University of Nebraska Press, 1982). The best overview of the women of this period is Patricia V. Stallard's *Glittering Misery: Dependents of the Indian Fighting Army* (Presidio Press, 1977). Also useful is Betty S. Alt and Bonnie D. Stone's *Camp Following: A History of the Military Wife* (Praeger, 1991), which places in the context of history their earlier book, *Uncle Sam's Brides: The World of Military Wives* (Walker and Co., 1990). Edward M. Coffman's *The Old Army: A Portrait of the American Army in Peacetime, 1784-1898* (Oxford University Press, 1986) includes a well-researched chapter on women and children. For an account of Navy wives of the period, Peter Karsten's *The Naval Aristocracy: The Golden Age of Annapolis and the Emergence of Modern American Navalism* (Free Press, 1972) is important but less complete.

In the study of military wives of the twentieth century, we have far less primary published material available. Alt and Stone's book (mentioned above) includes interviews from the post-World War II period, and Jane Metcalf's *Dowry of Uncommon Women* (X-Press, 1988) gives us many personal stories from the wives living at Air Force Village and their memories of Air Force life. Most of the accounts of the interwar years and World War II are out of print, but can be obtained through interlibrary loan at most post libraries. Katherine Tupper Marshall's *Together: Annals of an Army Wife* (Tupper and Lane, 1946) relates the much-told story that compares the Army wife to the tail of a kite, "the good appendage that steadies it." For additional insights on these years, see Helen Montgomery's *The Colonel's Lady* (New York, 1943); Eleanor Matthews Sliney's *Forward Ho!* (Vantage Press, 1960); and Maurre P. Mahin's *Life in the American Army from Frontier Days to Army Distaff Hall* (Baker-Webster, 1967). For valuable insights on a wife's view of the development of the Air Force, see Ruth Spaatz's *US. Air Force Oral History Interview* (USAF Oral History Program, March 1981). Carol M. Petillo, Boston College, will soon complete a book on Army wives prior to 1938.

The state of literature written about military wives in the second half of the twentieth century is remarkably underdeveloped. There are Mrs. Mark Clark's memoirs subtitled "Inside Story of an Army Wife," and her *Captain's Bride, General's Lady* (McGraw-Hill, 1956). The biographies of major Air Force leaders are generally silent on wives' roles and contributions. A beginning is Helen Maitland LeMay's *Oral History Interview* (Unpublished, October 1990). Many of the sources of the last forty years have been written for, rather than about, wives. An exception is the interesting booklet by Shauna Whitworth entitled "The Happy Contagion: or Story of the First Military Wives Club."

> BLUF
>
> Understand how we grew from Army Wives to Army Spouses while continuing the same strong support.

SOCIAL GUIDES FOR MILITARY SPOUSES

Social guides written specifically for military spouses by service spouses began to appear in formal publication at the time of World War II. Nancy Shea led the way with *The Army Wife* (Harper and Brothers, 1941); her purpose being, "As I realized the need of a guide book for Service wives..., I tried to write the type of book I needed as a bride, a book that a girl marrying into the Service might appreciate." There followed in succession: Nancy Shea and Anne Briscoe Pye's *The Navy Wife* (Harper and Brothers, 1942), and Nancy Shea's *The Air Force Wife* (Harper and Brothers, 1951). Ester Wier continued the process from the perspective of the 1950s with *Army Social Customs* (The Military Service Publishing Co., 1958); also *The Answer Book on Naval Social Customs* and *The Answer Book on Air Force Social Customs* (The Stackpole Co., 1959). The latter book was revised for the 1960s by Ester Wier into *What Every Air Force Wife Should Know* (Stackpole Books, 1966), and Helen LeMay (Mrs. Curtis E. LeMay) wrote in the forward, "We, as Air Force wives, have an obligation to our Country and to the Air Force, which is as important in our role as that of our husbands. Our participation... is essential for a successful Air Force Family." See also Betty Kinzer and Marion Leach's *What Every Army Wife Should Know* (Stackpole Books, 1966) and Silja Allen's *A Leading Lady* (By the Author, P.O. Box 1251, Vienna, VA 22180, 1978).

Though most guides written during this period were for the wives of all ranks and rates, their focus remained primarily on officers' wives. That noticeable gap was filled by Mary Preston Gross with *Mrs. NCO* (Beau Lac Publishers, 1969). This was followed by her other books: *Military Weddings and the Military Ball* (Beau Lac Publishers, 1974, 1983); *The Officer's Family Social Guide* (Beau Lac, 1977); and *The Noncommissioned Officer's Family Guide* (Beau Lac, 1985). The 1970s and 1980s brought a shift in focus from the prescriptive to specific, from personal guides to more official service etiquette. Combining both the personal perspective and military manners is Ann Crossley's *The Army Wife Handbook: A Complete Social Guide* (ABI Press, 1990, 2nd ed. 1994), and Ann Crossley's and Carol A. Keller's *The Air Force Wife Handbook: A Complete Social Guide* (ABI Press, 1992).

Most large posts and division-sized units have developed social guides for their spouse events, i.e., welcomes and farewells that are specific to their unit/post/camp/station. TRADOC senior spouses developed a social guide entitled *Training and Doctrine Command (TRADOC) Army Spouse Protocol and Social Guide.* Initially, the guide was for TRADOC units and Fort Eustis, but this guide has been shared and is useful for all Army spouses and command-team spouses. Part I "Protocol Customs and

Courtesies" was developed by the TRADOC Executive Services. Part II is the traditional "Planning Guide for Welcomes & Farewells," focused on TRADOC HQ/Ft Eustis. Part III is "Command Team Transitions"; and Part IV is "Army Spouse Vernacular Customs & Traditions A-Z." It is available at www.tradoc.army.mil/eso/pdfs/spousesguide.pdf.

SERVICE ETIQUETTE AND OFFICIAL PROTOCOL

The first edition of what has become the classic text on service etiquette appeared in 1958, for the naval service. The current edition is for all branches of the armed forces: Oretha D. Swartz, *Service Etiquette*, 5th ed. (Naval Institute Press). It contains an especially detailed section on weddings. For official protocol, see Mary Jane McCaffree and Pauline Innis's *Protocol: The Complete Handbook of Diplomatic and Social Usage* (Prentice-Hall, 1977 and 1985).

GENERAL ETIQUETTE AND MANNERS

Everyone who writes an etiquette book seeks to represent the needs of the present. Consequently, the rapid pace of change in America can account for the numerous guides recently written on the subject of manners. Some of the best authors have updated versions of previous best-sellers. Certainly, the wittiest author is Judith Martin in her *Miss Manners' Guide for the Turn-of-the-Millennium* (Pharos Books, 1989). The most convenient for quick reference is Marjabelle Young Stewart's *The New Etiquette: Real Manners for Real People in Real Situations an A-to-Z Guide* (St. Martin's Press, 1987). Two very personal books with different styles are: Charlotte Ford in *Etiquette: Charlotte Ford's Guide to Modern Manners* (Clarkson N. Potter, 1988), and Letitia Baldrige in *Letitia Baldrige's Complete Guide to the New Manners for the '90s* (Rawson Associates, 1990). For the traditionalist, there is the very fine book by Letitia Baldrige entitled *The Amy Vanderbilt Complete Book of Etiquette: A Guide to Contemporary Living* (Doubleday & Co., 1978). Also useful are *Emily Post on Entertaining*, and *Emily Post on Weddings* (Harper & Row, 1987), small easy-to-read booklets in a series written by Elizabeth L. Post. Also see Dorothea Johnson's *Entertaining and Etiquette: A Guide for Contemporary Living* (Acropolis Books Ltd., 1979). A great pamphlet for young spouses just entering their military walk of life is *The Once Over Lightly: A Practical Look at Protocol* by Bibs Reynard.

Another area of interest to today's Army spouse is business etiquette. Discussions about office manners and guides to business etiquette first began to appear in 1920. This type of book would naturally find great appeal in the 1980s and '90s. One of the best is *Letitia Baldrige's Complete Guide to Executive Manners* (Rawson Associates, 1985). Also helpful are *Emily Post on Business Etiquette* and *Emily Post on Invitations and Letters* (Harper & Row, 1990) by Elizabeth L. Post. For international business etiquette, look to *Kiss, Bow, or Shake Hands*, 2nd ed., 2006, by Terri Morrison and Wayne Conaway. This is a guide to proper international business protocol, along with practices, customs and philosophies of other countries.

ENTERTAINING

Entertaining's return to popularity in the 1980s was evident in an outpouring of books on this subject. There is every indication that the trend continues. Every post library has a good selection of guides for entertaining, as well as new, exciting cookbooks. The leading author in the field of entertaining is Martha Stewart, with over two million books sold; see *Entertaining* (Clarkson N. Potter, 1982). Two other popular authors are Alexandra Stoddard, whose books focus on lifestyles and decorating, and Mary Emmerling. In the specialty area is Michele Evans' *Fearless Cooking for Crowds: Beautiful Food for Groups of Eight through Fifty* (Times Books, 1986). Another useful series is the "Five-Minute" approach; see Jane Newdick, *The Five-Minute Centerpiece* (Breslich and Foss, 1991). A small book that's packed full of good suggestions for entertaining is *Entertaining Ideas for Military Wives* (Carlton Press, Inc., 1992). The author, Linda O. Anderson, is an Army wife who draws on her experiences and talents to help others add some zest to their entertaining.

SUPPORT RESOURCES

In the last decade, expanded research on military families has led to the publication of helpful support resource materials for military spouses and their families. Many informative pamphlets are published by each service. There is also significant work being done in this area by individual writers. Not to be missed is Kathleen O'Beirne's *Pass It On! How to Thrive in the Military Lifestyle* (Lifescape Enterprises, 1991). O'Beirne is the most-published author in the field of military families. *Pass It On!* is an anthology of all her articles from 1969-1991. It is exceptionally well indexed and a great asset for those who have clipped and saved O'Beirne's prose for years. Perry Smith focuses on life in the D.C. area, in his *Assignment Pentagon: The Insider's Guide to the Potomac Puzzle Palace*, 2nd ed., (Pergamon-Brassey's International Defense Publishers, Inc., 1993), as does the USO's *The Guide to Washington*, 5th ed.

Navy spouses are pioneers in the publication of support material for spouses. They created The Navy Wifeline Association in 1965, which is staffed five days a week by volunteers who manage the office and are responsible for its publications. Continuously revised and updated, these include "Sealegs," "Social Customs and Traditions of the Navy and Marine Corps," "Guidelines for the Spouses of Commanding and Executive Officers," and "Guidelines for the Spouses of Command Master Chiefs and Chiefs of the Boat." For information, write to The Navy Wifeline Association, Building 172, Washington Navy Yard, Washington, D.C. 20374.

Family Support Centers and ACS both have many good pamphlets and resource publications available. Additionally, other support resources not to be overlooked are the handouts frequently provided at military and military-related conferences and seminars. See Chaplain (Col) Hiram L. Jones' "Bibliography on Loss and Grief" (AWAG Convention handout, 1986), Chapter 28.

U.S. DEPARTMENT OF THE ARMY PUBLICATIONS

There are numerous Army regulations and publications that govern the customs and official protocol of the Army. They are available on post or may be obtained through inter-library loan at your post library. The ones used as resources for this book are as follows: *"A Guide to Protocol,"* Department of the Army Pamphlet 600-60; *"Salutes, Honors, and Visits of Courtesy,"* Army Regulation 600-25.

There are excellent pamphlets written by the students, wives, and staff at the Army's schools for professional development. Of special mention are: *The Leader's Link* (U.S. Army Command and General Staff College, 1986), written originally in 1982 as *The Commander's "Link,"* and *Choices & Challenges: A Guide for the Battalion Commander's Wife* (U.S. Army War College, 1991). The U.S. Army War College Spouses Academic Year 2011 published *Basics From the Barracks, Military Etiquette and Protocol*. A spouse's quick reference to the military's unique customs, courtesies, and traditions.

Though not a DA publication, *Army Officer's Guide*, 45th ed. (Stackpole Books, 1990) has served as an invaluable reference guide for Army officers for over sixty years. The current edition is no exception.

FAMILY/PARENTS' GUIDES

The Institute of Land Warfare Association of the United States Army published *Your Soldier Your Army, A Parents' Guide* by Vicki Cody. A wonderful guide for parents or parents-in-law of soldiers deploying.

VIDEOS

There are videos available that expand on the topics of this handbook. For manners and entertaining, see: "Miss Manners' on Weddings," and the "Martha Stewart's Secrets for Entertaining" series, Crown Video and WGBH Boston, 1988.

THE CHANGING WORLD OF MANNERS AND CUSTOMS

Even the most comprehensive guides to general manners and service etiquette need to be frequently updated. Just as we observed in our past, manners and customs are continually evolving to meet the needs of our ever-changing world and lifestyles. Fortunately, we have only to read our newspapers and magazines to keep in touch with the current thinking on manners.

Two of the most popular commentators on the subject of civilian manners are syndicated columnists. Judith Martin, as "Miss Manners," writes a daily column of humorous responses to some of the many letters she receives, over 100 letters weekly. Letitia Baldrige's weekly column on manners and gracious living, entitled "R.S.V.P.," appears in over 1,000 newspapers. They both answer questions sent to them by their readers, which means they are addressing the current issues and social concerns of today.

Wives' club magazines are another good source of information on protocol, with a particular emphasis on military-related issues. Many clubs feature a monthly column on this subject. The information you will find in these articles is generally most useful and written from the military perspective. However, true masters of military manners are very few. Most of us simply report on the world as we observe it.

Each Army spouse can become a skilled practitioner of "military manners" and enjoy the process. To do so, maintain a sense of humor, stay flexible, and approach the guidelines for proper etiquette as just that—guidelines, not rules. Following these guidelines will always be less important than showing genuine concern for the feelings of others, as courtesy is the only *essential* part of good manners.

* * * * * * * * * * * * *

"Back to the Future" — We have a solid foundation that has worked for over two centuries; it just always needs updating for the next generation.

Appendix

Army Spouses' Historical Chronology

1775-81—American Revolution
    Citizen soldiers and Regulars; wives considered "camp followers"
1815-60—National expansion; Frontier Army
    Wives and children called "companions of our exile"
1865-1900—No provision in Army regulations for wives and children
    Enlisted men "forbidden," then "discouraged" from marrying
1881—American Red Cross founded; serves military families to present
1898—War-time family allotments for enlisted men
1914-18—World War I
1917—Blue Star Flag tradition began
    War-time family allotments for officers
1941-45—World War II
1942—Public Law 490, dependency benefits provided
1948—Arlington Committee formed
    American Women's Activities Germany (AWAG) Conference
1950-53—Korean War
1950's—Military wives' clubs officially organized
1951—Defense Advisory Committee on Women in the Service formed
1961—Army Distaff Hall groundbreaking
1963-73—Vietnam War
1965—Army Community Service (ACS) founded
1966—CHAMPUS established
1969—National Military Wives Association (NMWA) founded
1972—Survivor Benefit Plan (SBP) initiated
1974—Public Law 42-659 garnishes military pay for child support and alimony
1979—Army Quality of Life Office created
1980—1st Army Family Symposium
1981—Army changes "dependent" to "family member"
1982—Army Family Liaison Office (FLO) opens
    Army publishes *The Commander's Link*
1983—Army White Paper, "The Army Family"
    Army Seminar for Wives; "command team training" initiated
    Public Law 97-252; right of divorcée to claim portion of retirement pay

       NMWA becomes National Military Family Association (NMFA)
       Military Spouse Priority Placement Program implemented
1984—May 23 proclaimed "National Military Spouse Day"
       DOD Family Policy Office created, military-wife director
1985—DOD adds spouse employment in Federal Women's Program
1986—Congress established Dependent Dental Program
       Exceptional Family Member Program became a full program
1987—DOD Authorization Act (hiring preference for spouses)
       Department of the Army issued a regulation implementing command-sponsored Family Support groups, now called Family Readiness Groups.
1991—Gulf War
1995—Operation Ready training materials for Family Assistance Centers
1996—Army National Guard Family Assistance Centers funded
1997—US Army Reserve Family Assistance Centers funded
1998—TRICARE established
1999—Military Child Education Coalition founded
2000—DOD allows limited unofficial information in Family Readiness Group newsletters
2001—Terrorist attack: World Trade Center, Pentagon, rural Pennsylvania
       War on Terrorism
       War in Afghanistan
2003-11—Iraq War
2005—Joint Federal Travel Regulation authorized transportation allowances to family members of seriously injured hospitalized servicemembers
2008—Higher Education Opportunity Act prohibits public institutions from charging-out-of-state tuition to members of the Armed Forces and their families whose domicile or permanent duty station is in that state and retains in-state tuition if the service member is reassigned outside state
       Officers' Spouse Clubs and Enlisted Spouse Clubs started combining into All Rank Spouse Clubs
2009—Spouse professional weight allowance authorized
       GI Benefits authorized to dependents of all ranks
2010—Operation New Dawn
2013—Gold Star Identification Card established authorizing unescorted installation access to Gold Star family members
2014 – Operation Inherent Resolve
2015 – Operation Freedom's Sentinel

A

Accidents, 104
Acronyms and Abbreviations, 372
ACS, 336
Addressing Envelopes, 73
Addressing Wedding Invitation Envelopes, 264
Advisors and Honorary President, 203
Advisory Positions, 297
Aides, 322
Arch of Sabers, 259
Army Family Team Building, 339
Army Song, 364
Army Structure, 365
Attaché Duty, 355
Avoid Family Separations, 379

B

Baby Showers, 194
Balls, 231
Bamboo Lap Trays, 170
Bartender, 171
Being a Good Sponsor, 333
Beverages, 154
Bridal Showers, 192
Buffet Dinners, 138
Buffet Manners, 105

C

Call Waiting, 50
Calling/Social Cards, 37
Cancellation or Postponement. *See* Invitations
Candle Wicks, 169
Candles, 141
Carrying the Umbrella, 287
Casual, 95
Centerpiece, 141
Chain of Command, 365
Change of Address, 34
Chapel Service at Military Funerals, 272
Clearing the Table, 143
Coffee Attendees, 180
Coffee Group Leader, 181
Coffee Hosts and Hostesses, 179
Coffees, 177
Command Performance, 286
Command Reflections, 304
Command Sergeants Major Spouses, 310
Commanders' Spouses, 291
Coping Skills for New Spouses, 324
Correspondence, 25
Corsage and Boutonniere, 222
Course Order, 148
Cremation, 274
Customs (Asia, Europe, Middle East), 347

D

Dealing with the Media, 383
Decorating Tips, 385
Dining-Out, 234
Dinner Partners, 155
Diplomatic and Civil Dignitaries, 17
Divorce Happens, 381
Dress for the Occasion, 93
Dress Term Chart, 95
Dress Terms, 93
Duty in Washington, D.C., 340
Duty with a Unified Combatant Command, 351
Duty with Another Service, 352

E

End of Command, 302
Ending Phone Calls, 48
Entertaining, 123
Entertaining Outside the Home, 172
Entertaining Tips, 167
Entertainment Record, 137
Escorts, 220
Etiquette at the Table, 113
Etiquette for Special Foods, 112
Examples of Invitations. *See* Invitations
Expressing Thanks, 115

F

Failure to Respond. *See* Responding to Invitations
Fallen Comrade Table, 232
Family Readiness Groups, 189
Farewell Coffee, 189
First Sergeants' Spouses, 309
Flag Etiquette, 282
Flowers, 170
Formal, 94
Formal Farewells, 233
Formal Invitations. *See* Invitations
FRG Leader, 191
Funeral Courtesies, 275

G

General Officers. *See* Addressing Envelopes
General Officers' Spouses, 317
Gifts, 300
Greeting Card Etiquette, 35
Guest Guidelines for Teas, 206
Guests' Responsibilities, 102

H

Head-Table Seating, 162
Helpful Books after a Death, 279
Honors at Military Funerals, 271
Hostess's Manners, 156
Houseguests, 107
How to Remember Names, 6
How You Can Help. *See* Military Funeral

I

Informal, 94
Informal Invitations. *See* Invitations
Informals, 26
Introducing and Addressing Military, 16
Introducing different groups, 10
Introducing Others, 15
Introductions, 10
Invitation Cover Sheet, 321
Invitation Reminders, 70
Invitation to the White House, 89
Invitations, 55
Ironing Linens, 168

J

Joint Assignments, 351

L

Ladder of Communication and Support, 312
Language Training, 349
Lap Trays, 140
Last-Minute Cancellation, 157
Late Guests, 157
Leadership Role, 295
Leadership Roles, 308
Leadership Skills, 294
Life after the Army, 388
Living on Post/Off Post, 329

M

Meeting People, 3
Memorial Service, 276
Military Chapel, 258
Military Doctors. *See* Addressing Envelopes
Military Education, 371
Military Funerals, 271
Military Time, 335
Military Weddings, 256
Moving to the Buffet, 142
Multiples of Four. *See* Seating Arrangements

N

Name Tags, 20
Names. *See* Meeting People
NCOs' Spouses, 306
New Year's Reception, 230
Newsletters, 187

O

On-Post Courtesies, 330
Oral Invitations and Responses, 85
Oral/Email Responses. *See* Responding to Invitations
Order of Precedence, 164
Ordering Guidelines for cards and informals, 43
Overseas Living, 346

P

Parades General Guidance, 248
Parades, Changes of Command, Changes of Responsibility, Retirements, 241
Party Checklist, 135
Party Preparations, 132
Pitfalls to Guard Against, 318
Place Cards, 145
Place Plates/Chargers, 150
Place Setting, 150
Places of Honor. *See* Seating Arrangements
Planning a Tea, 210
Pourers Guidelines, 207
Pre-Party Schedule, 132
Prologue-Legacies, xviii
Promotion and Award Ceremonies, 284
Promotion Lists Require Kindness, 380
Public Speaking, 298

R

Rank and Initials. *See* Addressing Envelopes
Rank Insignia Chart, 357
Rank Structure, 369
Receiving Line, 227
Receiving-Line Guidance, 228
Reception Attire and Arrival, 223
Reception Line, 224
Reception/Receiving Line Diagram, 229
Receptions and Receiving Lines, 223
Reciprocate, 107
Red Carpet, 227
Responding to Invitations, R.s.v.p., 83
Responding to Wedding Invitation, 263
Response Complications for Invitations, 85
Response Indicators. *See* Responding to Invitations
Response Time. *See* Responding to Invitations
Retirement Parades, 254
Revolving Door Courtesy, 287
Round Table Seating, 162

S

Seated Dinners, 145
Seated-Dinner Manners, 105
Seating Arrangements, 158
Seating at Parades, 246
Service Customs, 353
Serving Wines, 170

Setting Up a Receiving Line, 227
Signatures, 33
Silver-Polishing Gloves, 169
Sir and Ma'am, 19
Sister Services, 352
Social (Official) Calendar, 319
Social and Business Correspondence, 28
Social Guides for Military Spouses, 394
Sock Color, 98
Spills, 169
Spiritual Health, 382
Spouse Club Activities, 200
Spouse Club Board, 201
Spouse Club Courtesies, 201
Spouse Club Elections, 202
Spouse Clubs, 197
Spouse Employment, 338
Staff Car Seating, 284
Standing to Talk, 14
Stationery, 25
Stewardship of Quarters, 323
Substitute Leading Spouse, 299

T

Table Appearance, 153
Table Manners, 109
Tea Committee, 213
Tea Invitations, 211
Teas, 205
Telephone Manners, 46
Thank-You Note Formula, 118
Titles, 33
Toasting, 236
Two Invitations for the Same Time, 87
Tying a Bow Tie, 98

U

Uniforms at Military Weddings, 257
Unit Structure, 368
Ushers and Saber Bearers, 267
Using First Names, 18

V

Voicemail and Answering Machines, 49
Volunteerism, 337

W

Warrant Officer History, 370

Wearing Special Awards, 98
Wedding Announcements, 265
Wedding Invitations, 260
Wedding Receptions, 269
Wedding Rehearsal and Dinner, 266
Wedding, Birth, and Adoption
    Announcements, 195
Welcome Coffee, 187
Welcome Courtesies, 332
Welcoming Committee, 186
What to Say and Not to Say, 278
When Commander is Relieved, 381
When No Response Is Received, 87
When to Stand, 285
When to Stand at Parades, 250
Which Fork to Use, 110
Working Spouses, 314
Written Thanks, 116

About the Authors

Ann Crossley has a background better suited than most to write for an Army spouse audience. She grew up as the daughter of an Army NCO who went on to become a warrant officer. During her childhood, she lived with her family in Japan, France, and Germany, and spent her high-school years at American military boarding schools in Europe. Ann earned a home economics degree from the University of Georgia (Phi Kappa Phi) and then married her husband, Bill, the day after he was commissioned as a second lieutenant. She "served" with him for 28 years, until he retired from the Army in 1988 as a brigadier general.

Ann's professional experience in matters of protocol, etiquette, and entertaining includes courses on these topics at the Foreign Service Institute. She has spoken and written for a wide variety of audiences, including officer and NCO spouses, spouse clubs, Army teenagers, officer professional development classes, and American Women's Activities in Germany (AWAG) Conference delegates. When her husband retired, she turned to the broader audience and authored *The Army Wife Handbook* and *The Air Force Wife Handbook*. Both books enjoyed great success, and this new book, *The Army Spouse Handbook*, is intended to bring updated 21st-century guidance to today's Army spouses.

Ann and her husband lived in Atlanta for the past twenty years, and are now moving to Florida to settle into a retirement community a few blocks from the Jacksonville Beach.

Ginger Perkins has thrived as an Army spouse for 38 years of active duty that included over 20 moves around the world and multiple combat deployments as the spouse of the commanding officer until he retired from the Army in 2018 as a four-star general. She is also the mother of soldiers with a daughter who is a Blackhawk helicopter pilot with two combat tours to Afghanistan and a son who is a West Point graduate and engineer.

Ginger's experience and dedication have been forged through the trials of combat and sacrifice. Ginger knew the Army spouses and families needed encouragement, support, and inspiration. She has had the honor to lead multiple spouse groups and community organizations, each time gaining insights into how important history and traditions are to understand why we serve this great Nation. Ginger has published several Army spouse advisory and unit social guides, and most recently the *Training and Doctrine Command (TRADOC) Army Spouse Protocol and Social Guide*. For several years Ginger was part of the Army's Professional Military Education (PME) program teaching Army spouse customs and traditions to battalion, brigade, and general officer spouses.

Ginger currently lives with her husband in New England and continues to advocate for the Army Spouse.

Comments from Committee Members

"As a new Army spouse preparing for our first PCS move, I was purchasing books about Germany and also bought a copy of *The Army Wife Handbook*. I was hungry for information, anything that could help to prepare me for a military and Army life. This book has traveled with me over the course of 22 years, during 11 PCS moves in three countries and 7 different states. My original copy is dog-eared, underlined, marked up, and has notes scribbled in the margins. I used this book as a guide, and along the way it helped me to view each PCS as an invitation to start anew, an opportunity to explore the world, to make lifelong friends, to give of myself for a greater good, and has enabled me to thrive in this military life and truly embrace being an Army wife. With a fresh update, it is the perfect guide for today's Army Spouse."
—*Aimee Randazzo, Army Spouse 22 years and happily still counting*

"Mrs. Crossley's handbook served as one of my favorite reference materials during my wonderful military journey. In addition to still owning both of her previous editions, I incorporated the tradition of purchasing and gifting to newly wedded spouses I encountered, just as I had received as a new Army spouse, along with the personal message that this was a strong source of my comfort and foundation for military life. It was an honor to assist her in this newest edition."
—*Linda Via, Army Spouse for 34* years

"Being an Army spouse is hard...and it doesn't always get easier. However, you do get stronger and you learn to survive and thrive through a lot of things you didn't think you ever would. *The Army Spouse Handbook* will be an invaluable compass as you navigate this journey and guide you to be the strongest version of you."
—*Connie Roy, Army Spouse, Army Civilian, 29 years*

"Thank you for this wonderful guide to life in a very specialized culture called the Army. But, when push comes to shove, and you are at a loss for what to do, do what Army spouses have done since the beginning of time. Join ranks and be yourselves!"
— *Carol Brooks, Army Brat, and Army Spouse of 36 years*

"This book touches on everything and has been a great resource for me at every step of my military journey—Army Brat, ROTC Cadet, Soldier, and now as a newlywed Military Spouse." —*MAJ Cassandra Perkins*

"This is a great guide to help every Spouse 'maneuver' in their journey and Army adventure! Stay true to yourself. Do the right think and do things right."
—*Angel Mangum, Army Spouse 28 years*

To order this book, please go to sales@armyspousehandbook.com. Price is $27 per book, plus shipping and handling. Reduced prices are available for spouse clubs when buying in bulk.